Advances in Delays and Dynamics

Volume 12

Delay systems are largely encountered in modeling propagation and transportation phenomena, population dynamics and representing interactions between interconnected dynamics through material, energy and communication flows. Thought as an open library on delays and dynamics, this series is devoted to publish basic and advanced textbooks, explorative research monographs as well as proceedings volumes focusing on delays from modeling to analysis, optimization, control with a particular emphasis on applications spanning biology, ecology, economy and engineering. Topics covering interactions between delays and modeling (from engineering to biology and economic sciences), control strategies (including also control structure and robustness issues), optimization and computation (including also numerical approaches and related algorithms) by creating links and bridges between fields and areas in a delay setting are particularly encouraged.

Indexed by zbMATH.

More information about this series at https://link.springer.com/bookseries/11914

Giorgio Valmorbida · Wim Michiels ·
Pierdomenico Pepe
Editors

Accounting for Constraints in Delay Systems

Springer

Editors
Giorgio Valmorbida
Laboratoire des Signaux et Systèmes
CentraleSupélec
Gif-sur-Yvette, France

Wim Michiels
Department of Computer Science
KU Leuven
Heverlee, Belgium

Pierdomenico Pepe
Department of Information Engineering,
Computer Science, and Mathematics
University of L'Aquila
L'Aquila, Italy

ISSN 2197-117X ISSN 2197-1161 (electronic)
Advances in Delays and Dynamics
ISBN 978-3-030-89013-1 ISBN 978-3-030-89014-8 (eBook)
https://doi.org/10.1007/978-3-030-89014-8

This Springer imprint is published by the registered company Springer Nature Switzerland AG
The registered company address is: Gewerbestrasse 11, 6330 Cham, Switzerland

Preface

The last decade has witnessed a growing research activity on methods and applications of time-delay systems. In science and engineering, time-delay models naturally appear in the description of dynamical systems because the transfer of material, energy, or information is mostly not instantaneous. For example, delay elements in mathematical models may describe heredity in population dynamics, feedback delays in control loops, coupling delays in multi-agent systems and networks, and regenerative effects in machining applications. Time-delay models also arise as approximate models of other classes of distributed parameter systems (e.g., systems described by partial differential equations) or in the context of approximating control laws (e.g., replacing the derivative action in a PID control law by a finite difference to ease the implementation or to reduce the sensitivity to sensor noise).

Time-delay systems constitute a class of infinite-dimensional systems. Consequently, surprisingly rich dynamics can be observed, and many challenges appear from a control design point of view. The purpose of this volume in the series *Advances in Delays and Dynamics* (ADD@S) is to carefully select recent results on the analysis and control synthesis for time-delay systems, thereby following two transversal axes.

The first axis concerns fundamental problems in the field, including stability analysis and stabilization, characterizing the region of attraction, relative stability, and synchronization problems. Models in terms of delay differential equations, convolution models in the context of supply networks, and models described by coupled ordinary and partial differential equations are studied. The second axis is concerned with advances in mathematical and computational methods for analysis and design. The discussed methods focus on strategies rooted in Lyapunov's second method and convex optimization, possibly combined with polynomial representations or approximations, methods exploiting internally positive representations, and the robust stability analysis via the concept of set-invariance. Besides, the computation of characteristic roots for stability analysis and polynomial chaos expansions for uncertainty quantification are visited, and algebraic tools relying on interpretations of differential operators in terms of Ore extensions are studied.

This volume particularly highlights how the control design can consider fundamentally different constraints. These constraints consist of structural constraints on the controller (e.g., imposing a PID controller structure), input and state constraints in the mathematical model of the system or the specification of the control task, or constraints related to the implementation such as the sampled-data nature inherent to digital communication, the robustness to disturbances or sensor noise, or saturation in the actuation.

The original idea for this volume appeared at a specialized workshop on November 22–24, 2017, in the Laboratoire de Signaux et Systèmes, Gif-sur-Yvette, France. This workshop was the first one in a series of meetings of the International Scientific Coordination Network on Delay and Constraints in Distributed Parameter Systems (SPa-DisCo), funded by the French Center for Scientific Research (CNRS).

The contents of the book are organized into two parts:

Analytical Methods for Analysis and Design

In chapter "Stabilizing Systems with Uncertain Time Delay via PID Control: How Good Is It?", contributed by Dan Ma and Jie Chen, the maximal time delay in the feedback loop, which can be managed by a single PID regulator, is investigated for linear time-invariant systems. Explicit expressions of the achievable delay margin and tractable bounds are provided for low order unstable systems with uncertain constant time-delay.

In chapter "Stability Analysis of Linear Delay Systems via Internally Positive Representations: An Overview", contributed by Vittorio De Iuliis, Alfredo Germani, and Costanzo Manes, stability techniques for positive systems are exploited for the analysis of general (nonpositive) linear time-delay systems by the use of an internally positive representation. The methodology is applied to systems described by difference equations, differential equations, and coupled difference differential equations with time-varying delays.

In chapter "Control by Set Invariance of a Class of Convolution Systems", contributed by Berna Bou Farraa, Rosa Abbou, and Jean Jacques Loiseau, the control of convolution systems subject to constraints is investigated. An invariance property of polytope sets of continuous-time systems is investigated. The problem is stated in terms of a controlled input delay system, with constraints and bounded disturbances. Applications to supply chains are investigated.

In chapter "Sub-platooning via Agent Separation for Improved Traffic Flow Metrics in a Car-Following Model", contributed by Min Koh, Adrián Ramírez, and Rifat Sipahi, the formation of traffic jams is investigated. Also, a control strategy is proposed, which consists of disconnecting vehicles if a constraint is violated. This strategy involves the separation of a platoon into sub-platoons and the choice of suitable leaders. A so-called optimal velocity model, involving the driver's sensitivity and reaction time delay is exploited in the analysis and design.

In chapter "On the Ore Extension Ring of Differential Time-Varying Delay Operators", contributed by Alban Quadrat and Rosane Ushirobira, linear systems with time-varying delays are studied within the framework of algebraic methodologies. By exploiting the notion of skew polynomial rings, an effective construction of the ring of differential time-varying delay operators as an Ore extension is provided. Essential properties, such as controllability, parametrizability, and flatness, are then investigated.

In chapter "Practical Stability Preservation Under Sampling, Actuation Disturbance and Measurement Noise, for Globally Lipschitz Time-Delay Systems", contributed by Mario Di Ferdinando, Pierdomenico Pepe, and Emilia Fridman, practical stability preservation under sampling for globally Lipschitz time-delay systems is studied. It is proved that such a property holds in the presence of bounded actuation disturbances and measurement noise when globally Lipschitz state feedback, globally exponentially stabilizing in continuous-time, is emulated with suitably fast sampling.

Numerical Methods for Analysis and Design

In chapter "15 Years or So of Pseudospectral Collocation Methods for Stability and Bifurcation of Delay Equations", contributed by Alessia Andò, Dimitri Breda, Davide Liessi, Stefano Maset, Francesca Scarabel, and Rossana Vermiglio, numerical methods based on pseudo-spectral collocation techniques for the stability and bifurcation analysis of coupled differential and difference time-delay systems are overviewed. Suitable approximations of the solution operator or its infinitesimal generators are exploited. The approximation of nonlinear delay differential equations by a system of ordinary differential equations is addressed as well.

In chapter "Sensitivity Analysis for Stability of Uncertain Delay Differential Equations Using Polynomial Chaos Expansions", contributed by Rossana Vermiglio and Andrea Zamolo, the stability analysis of linear(ized) systems with delays, affected by uncertain parameters modeled as random variables, is investigated. A numerical method is provided for computing the polynomial chaos expansion of the spectral abscissa function. The computational cost of evaluating the integrals is reduced by an innovative integration scheme relying on interpolation in Padua points.

In chapter "Lyapunov Stability of a Coupled Ordinary Differential System and a String Equation with Polytopic Uncertainties.", contributed by Matthieu Barreau, Alexandre Seuret, and Frédéric Gouaisbaut, the robust stability analysis of systems described by coupled ODE/PDE equations, subject to polytopic uncertainties, is investigated. By invoking Lyapunov theory, conditions in terms of linear matrix inequalities are obtained. The analysis methods are validated by means of an example, where the mathematical model describes regenerative chatter in the particular case of modeling the wave as a pure delay.

In chapter "Regional Stability of Nonlinear Sampled-Data Controlled Systems Under Actuator Saturation: A Quasi-LPV Approach", contributed by Alessandra

Helena Kimura Palmeira, João Manoel Gomes da Silva Jr., and Jeferson Vieira Flores, sampled-data state feedback control with saturated actuators is investigated for control-affine nonlinear systems. An approach based on local quasi-LPV models is adopted. Convex optimization problems allow computing the estimates of the region of attraction or the maximal inter-sampling time using stability conditions provided in terms of linear matrix inequalities.

In chapter "A Method for Computation and Analysis of Partial Synchronization Manifolds of Delay Coupled Systems", contributed by Libo Su, Wim Michiels, Erik Steur, and Henk Nijmeijer, a systematic method for the computation and analysis of partial synchronization manifolds of networks of diffusive delay-coupled nonlinear systems is provided. In addition, a systematic procedure for separating the dynamics on the synchronization manifold and the synchronization error dynamics is proposed, which is a pre-requisite for characterizing the stability of partial synchronization manifolds. The numerical methods are implemented in open-source MATLAB software.

In chapter "Analysis of Linear Partial Differential Equations Using Convex Optimization", contributed by Aditya Gahlawat and Giorgio Valmorbida, the stability of systems described by PDEs is investigated. The approach does not make use of approximations into ordinary or delay differential equations but exploits the Lyapunov theory with suitable functionals on chosen Hilbert spaces. By using sum-of-squares relaxations, the analysis is reduced to checking the feasibility of linear matrix inequalities subject to affine constraints, which is a convex optimization problem.

Gif-sur-Yvette, France	Giorgio Valmorbida
Leuven, Belgium	Wim Michiels
L'Aquila, Italy	Pierdomenico Pepe
February 2021	

Contents

Acronyms

BPC	Pitchfork Bifurcation
CE	Characteristic Equation
DDE	Delay Differential Equations
DTD	Differential Time-Delay
DTVD	Differential Time-Varying Delay
GES	Globally Exponential Stablility
H	Hopf (bifurcation)
IG	Infinitesimal Generator
IPR	Internally Positive Representation
LF	Lyapunov Functional
LMI	Linear Matrix Inequality
LPV	Linear Parameter Varying
LTI	Linear-Time-Invariant
MAS	Multi-Agent System
MC	Monte Carlo (simulations)
MIMO	Multi-Input Multi-Output
NISP	Non-Intrusive Spectral Projection
ODE	Ordinary Differential Equation
OVF	Optimal Velocity Function
OVM	Optimal Velocity Model
PCE	Polynomial Chaos Expansion
PD	Proportional Derivative, Period Doubling
PDE	Partial Differential Equation
PI	Proportional Integral
PID	Proportional Integral Derivative
RFDE	Retarded Functional Differential Equation
SA	Sensitivity Analysis
SIG	Stochastic Infinitesimal Generator
SISO	Single-Input Single-Output
SOS	Sum-Of-Squares
T-S	Takagi-Sugeno

UDDS Uncertain Delay Differential Equations
UQ Uncertainty Quantification

Analytical Methods for Analysis and Design

Stabilizing Systems with Uncertain Time Delay via PID Control: How Good Is It?

Dan Ma and Jie Chen

Abstract This chapter concerns the delay margin achievable using PID controllers for linear time-invariant (LTI) systems subject to variable, unknown time delays. The basic issue under investigation addresses the question: What is the largest range of time delay so that there exists a single PID controller to stabilize the delay plants within the entire range? Delay margin is a fundamental measure of robust stabilization against uncertain time delays and poses a fundamental, longstanding problem that remains open except in simple, isolated cases. In this chapter we present explicit expressions of the exact delay margin and its upper bounds achievable by a PID controller for low-order delay systems, notably the first- and second-order unstable systems with unknown constant delays. While furnishing the fundamental limits of delay within which a PID controller may robustly stabilize a delay process, our results should also provide useful guidelines in tuning PID parameters and in the analytical design of PID controllers.

1 Introduction

Delay margin is a fundamental measure of robust stabilization against uncertain time delays and it addresses a central issue in the study of feedback stabilization of time-delay systems: What is the largest range of delay so that there exists a single controller that can robustly stabilize the delay plant within that entire range? Much

The authors gratefully acknowledge the support of Hong Kong RGC under the projects CityU 11200415, CityU 111613, and the support of Natural Science Foundation of China under grant numbers 61603079, 61773098 and the Fundamental Research Funds for the Central Universities under Grant N150404019.

D. Ma (✉)
College of Information Science and Engineering, Northeastern University, Shenyang, China
e-mail: madan@mail.neu.cn.edu

J. Chen
Department of Electronic Engineering, City University of Hong Kong, Hong Kong, China
e-mail: jichen@cityu.edu.hk

G. Valmorbida et al. (eds.), *Accounting for Constraints in Delay Systems*, Advances in Delays and Dynamics 12,
https://doi.org/10.1007/978-3-030-89014-8_1

of the past work on the delay margin problem can be seen as a tradeoff between the complexity of control laws and the delay margin achievable. When allowed more sophisticated control laws, such as linear periodic controllers [13], nonlinear periodic controllers [5], and nonlinear adaptive controllers [4, 8], the delay margin can be made infinite; in other words, a LTI delay plant can be stabilized by these complex, non-LTI controllers for arbitrarily long uncertain delays. On the other hand, when confined to LTI controllers, the delay margin can only be finite [10] (see also, e.g., [12]), and the achievable margins vary. More specifically, for general LTI systems, upper bounds on the delay margin were obtained in [6, 12, 16], which establish limits beyond which no finite-dimensional LTI output feedback controllers may exist to robustly stabilize the delay plant family within the margin. The bounds, however, are tight generally for systems with no more than one unstable pole and one nonminimum phase zero, for which they were found to be exact; otherwise, for systems containing more unstable poles, the bounds tend to be crude and pessimistic. In contrast, lower bounds were derived in [15] for systems with an arbitrary number of unstable poles and nonminimum phase zeros, wherein it was also shown that a LTI controller can be synthesized using $\mathscr{H}_\infty$ optimal control techniques, which guarantees closed-loop robust stabilization when the delay value falls into the range. Other bounds were also made available under more specialized circumstances by employing finite- or infinite-dimensional LTI controllers based on a predictor feedback mechanism (see, e.g., [10, 20, 23–25]), and a numerical method was proposed in [10] with the goal to increase the delay margin in an iterative manner. Additionally, problems similar to that of delay margin have been studied in the context of multi-agent consensus, in, e.g., [9, 14, 21].

In this chapter we study the delay margin achievable by LTI controllers of further restricted structure and complexity, i.e., those of PID type [1, 22]. PID control is favored for its ease of implementation, and has proven to be the most popular means in controlling industrial processes; a recent expert survey shows the remarkable, widespread approval of PID controllers [17]. Earlier work on the delay margin by PID control includes [10] (see also [11] (p. 154)), where the exact delay margin was found for first-order systems achievable by proportional static feedback. In the comprehensive developments of [18, 19], likewise, the delay margin problem was addressed using full PID controllers, wherein it was shown that for first-order systems, PID controllers can double the delay margin achievable solely by proportional feedback, a margin that was also found to be achievable by a predictor-type observer feedback [10]. The analysis in [18, 19] was carried out on the closed-loop quasipolynomial resulted from the delay plant, by a rather elaborate derivation based on an extended *Hermite-Biehler theorem* for quasipolynomials. Despite the considerable effort and lengthy derivations in [18] and elsewhere, however, the delay margin was found only for the first-order system. It appears difficult to extend this approach to higher-order plants, and the highly intricate nature of the approach tends to obscure the insight much sought after.

From a rather different perspective, we analyze the open-loop frequency response of a delay system. The idea is to identify the critical frequency where a stable delay-free closed-loop system loses its stability due to time delay. We consider a number of first- and second-order delay plants, with one or two unstable poles, correspondingly. In each case we derive exact delay margin and its bounds, which in turn avail two sets of results, dependent on the PID coefficients or independent of the controller, respectively. As such, the former set of results provide useful guidelines in tuning PID parameters and in designing analytically a PID controller, while the latter give the intrinsic limits in robustly stabilizing delay plants by a PID controller. More specifically, for systems with one unstable pole, we find analytical expressions of the delay margin, while for systems with more than one unstable pole, we find that the delay margin can be computed by finding the maximum of an explicit function of two real variables. The results consequently reduce considerably the complexity in determining the exact margin, and make it possible to numerically compute the margin; otherwise, the computation can be intensive whether by a brute-force search or using the Hermite-Biehler theorem. Indeed, in the latter vein, even with a fixed delay, finding a set of stabilizing PID coefficients is by no means an easy task. Moreover, the upper bounds provide a priori the limits independent of design, which, other than their accuracy, shed much insight desired; for example, the results show how the delay margin may be constrained by the plant unstable poles and nonminimum phase zeros. Note also that in using PID control, it is prevalent to consider low-order plants, and in fact, in many instances first-order plants, partly because industrial processes are often modeled by first- and second-order systems, partly due to the limitation of PID controllers in controlling high-order dynamics; indeed, in general it is not guaranteed that PID controllers can stabilize even a high-order system free of delay, lest that the system may contain delays. Thus, in practice it appears adequate and non-restrictive to consider systems up to the second-order.

Our contribution can be summarized as follows. In Sect. 2, we introduce the delay margin problem and its variant defined with PID controllers. In Sect. 3, we consider first-order unstable delay plants, for which we derive exact expressions of the delay margin with fixed PID controller coefficients and accordingly the maximal delay margin achievable by optimizing the PID controller. The result coincides with that of [18], and interestingly, that of [10, 12] in a more general setting, thus showing that in robustly stabilizing a first-order delay plant, a PID controller is in fact among the optimal; indeed, further to this point, our constructive proof reveals that it suffices to employ a PD controller only. The development, which arguably appears much simpler than those based on the Hermite-Biehler theorem, also yields useful insight: it shows that the delay margin is achieved mainly by the action of derivative control, while the integral control does no more than to undermine robust stabilization, a conflict seen as the tradeoff between control performance and robustness consistent with long-held design heuristics. Yet further, this insight and technical simplicity carries us forward to Sect. 4, where we derive exact expressions and bounds for several second-order unstable plants. Systems with two or more unstable poles pose a significant technical difficulty; as we noted above, no exact result has been found for such plants with general LTI controllers or PID controllers. We demonstrate, however, that the delay margin can be computed exactly by searching for the maximum of an explicit function of two variables. Other than the exact results, the bounds obtained herein are significantly tighter than those known, e.g., in [6, 12]. Numerical examples are provided in Sect. 5, and the chapter concludes in Sect. 6.

Due to space consideration we shall omit all proofs except that of Theorem 1. The latter serves to illustrate the essential flavor of our technical development throughout the entire paper.

2 Preliminaries and Problem Formulation

We consider the feedback system depicted in Fig. 1, in which $P_\tau(s)$ denotes a plant to be controlled subject to an unknown delay τ, whose transfer function is given by

$$P_\tau(s) = P_0(s)e^{-\tau s}, \quad \tau \geq 0, \tag{1}$$

where $P_0(s)$ is a finite dimensional delay-free plant. Suppose that $P_0(s)$ can be stabilized by a certain finite-dimensional LTI controller $K(s)$. How large may τ be, before the closed-loop system loses its stability? This question concerns the computation of the so-called delay margin, defined as

$$\bar{\tau} = \sup\{\mu \geq 0 : \text{ There exists some } K(s) \text{ that stablizes } P_\tau(s),\ \forall \tau \in [0, \mu)\}.$$

Stated in words, the problem is to determine the largest delay range within which $P_\tau(s)$ can be robustly stabilized by a finite-dimensional LTI controller $K(s)$. The delay margin problem has been under scrutiny for some time. In [12], the exact delay margin was found for LTI plants containing one unstable pole and one nonminimum phase zero, achievable by general LTI controllers of a possible high-order.

Of particular interest in this chapter is the delay margin achievable by LTI controllers of a specific structure, that is, those of the PID structure:

$$K_{PID}(s) = k_p + \frac{k_i}{s} + k_d s. \tag{2}$$

In other words, we are concerned with

$$\bar{\tau}_{PID} = \sup\{\mu \geq 0 : \text{ There exists some } K_{PID}(s) \text{ that stablizes } P_\tau(s),\ \forall \tau \in [\,0, \mu)\}.$$

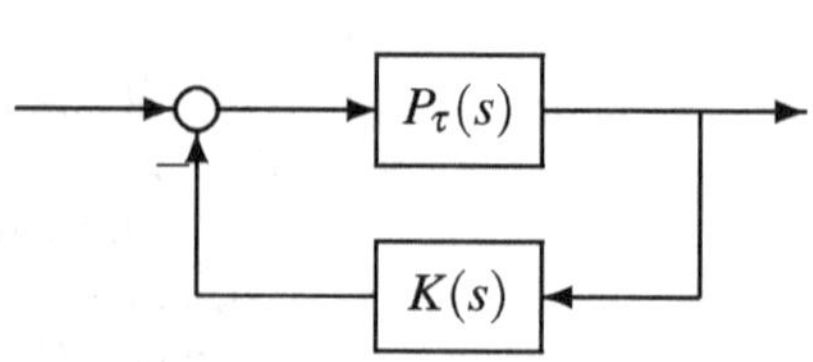

Fig. 1 Feedback control of a delay plant $P_\tau(s)$

This amounts to determining the largest range of delay tolerable by a PID controller. Also of interest is the delay margin achievable by a specific PID controller, with a fixed set of the controller parameters k_p, k_i, and k_d. We define the latter by

$$\bar{\tau}_{PID}(K_{PID}) = \sup\{\mu \geq 0 : K_{PID}(s) \text{ stabilizes } P_\tau(s), \ \ \forall \tau \in [\,0, \mu)\}.$$

Evidently,

$$\bar{\tau}_{PID} = \sup\{\bar{\tau}_{PID}(K_{PID}) : \ K_{PID}(s) \text{ stabilizes } P_\tau(s), \ \forall \tau \in [\,0, \mu)\}.$$

In designing PID controllers, it is rather customary to consider low-order, and in fact first-order systems. This is consistent with control of industrial processes, where the process dynamics is often modeled by a first-order plant. Otherwise, one attempts to control the dominant modes of higher-order systems. It is worth noting that for a stable plant, one can trivially stabilize it robustly independent of delay; that is, the delay margin is infinite.[1] For this reason, we follow the suit by considering first- and second-order unstable plants. Consider the open-loop transfer function

$$L_0(s) = P_0(s) K_{PID}(s). \tag{3}$$

Throughout this chapter, we impose the following assumption.

Assumption 1 (i) $|L_0(0)| > 1$,(ii) $|L_0(\infty)| < 1$.

We note that the assumption (i) is necessary for a system to achieve disturbance attenuation, while the assumption (ii) simply means that the open-loop gain must roll off at high frequencies. Both are required of any realistic control design; indeed, it is typically required that $|L_0(j\omega)| >> 1$ at low frequencies and $|L_0(j\omega)| << 1$ at high frequencies.

In the following lemma we collect some useful properties of the $\tan^{-1}$ function, which can either be easily verified or found from, e.g., [2, 3]. The inequality in Lemma 1 (iii) is referred to as the *Shafer inequality* [3].

Lemma 1 *Suppose that $\xi \geq 0$, $\eta \geq 0$. Then,*

(i) $\tan^{-1}\xi$ *is monotonically increasing with* ξ, *and* $\tan^{-1}\xi/\xi$ *is monotonically decreasing with* ξ.

(ii)

$$\frac{\xi}{1+\xi^2} \leq \tan^{-1}\xi \leq \min\{\xi, \ \pi/2\}.$$

(iii)

$$\tan^{-1}\xi \leq \frac{\pi\xi}{\pi - 2 + 2\sqrt{1+\xi^2}}.$$

[1] It should be noted, nonetheless, that the tuning of PID controllers remains a delicate task even for stable plants. The stabilization of stable delay plants, while not a subject of our present study, is extensively discussed in, e.g., [18, 19].

(iv)

$$\left|\tan^{-1}\xi - \tan^{-1}\eta\right| \leq |\xi - \eta| .$$

(v)

$$\tan^{-1}\xi + \tan^{-1}\eta = \begin{cases} \tan^{-1}\frac{\xi+\eta}{1-\xi\eta}, & \xi\eta < 1, \\ \tan^{-1}\frac{\xi+\eta}{1-\xi\eta} + \pi. & \xi\eta > 1. \end{cases}$$

3 First-Order Unstable Systems

In Silva et al. [18], the first-order unstable delay plant

$$P_\tau(s) = \frac{1}{s-p}e^{-\tau s}, \quad p > 0 \tag{4}$$

was considered, where it was found that

$$\bar{\tau}_{PID} = \frac{2}{p}. \tag{5}$$

Independently, this bound was also found in [10] as the maximum delay range within which the system (4) can be stabilized by an observer-based delay feedback. Furthermore, in light of [12], it is clear at once that for this system

$$\bar{\tau} = \bar{\tau}_{PID}.$$

In other words, for a first-order unstable plant, PID controllers are in fact among the optimal to achieve the largest possible delay margin. In this section, we provide an alternative, constructive proof for this result, which, unlike the elaborate analysis in [18] based on the Hermite-Bieler stability criterion, appears conceptually simple. The proof also lends an launching pad toward our investigation of more sophisticated plants of a higher order. Furthermore, the approach leads to the exact margin $\bar{\tau}_{PID}(K_{PID})$ for any given stabilizing PID controller $K_{PID}(s)$, and shows how in the limit it may be constructed to achieve asymptotically the delay margin $\bar{\tau}_{PID}$.

Theorem 1 *Let $P_\tau(s)$ be given by (4). Then for $K_{PID}(s)$ to stabilize $P_0(s)$ and satisfy Assumption 1, it is necessary and sufficient that $k_p > p$, $k_i > 0$, and $|k_d| < 1$. Under these conditions,*

$$\bar{\tau}_{PID}(K_{PID}) = \frac{\tan^{-1}\frac{\omega_0}{p}}{\omega_0} + \frac{\tan^{-1}\frac{k_d\omega_0 - \frac{k_i}{\omega_0}}{k_p}}{\omega_0}, \tag{6}$$

where $\omega_0 > 0$ is given by

$$\omega_0^2 = \frac{\frac{k_p{}^2 - 2k_d k_i - p^2}{1-k_d{}^2} + \sqrt{(\frac{k_p{}^2 - 2k_d k_i - p^2}{1-k_d{}^2})^2 + 4\frac{k_i{}^2}{1-k_d{}^2}}}{2}.$$

Furthermore,

$$\bar{\tau}_{PID}(K_{PID}) \le \frac{1}{p} + \frac{k_d}{k_p}, \tag{7}$$

$$\bar{\tau}_{PID} = \frac{2}{p}. \tag{8}$$

Proof We first determine the range of the triplet $(k_p,\ k_i,\ k_d)$ so that $K_{PID}(s)$ stabilizes $P_0(s)$. This concerns the closed-loop characteristic equation

$$(1 + k_d)s^2 + (k_p - p)s + k_i = 0.$$

It follows from the Routh-Hurwitz criterion that for this second-order polynomial to be stable, it is both necessary and sufficient that all the coefficients have the same sign, i.e.,

$$(i) \begin{cases} k_p > p \\ k_i > 0 \\ k_d > -1, \end{cases} \quad \text{or} \quad (ii) \begin{cases} k_p < p \\ k_i < 0 \\ k_d < -1. \end{cases}$$

Note, however, that $|L_0(\infty)| = |k_d|$. In light of Assumption 1, Case (ii) is excluded. Next, we examine the open-loop frequency response

$$L_0(j\omega) = \frac{1}{-p + j\omega}\left(k_p + \frac{k_i}{j\omega} + jk_d\omega\right), \tag{9}$$

which gives rise to

$$|L_0(j\omega)|^2 = \frac{(k_p{}^2 - 2k_d k_i) + k_d{}^2\omega^2 + \frac{k_i^2}{\omega^2}}{\omega^2 + p^2}.$$

Setting $|L_0(j\omega_0)| = 1$ yields the unique solution

$$\omega_0^2 = \frac{\frac{k_p{}^2 - 2k_d k_i - p^2}{1-k_d{}^2} + \sqrt{(\frac{k_p{}^2 - 2k_d k_i - p^2}{1-k_d{}^2})^2 + 4\frac{k_i{}^2}{1-k_d{}^2}}}{2}.$$

In other words, there exists a unique $\omega_0 > 0$ such that $|L_0(j\omega_0)| = 1$. At $\omega = \omega_0$, we have

$$\angle L_0(j\omega) = \pi + \tan^{-1}\frac{\omega_0}{p} + \tan^{-1}\frac{k_d\omega_0 - \frac{k_i}{\omega_0}}{k_p}.$$

Since

$$L_0(j\omega_0) = e^{j\angle L_0(j\omega_0)} = -e^{j(\tan^{-1}\frac{\omega_0}{p} + \tan^{-1}\frac{k_d\omega_0 - \frac{k_i}{\omega_0}}{k_p})},$$

we can match the phase of $L_0(j\omega_0)$ with that of the delay, by setting

$$\tau_0\omega_0 = \tan^{-1}\frac{\omega_0}{p} + \tan^{-1}\frac{k_d\omega_0 - \frac{k_i}{\omega_0}}{k_p}, \tag{10}$$

for some $\tau_0 \geq 0$. Evidently,

$$1 + P_0(j\omega_0)K_{PID}(j\omega_0)e^{-j\tau_0\omega_0} = 0,$$

and for any $\tau < \tau_0$, $1 + P_0(j\omega)K_{PID}(j\omega)e^{-j\tau\omega} \neq 0$ for all $\omega \geq 0$; that is, the system is stable for all $\tau < \tau_0$. Consequently,

$$\bar{\tau}_{PID}(K_{PID}) = \tau_0 = \frac{\tan^{-1}\frac{\omega_0}{p}}{\omega_0} + \frac{\tan^{-1}\frac{k_d\omega_0 - \frac{k_i}{\omega_0}}{k_p}}{\omega_0}.$$

This proves (6). To establish (7), we note that if $k_d\omega_0 \geq \frac{k_i}{\omega_0}$, then from Lemma 1 (ii),

$$\tau_0 \leq \frac{1}{p} + \frac{k_d - \frac{k_i}{\omega_0{}^2}}{k_p} \leq \frac{1}{p} + \frac{k_d}{k_p}.$$

On the other hand, if $k_d\omega_0 < \frac{k_i}{\omega_0}$, then according to Lemma 1 (iv), we have

$$\begin{aligned}
\tau_0 &= \frac{\left|\tan^{-1}\frac{\omega_0}{p} - \tan^{-1}\frac{\frac{k_i}{\omega_0} - k_d\omega_0}{k_p}\right|}{\omega_0} \\
&\leq \frac{\left|\frac{\omega_0}{p} - \frac{\frac{k_i}{\omega_0} - k_d\omega_0}{k_p}\right|}{\omega_0} \\
&= \frac{1}{p} + \frac{k_d}{k_p} - \frac{k_i}{k_p}\frac{1}{\omega_0{}^2} \\
&\leq \frac{1}{p} + \frac{k_d}{k_p}.
\end{aligned}$$

Thus, (7) is proved. Finally, to establish (8), we note that for the ranges of $|k_d| < 1$ and $k_p > p$,

$$\bar{\tau}_{PID} \leq \frac{1}{p} + \frac{1}{p} = \frac{2}{p}. \tag{11}$$

The upper bound in (11), however, can be achieved asymptotically by selecting

$$\begin{cases} k_d = 1 - \varepsilon, \\ k_p = p + \varepsilon^2, \\ k_i = \varepsilon^3, \end{cases} \tag{12}$$

for a sufficiently small $\varepsilon > 0$. Indeed, with k_d, k_p and k_i given in (12), we have $\omega_0^2 = p\varepsilon + o(\varepsilon) \to 0$ as $\varepsilon \to 0$. Consequently, $\bar{\tau}_{PID}(K_{PID}) \to 2/p$. This proves (8), and hence completes the proof. ■

Remark 1 From the above construction of PID gains (k_p, k_i, k_d), it is clear that only k_d and k_p contribute to the robust stabilization of $P_\tau(s)$ for $\tau \in [0, \; \bar{\tau}_{PID})$, while k_i is immaterial. On the contrary, it is clear from the proof of Theorem 1 that k_i or alternatively, the integral control tends to make the delay margin smaller. This is consistent with one's intuition, since integral control is generally introduced to achieve performance objectives such as asymptotic tracking, which necessarily is in conflict with and is therefore seen as a tradeoff to stabilization. Indeed, it is not difficult to see that the delay margin in (8) can be achieved solely by the use of a PD controller. To this end, we note that with $k_i = 0$, the closed-loop characteristic equation becomes $(1 + k_d)s + (k_p - p) = 0$. The crossover frequency $\omega_0 \to 0$ when $k_p \to p$. In light of Lemma 1 (i), both $\left(\tan^{-1}(\omega_0/p)\right)/\omega_0$ and $\left(\tan^{-1}(k_d\omega_0/k_p)\right)/\omega_0$ are monotonically decreasing functions of ω_0, which implies that

$$\bar{\tau}_{PID} = \lim_{\substack{k_p \to p, \\ k_d \to 1}} \frac{\tan^{-1}\frac{\omega_0}{p}}{\omega_0} + \frac{\tan^{-1}\frac{k_d\omega_0}{k_p}}{\omega_0} = \frac{2}{p}.$$

■

On the other hand, when in the absence of derivative control, the delay margin is reduced. We summarize this observation in the following corollary, which states that with PI control or static feedback alone, the delay margin reduces to $1/p$. Note that the case of static feedback was established previously in [7, 10, 11].

Corollary 1 *Let $P_\tau(s)$ be given by (4), and $k_d = 0$. Then*

$$\bar{\tau}_{PID} = \frac{1}{p}. \tag{13}$$

In light of Remark 1 and Corollary 1, we assert that PD controllers are essential for stabilization, while the integral control will invariably reduce the delay margin. Hence, throughout the rest of the chapter, we shall consider PD controllers only.

Remark 2 In practice, it is typical to implement the derivative control in conjunction with a low-pass filter [1, 22], so that the PID controller possesses the form of

$$K_{PID}^{T_f}(s) = k_p + \frac{k_i}{s} + \frac{k_d s}{1 + T_f s}, \tag{14}$$

where $T_f > 0$. By mimicking the above derivation, it is possible to estimate the delay margin achievable by $K_{PID}^{T_f}(s)$ as well, which helps demonstrate the tradeoff between the achievable margin and the practical implementation. The following corollary is presented to this effect. The result shows that the inclusion of the filter will reduce the delay margin achievable; intuitively, the filter will reduce the bandwidth of the derivative control and hence the reduction will depend on the time constant T_f relative to the unstable pole p. The effect, however, can be made negligible by selecting T_f such that $T_f p << 1$. In particular, when $T_f \to 0$, the margin approaches the ideal $\bar{\tau}_{PID}$, while with $T_f \to \infty$, it reduces to that achievable by proportional feedback. ∎

Corollary 2 *Let $P_\tau(s)$ be given by (4), and $k_i = 0$. Denote by $\bar{\tau}_{PID}(K_{PID}^{T_f})$ the delay margin achievable by the controller $K_{PID}^{T_f}(s)$. Then for $K_{PID}^{T_f}(s)$ to stabilize $P_0(s)$, it is necessary and sufficient that $k_p > p$ and $k_d + T_f(k_p - p) + 1 > 0$. Under these conditions,*

$$\bar{\tau}_{PID}(K_{PID}^{T_f}) = \frac{\tan^{-1}\frac{\omega_0}{p}}{\omega_0} + \frac{\tan^{-1}\frac{\frac{k_d}{k_p}\omega_0}{1+T_f\left(T_f+\frac{k_d}{k_p}\right)\omega_0^2}}{\omega_0}, \tag{15}$$

where $\omega_0 > 0$ is given by

$$\omega_0^2 = \frac{(k_d + k_p T_f)^2 - (1 + T_f^2 p^2)}{2T_f^2} + \frac{\sqrt{\left[(k_d + k_p T_f)^2 - (1 + T_f^2 p^2)\right]^2 + 4T_f^2(k_p^2 - p^2)}}{2T_f^2}.$$

With k_p and k_d given by (12), where $\varepsilon \to 0$,

$$\bar{\tau}_{PID}(K_{PID}^{T_f}) \leq \frac{1}{p}\left(1 + \frac{1}{1 + 2T_f p}\right). \tag{16}$$

We conclude this section with a discussion on possible extensions to high-order systems. At the outset, we note that it is generally difficult, if ever possible, to stabilize high-order delay systems using PID control. Indeed, a simple argument based on the root locus or Routh-Hurwitz criterion reveals that a delay-free triple integrator system, or one with three real unstable poles, can in no event be stabilized by PID control, lest that the system may contain delays. Notwithstanding this fundamental limitation of PID control, it remains possible to stabilize high-order systems by augmented, PID-based controllers. To illustrate, consider, for example, the plant

$$P_\tau(s) = \frac{P_m(s)}{s-p}e^{-\tau s}, \tag{17}$$

where $P_m(s)$ is a stable and minimum phase transfer function. By employing the controller

$$K(s) = \frac{P_m^{-1}(s)}{(1+\delta s)^n}K_{PID}(s),$$

where n is an integer so that $K(s)$ remains proper, one can readily show that with a sufficiently small $\delta > 0$,

$$\bar{\tau} = \frac{2}{p}.$$

In other words, using an appropriately augmented PID controller, one can still attain the delay margin, despite that the plant may be of a high order and contain an unstable pole.

4 Second-Order Unstable Plants

In this section we study second-order unstable systems. Delay margin for second-order delay systems generally presents more difficult problems, which has barely been addressed in the previous work. At present, no exact delay margin has been found for second-order plants, either with general LTI controllers or PID controllers.

4.1 Real Poles

We first consider plants that contain a pair of distinct real unstable poles p_1 and p_2:

$$P_\tau(s) = \frac{1}{(s-p_1)(s-p_2)}e^{-\tau s}, \quad p_1 > 0, \quad p_2 > 0. \tag{18}$$

Both the exact delay margin and explicit bounds are given in the following theorem.

Theorem 2 *Let $P_\tau(s)$ be given by (18), and $k_i = 0$. Then for $K_{PID}(s)$ to stabilize $P_0(s)$ and satisfy Assumption 1, it is necessary and sufficient that $k_p > p_1 p_2$ and $k_d > p_1 + p_2$. Under these conditions,*

$$\bar{\tau}_{PID}(K_{PID}) = \frac{\tan^{-1}\frac{\omega_0}{p_1}}{\omega_0} + \frac{\tan^{-1}\frac{\omega_0}{p_2}}{\omega_0} + \frac{\tan^{-1}\frac{k_d\omega_0}{k_p}}{\omega_0} - \frac{\pi}{\omega_0}, \tag{19}$$

where $\omega_0 > 0$ *is given by*

$$\omega_0{}^2 = \frac{k_d{}^2 - (p_1^2 + p_2^2)}{2} + \frac{\sqrt{\left(k_d{}^2 - (p_1^2 + p_2^2)\right)^2 + 4(k_p^2 - p_1^2 p_2^2)}}{2}. \tag{20}$$

Furthermore,

$$\bar{\tau}_{PID}(K_{PID}) \leq \frac{k_d}{k_p}, \tag{21}$$

$$\bar{\tau}_{PID} = \sup\left\{\bar{\tau}_{PID}(K_{PID}) : \, k_p > p_1 p_2, \, k_d > p_1 + p_2\right\}, \tag{22}$$

and

$$\bar{\tau}_{PID} \leq \frac{\tan^{-1}\left(\sqrt{\frac{2p_2}{p_1}} + \sqrt{\frac{2p_1}{p_2}}\right)}{\sqrt{2p_1 p_2}}. \tag{23}$$

For a given $P_0(s)$, the exact expression (22) in Theorem 2 shows that the delay margin $\bar{\tau}_{PID}$ can be found by a search of maximum over two variables k_p and k_d, with $\bar{\tau}_{PID}(K_{PID})$ given explicitly as a function of k_p and k_d in terms of (19) and (20). This result thus simplifies considerably the complexity in determining the exact delay margin than done by a brute-force method; in the latter vein, one should note that even with a fixed delay, finding a set of stabilizing PID coefficients is a nontrivial task. In contrast, the explicit bound in (23) gives a fundamental limit showing how the delay margin may be constrained by the unstable poles.

Remark 3 In the earlier work [6], allowing the use of general LTI controllers, it was shown that the delay margin $\bar{\tau}$ satisfies the inequality

$$\bar{\tau} \leq \frac{2\left(\frac{1}{p_1} + \frac{1}{p_2}\right)}{1 + \frac{p_1}{p_2} + \frac{p_2}{p_1}}. \tag{24}$$

From Lemma 1 (iii), we have

$$\begin{aligned}
\tan^{-1}\left(\sqrt{\frac{2p_2}{p_1}} + \sqrt{\frac{2p_1}{p_2}}\right) &\leq \frac{\pi\left(\sqrt{\frac{2p_2}{p_1}} + \sqrt{\frac{2p_1}{p_2}}\right)}{\pi - 2 + 2\sqrt{5 + \frac{2p_2}{p_1} + \frac{2p_1}{p_2}}} \\
&\leq \frac{\pi\left(\sqrt{\frac{2p_2}{p_1}} + \sqrt{\frac{2p_1}{p_2}}\right)}{\pi + 4},
\end{aligned}$$

where the latter inequality follows from the fact that $\frac{p_2}{p_1} + \frac{p_1}{p_2} \geq 2$. It then follows from (23) that

$$\bar{\tau}_{PID} \leq \frac{\pi}{\pi + 4} \left(\frac{1}{p_1} + \frac{1}{p_2} \right),$$

which can be shown to improve (24) whenever

$$\frac{p_2}{p_1} + \frac{p_1}{p_2} \leq 1 + \frac{8}{\pi}.$$

Remark 4 It can be shown analogously that the inclusion of the integral term, i.e., $k_i \neq 0$, will lead to the reduction of the delay margin, thus reaffirming the consideration of PD control only. It is also easy to see from (20) that at $k_d = p_1 + p_2$, $\bar{\tau}_{PID}(K_{PID}) = 0$. This implies that the delay margin will never be reached at the boundary of the PD parameter space, and it highlights again the critical role of the derivative control in stabilization. On the other hand, by setting $k_p = p_1 p_2$, we obtain a lower bound on the delay margin, which can be computed by a line search method. This observation is given in the following corollary. ■

Corollary 3 *Let $P_\tau(s)$ be given by (18), $k_i = 0$ and define*

$$\underline{g}(k_d) = \frac{\tan^{-1} \frac{k_d \sqrt{k_d^2 - (p_1^2 + p_2^2)}}{p_1 p_2} - \tan^{-1} \frac{(p_1 + p_2)\sqrt{k_d^2 - (p_1^2 + p_2^2)}}{k_d^2 - (p_1^2 + p_2^2 + p_1 p_2)}}{\sqrt{k_d^2 - (p_1^2 + p_2^2)}}.$$

Then,

$$\bar{\tau}_{PID} \geq \sup \left\{ \underline{g}(k_d) : k_d > p_1 + p_2 \right\}. \tag{25}$$

In a limiting case, it is of interest to consider plants with double unstable poles, that is,

$$P_\tau(s) = \frac{1}{(s - p)^2} e^{-\tau s}, \quad p > 0. \tag{26}$$

The result given below follows as a simple corollary of Theorem 2.

Corollary 4 *Let $P_\tau(s)$ be given by (26), and $k_i = 0$. Then for $K_{PID}(s)$ to stabilize $P_0(s)$ and satisfy Assumption 1, it is necessary and sufficient that $k_p > p^2$ and $k_d > 2p$. Under these conditions,*

$$\bar{\tau}_{PID}(K_{PID}) = \frac{2 \tan^{-1} \frac{\omega_0}{p}}{\omega_0} + \frac{\tan^{-1} \frac{k_d \omega_0}{k_p}}{\omega_0} - \frac{\pi}{\omega_0}, \tag{27}$$

where $\omega_0 > 0$ is given by

$$\omega_0^2 = \frac{k_d{}^2 - 2p^2 + \sqrt{(k_d{}^2 - 2p^2)^2 + 4(k_p{}^2 - p^4)}}{2}. \tag{28}$$

Furthermore,

$$\bar{\tau}_{PID}(K_{PID}) \le \frac{k_d}{k_p}, \tag{29}$$

$$\bar{\tau}_{PID} = \sup\left\{\bar{\tau}_{PID}(K_{PID}) : \, k_p > p^2, \, k_d > 2p\right\}, \tag{30}$$

and

$$\bar{\tau}_{PID} \le \left(\frac{\tan^{-1}(2\sqrt{2})}{\sqrt{2}}\right)\frac{1}{p}. \tag{31}$$

Remark 5 In [6], with general LTI controllers, an explicit upper bound on the delay margin of the plant (26) was found as

$$\bar{\tau} \le \left(\frac{4}{3}\right)\frac{1}{p}.$$

The bound in (31) is approximately equal to $0.87/p$, which is significantly less than, and certainly consistent with that in [6]. ■

Also of interest is the case when the plant contains an integrator, in addition to an unstable pole. The preceding results (Theorem 1, Corollary 4) indicate that in the presence of a pure integrator, the delay margin is infinite, seemingly pointing to the contention that an integrator has no effect on the delay margin. The following result, however, shows otherwise, that together with an unstable pole, the presence of an integrator will reduce the delay margin.

Corollary 5 *Let*

$$P_\tau(s) = \frac{1}{s(s-p)}e^{-\tau s}, \qquad p > 0.$$

Then,

$$\bar{\tau}_{PID} = \frac{1}{p}. \tag{32}$$

In comparison to Theorems 1, 2 and Corollary 4 both point to the fact that it is considerably more difficult to robustly stabilize a delay plant with more than one unstable pole, a difficulty that may manifest itself through the technical difficulty as well in obtaining an explicit expression of the exact delay margin. It further exhibits the limitation of LTI controllers in robustly stabilizing systems of a higher order.

4.2 Complex Conjugate Poles

A second-order delay plant with a pair of complex conjugate poles is described by

$$P_\tau(s) = \frac{1}{(s-p)(s-\bar{p})} e^{-\tau s}, \tag{33}$$

where $p = \alpha + j\beta$, $\mathrm{Re}(p) = \alpha > 0$, and $\bar{p}$ denotes the complex conjugate of p. As stipulated in the preceding section, we consider as well a PD controller given by $K_{PID}(s) = k_p + k_d s$. Likewise, we show below that the exact delay margin can be determined by finding the maximum of a function of two variables, and an explicit upper bound can be established.

Theorem 3 *Let $P_\tau(s)$ be given by (33), and $k_i = 0$. Then for $K_{PID}(s)$ to stabilize $P_0(s)$ and satisfy Assumption 1, it is necessary and sufficient that $k_p > p^2$ and $k_d > 2\alpha$. Under these conditions,*

$$\bar{\tau}_{PID}(K_{PID}) = \frac{\tan^{-1}\frac{\omega_0-\beta}{\alpha}}{\omega_0} + \frac{\tan^{-1}\frac{\omega_0+\beta}{\alpha}}{\omega_0} + \frac{\tan^{-1}\frac{k_d\omega_0}{k_p}}{\omega_0} - \frac{\pi}{\omega_0}, \tag{34}$$

where $\omega_0 > 0$ is given by

$$\omega_0^2 = \frac{k_d^2 + 2\beta^2 - 2\alpha^2}{2} + \frac{\sqrt{(k_d^2 + 2\beta^2 - 2\alpha^2)^2 + 4(k_p^2 - |p|^4)}}{2}. \tag{35}$$

Furthermore,

$$\bar{\tau}_{PID}(K_{PID}) \le \frac{k_d}{k_p}, \tag{36}$$

$$\bar{\tau}_{PID} = \sup\left\{\bar{\tau}_{PID}(K_{PID}) : \, k_p > |p|^2, \, k_d > 2\alpha\right\}, \tag{37}$$

and

$$\bar{\tau}_{PID} \le \begin{cases} \dfrac{\tan^{-1}\frac{2\sqrt{2}\alpha}{|p|}}{\sqrt{2}|p|}, & \text{if } \beta < \alpha. \\ \left(\dfrac{1}{|p|}\right)\dfrac{(\pi/4)}{\sqrt{\frac{\pi}{2} - \left(\frac{\sqrt{2}\alpha}{|p|}\right)^2}}. & \text{if } \beta \ge \alpha. \end{cases} \tag{38}$$

4.3 Oscillatory Poles

Systems with imaginary poles exhibit rather different characteristics in robust stabilization against time delays. As demonstrated in [12], such poles generally make the delay margin larger, compared to their real counterparts with the same magnitudes. This is consistent with the intuition that systems with purely imaginary poles are less difficult to control than those with strictly unstable poles. Consider the plant

$$P_\tau(s) = \frac{1}{s^2 + \omega_c^2} e^{-\tau s}, \qquad \omega_c \geq 0, \tag{39}$$

where $P_0(s)$ contains a pair of imaginary poles at $\pm j\omega_c$. In this case, we give below a more computable expression of the exact delay margin, which reduces the search over two variables to one involving a single variable.

Theorem 4 *Let $P_\tau(s)$ be given by (39), and $k_i = 0$. Then for $K_{PID}(s)$ to stabilize $P_0(s)$ and satisfy Assumption 1, it is necessary and sufficient that $k_p > \omega_c^2$ and $k_d > 0$. Under these conditions,*

$$\bar{\tau}_{PID}(K_{PID}) = \frac{\tan^{-1} \frac{k_d \omega_0}{k_p}}{\omega_0}, \tag{40}$$

where $\omega_0 > 0$ is given by

$$\omega_0^2 = \frac{k_d^2 + 2\omega_c^2 + \sqrt{(k_d^2 + 2\omega_c^2)^2 + 4(k_p^2 - \omega_c^4)}}{2}. \tag{41}$$

Furthermore,

$$\bar{\tau}_{PID}(K_{PID}) \leq \frac{k_d}{k_p}, \tag{42}$$

and

$$\bar{\tau}_{PID} = \frac{1}{\omega_c} \sup_{k_d > 0} \frac{\tan^{-1}(k_d/\omega_c)\sqrt{(k_d/\omega_c)^2 + 2}}{\sqrt{(k_d/\omega_c)^2 + 2}}, \tag{43}$$

$$\left(\frac{\tan^{-1}(\sqrt{7}/3)}{\sqrt{(7/3)}} \right) \frac{1}{\omega_c} \leq \bar{\tau}_{PID} \leq \left(\frac{\sqrt{\pi}}{2\sqrt{2}} \right) \frac{1}{\omega_c}. \tag{44}$$

Finally, $\bar{\tau}_{PID} = \infty$ when $\omega_c = 0$.

It is clear from Theorem 4 that the exact delay margin for systems with one pair of imaginary poles can be computed by employing, e.g., a line search method. The upper bound in (44) is significantly smaller than that found in [12], which concerns the delay margin achievable using general LTI controllers. Theorem 4 also corroborates with earlier results in [20, 24, 25], where it was shown that for systems with unstable poles solely at the origin, an arbitrarily large delay margin can be achieved by finite-dimensional state feedback [24], and that for systems with non-zero imaginary poles, only finite delay margin can be attained [20].

5 Examples

We now use a number of numerical examples to illustrate our results. In all cases, we set $k_i = 0$ in the PID controller $K_{PID}(s)$.

Example 1 We first consider the second-order plant

$$P_\tau(s) = \frac{1}{(s - p_1)(s - p_2)} e^{-\tau s},$$

where p_1 and p_2 are both unstable poles. Different combinations of these poles are examined in the sequel.

Distinct Real Poles: $p_1 > 0,\ p_2 > 0$. In this case, we take three pairs $(p_1,\ p_2) = (0.6,\ 0.8),\ (1,\ 1.2)$, and $(0.4,\ 2)$, taking into account the proximity of the poles from the imaginary axis and their relative locations. Figure 2 shows the 3-D surfaces of $\bar{\tau}_{PID}(K_{PID})$ varying with the parameters k_p and k_d. The computation gives the maximum of $\bar{\tau}_{PID}(K_{PID})$ as 0.3960, 0.2513 and 0.2595, respectively, while the upper bound in (23) yields 1.2596, 0.7954 and 1.0381. In spite of this seemingly significant discrepancy, Fig. 3, which plots the upper bound in (23) versus the available lower and upper bounds achievable by general LTI controllers by fixing $p_1 = 0.2$ and allowing p_2 to vary, demonstrates that for an appreciable range of the unstable poles, the upper bound in (23) remains to offer a sizable improvement.

Complex Conjugate Poles: $p_1 = p = \alpha + j\beta,\ p_2 = \bar{p},\ \alpha > 0$. Figure 4 shows the 3-D surfaces of $\bar{\tau}_{PID}(K_{PID})$ for three different pairs of complex conjugate poles with $p = 24 + j3.2,\ 0.006 + j0.8,\ 1 + j$, which exhibit a similar trait as in the case of distinct real poles. The computation gives the maximum of $\bar{\tau}_{PID}(K_{PID})$, that is, the exact delay margin $\bar{\tau}_{PID}$, as 0.0115, 0.7675 and 0.2329, respectively. The upper bound in (38), instead, is found to be 0.0359, 0.7833 and 0.7351. It is of interest to note that the first two cases correspond to the conditions $\alpha >> \beta$ and $\alpha << \beta$, respectively, and both are consistent with the assertions that Corollary 4 and Theorem 4 provide improved upper bounds on $\bar{\tau}_{PID}$ under the corresponding circumstances.

Oscillatory Poles: $p_1 = j\omega_c,\ p_2 = -j\omega_c,\ \omega_c > 0$. From the proof of Theorem 4, we know that $\bar{\tau}_{PID}(K_{PID})$ is monotonically decreasing with $k_p > \omega_c^2$. Figure 5 shows the 3-D surfaces of $\bar{\tau}_{PID}(K_{PID})$ varying with the parameters k_p and k_d. The analytical expression of the exact delay margin in (43) is a smooth function of the variable k_d, whose maximum can be found using a line search method. Figure 6 shows the univariate function in (43) and the upper bound in (44). The maximum in these cases is found to be 0.6198, 0.3099 and 0.1550, while the upper bound is given by 0.6267, 0.3133 and 0.1567, demonstrating the striking accuracy of the bound.

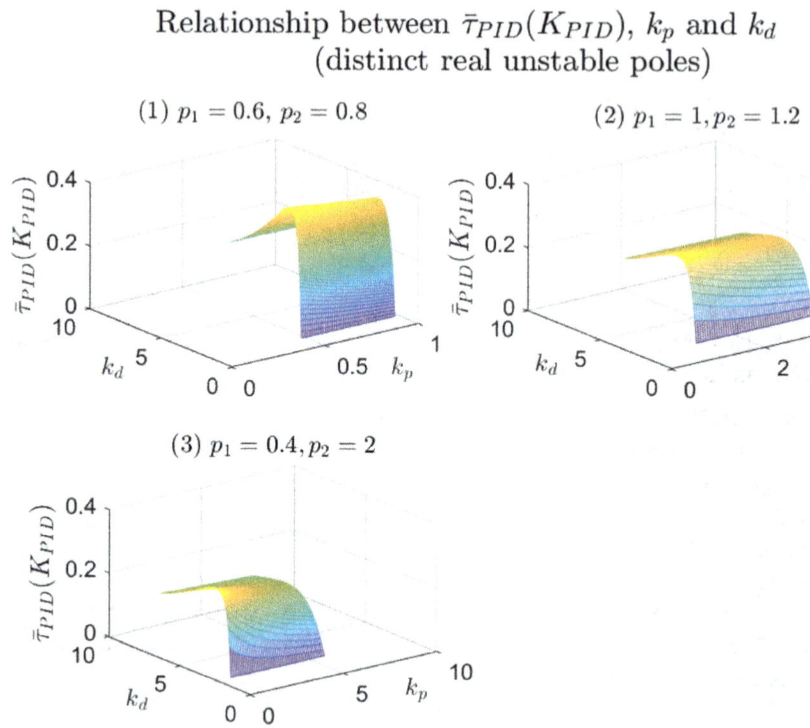

Fig. 2 3-D surface of the delay margin varying with PD parameters: Distinct real unstable poles

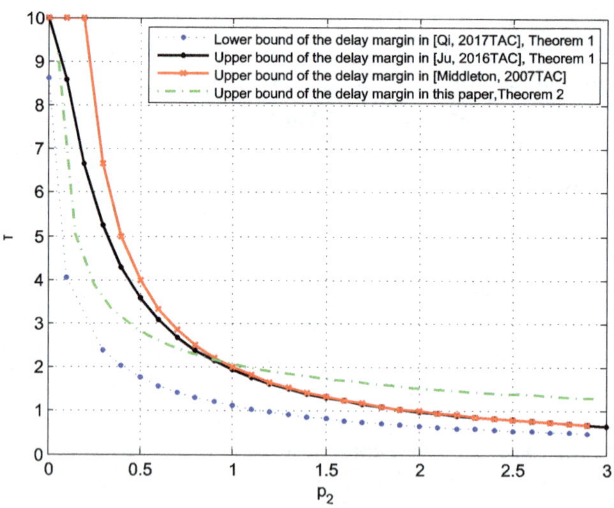

Fig. 3 Comparison of bounds: PID controller versus general LTI controllers

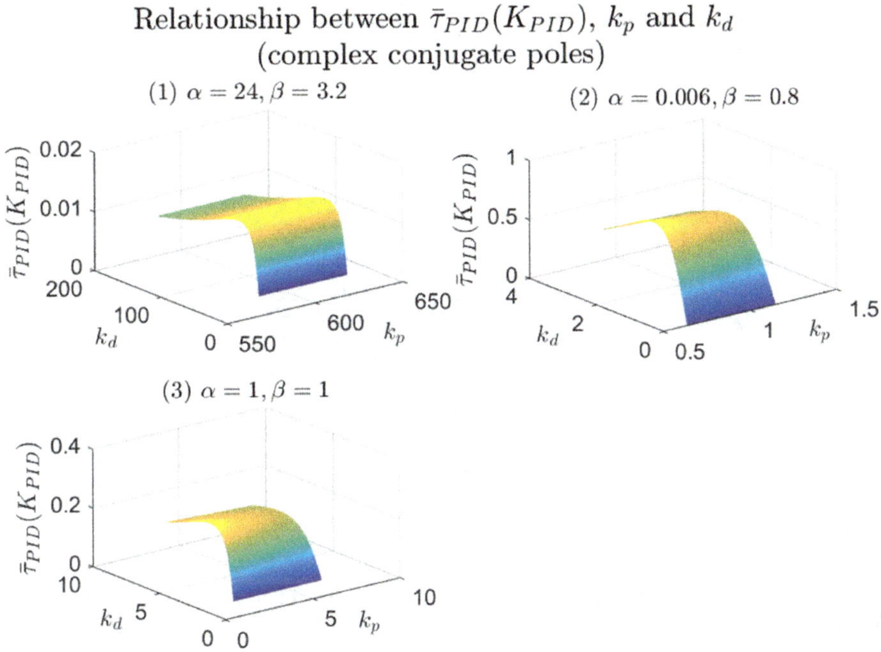

Fig. 4 3-D surface of the delay margin varying with PD parameters: complex conjugate poles

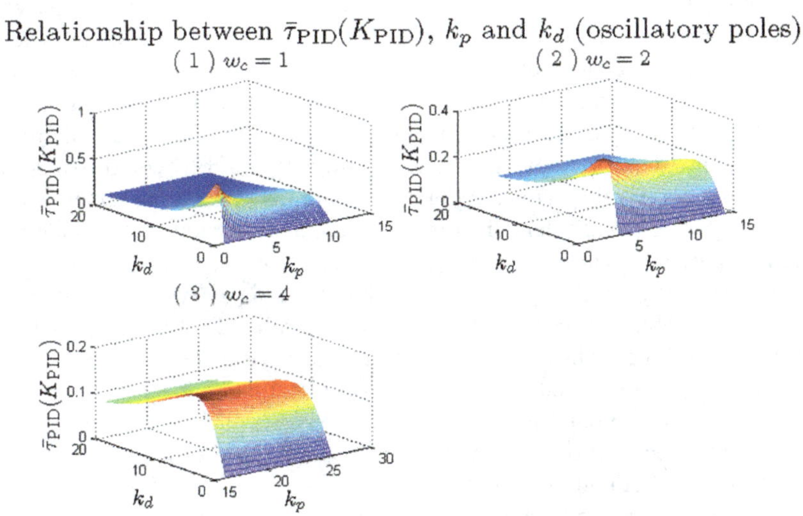

Fig. 5 3-D surface of the delay margin varying with PD parameters: Oscillatory poles

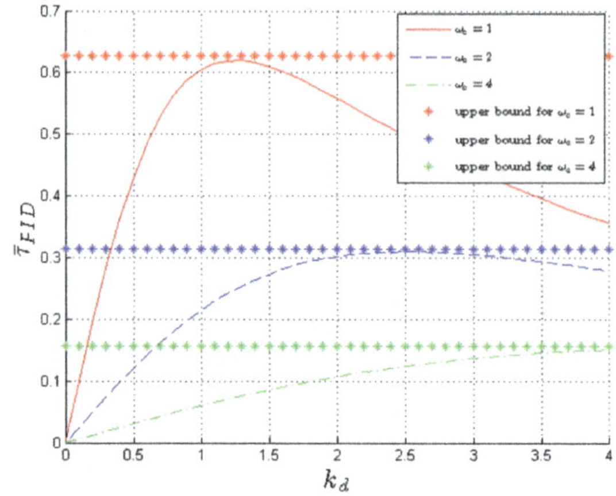

Fig. 6 Delay margin varying with k_d: Oscillatory poles

6 Conclusion

In this chapter we have studied the delay margin of LTI delay systems achievable using PID controllers. We derived the exact delay margin and its bounds for first- and second-order delay plants, both dependent on and independent of PID controller parameters. Other than the exact results, the bounds obtained herein are significantly tighter than those known; in other words, when using PID control, the earlier bounds obtained elsewhere will be overly pessimistic compared to the ones derived herein. From both a conceptual and technical standpoint, it appears, arguably, that our analysis is much simpler than those based on the Hermite-Biehler theorem for quasipolynomial analysis, and sheds much light into the limit of robust stabilization of unstable time-delay systems, as well as the limitation of PID controllers in stabilizing such plants. While giving the fundamental limits in the robust stabilization of delay plants by PID controllers, the results should also provide useful guidelines in tuning PID parameters and in the analytical design of PID controllers.

Notwithstanding the ability in stabilizing low-order delay systems, with only two parameters available for design of a stabilizing controller, PID control is inherently limited in controlling high-order systems. While we have analyzed cases where such an extension is possible, fundamental challenges remain. For this purpose, it appears more advantageous to employ more sophisticated control design methods such as those based on the $\mathcal{H}_\infty$ optimal control in [15] and those based on the predictor feedback in [20, 25], though each of the latter methods suffers from its own limitations and expenses. Finally, while in this chapter we have focused exclusively on unstable delay systems, the design of PID controllers for stable delay plants remain to be of practical interest, where likely the tuning of the PID coefficients may override the pursuit of a larger delay margin; indeed, sheer for maximizing the range of stabilization, the delay margin for a stable plant can be made arbitrarily large.

References

1. K. Aström, T. Hagglund, *PID Controllers: Theory, Design, and Tuning* (2nd edn., ISA, 1995)
2. G. Boros, V.H. Moll, Sums of arctangents and some formulas of Ramanujan. SCIENTIA Ser. A: Math. Sci. **11**, 13–24 (2005)
3. C. Chen, W. Cheung, W. Wang, On Shafer and Carlson inequalities. J. Inequal. Appl. **2011**, 1–11 (2011)
4. B.P. Delphine, C. Jonathan, P. Nicolas, Adaptive control scheme for uncertain time-delay systems. Automatica **48**, 1536–1552 (2012)
5. D.L. Gaudette, D.E. Miller, Adaptive stabilization of a class of time-varying systems with an uncertain delay. Math. Control Signals Syst. **28**(3) (2016)
6. P. Ju, H. Zhang, Further results on the achievable delay margin using LTI control. IEEE Trans. Autom. Control **61**(10), 3134–3139 (2016)
7. V.L. Kharitonov, S.I. Niculescu, J. Moreno, W. Michiels, Static output feedback stabilization: necessary conditions for multiple delay controllers. IEEE Trans. Autom. Control **50**(1), 82–86 (2005)
8. N.B. Liberis, M. Krstic, *Nonlinear Control Under Nonconstant Delays* (SIAM Press, Philadelphia, PA, 2013)
9. U. Münz, A. Papachristodoulou, F. Allgöwer, Delay robustness in consensus problems. Automatica **46**(8), 1252–1265 (2010)
10. W. Michels, K. Engelborghs, P. Vansevenant, D. Roose, Continuous pole placement for delay equations. Automatica **38**, 747–761 (2002)
11. W. Michiels, S.I. Niculescu, *Stability and Stabilization of Time-Delay Systems: An Eigenvalue-Based Approach* (SIAM Press, Philadelphia, PA, 2007)
12. R.H. Middleton, D.E. Miller, On the achievable delay margin using lti control for unstable plants. IEEE Trans. Autom. Control **52**, 1194–1207 (2007)
13. D.E. Miller, D.E. Davison, Stabilization in the presence of an uncertain arbitrarily large delay. IEEE Trans. Autom. Control **50**(8), 1074–1089 (2005)
14. T. Qi, L. Qiu, J. Chen, Multi-agent consensus under delayed feedback: fundamental constraint on graph and fundamental bound on delay, in *Proceedings of 2013 American Control Conference*, Washington D.C., USA, June 2013, pp. 952–957
15. T. Qi, J. Zhu, J. Chen, Fundamental limits on uncertain delays: when is a delay system stabilizable by LTI controllers? IEEE Trans. Autom. Control **62**(3), 1314–1328 (2017)
16. T. Qi, J. Zhu, J. Chen, On delay radii and bounds of MIMO systems. Automatica **77**, 214–218 (2017)
17. T. Samad, A survey on industry impact and challenges thereof. IEEE Control Syst. Mag. **37**(1), 17–18 (2017)
18. G.J. Silva, A. Datta, S.P. Bhattacharyya, New results on the synthesis of PID controllers. IEEE Trans. Autom. Control **47**(2), 241–252 (2002)
19. G.J. Silva, A. Datta, S.P. Bhattacharyya, *PID Controllers for Time-Delay Systems* (Birkhuster, Basel, 2005)
20. Y. Wei, Z. Lin, Maximum delay bounds of linear systems under delay independent truncated predictor feedback. Automatica **83**(9), 65–72 (2017)
21. J.J. Xu, H.S. Zhang, L.H. Xie, Input delay margin for consensusability of multi-agent systems. Automatica **49**(6), 1816–1820 (2013)
22. D.Y. Xue, Y.Q. Chen, D.P. Atherton, *Linear Feedback Control Analysis and Design with MATLAB* (Springer, 2002)
23. S.Y. Yoon, Z. Lin, Truncated predictor feedback control for exponentially unstable linear systems with time-varying input delay. Syst. Control Lett. **62**(10), 837–844 (2013)
24. B. Zhou, Z. Lin, G.-R. Duan, Properties of the parametric lyapunov equation-based low-gain design with applications in stabilization of time-delay systems. IEEE Trans. Autom. Control **54**(7), 1698–1704 (2009)
25. B. Zhou, Z. Lin, G.-R. Duan, Truncated predictor feedback for linear systems with long time-varying input delays. Automatica **48**(10), 2387–2399 (2012)

Stability Analysis of Linear Delay Systems via Internally Positive Representations: An Overview

Vittorio De Iuliis, Alfredo Germani, and Costanzo Manes

Abstract The stability analysis of time-delay systems is a tough task even in the linear case, especially when time-varying delays are considered. In recent years, for the class of positive delay systems, a number of insightful works appeared in the literature, highlighting their peculiar stability properties, which lead to simple and easy to check necessary and sufficient conditions for the delay-independent stability. This fact poses the question whether some results that hold for positive systems can be exported to arbitrary systems. In some recent papers we gave an answer to this question by extending to the time-delay setting the Internally Positive Representation technique, a method that allows to systematically construct positive representations of arbitrary systems. In this work, after a survey on recent stability results for positive delay systems, an overview of the Internally Positive Representation technique and of its implications on the stability analysis of arbitrary systems is reported. Three classes of linear systems are considered: delay differential, delay difference, and coupled delay differential-difference systems.

1 Introduction

Positive systems are systems whose trajectories stay nonnegative for all future times if started from nonnegative initial conditions and excited by nonnegative inputs. Since positivity is a fundamental property of a huge amount of physical systems, it is not surprising that positive models have been extensively studied in the last decades, highlighting a remarkable set of properties and applications particularly in the linear

V. De Iuliis · A. Germani · C. Manes (✉)
Dipartimento di Ingegneria e Scienze dell'Informazione, e Matematica, Università degli Studi dell'Aquila, Via Vetoio, 67100 Coppito, AQ, Italy
e-mail: costanzo.manes@univaq.it

V. De Iuliis
e-mail: vittorio.deiuliis@univaq.it

A. Germani
e-mail: alfredo.germani@univaq.it

G. Valmorbida et al. (eds.), *Accounting for Constraints in Delay Systems*, Advances in Delays and Dynamics 12,
https://doi.org/10.1007/978-3-030-89014-8_2

case [20, 40]. One of the strength of linear positive systems relies in the simplified stability analysis, which proves to be a strong advantage in the time-delay setting. Indeed, while studying the stability of linear delay systems can be a hard task (see e.g. [23, 26]), the positivity assumption vastly simplifies the analysis, leading to easy to check necessary and sufficient conditions on the system matrices, as recently shown in several works on some classes of positive linear systems with delays [1, 19, 29, 36, 48, 61]. The nice properties of positive delay systems gave us the main motivation for a series of works, which we survey in this chapter. Indeed, the key idea of our work is providing a method to systematically construct a positive representation of an arbitrary (i.e. not necessarily positive) delay system, thus enabling to apply to the latter properties that only hold for its positive representation.

We first introduce the technique and its consequences on stability for two classes of linear delay systems: differential systems with multiple time-varying delays, i.e. systems driven by

$$\dot{x}(t) = A_0 x(t) + \sum_{k=1}^{m} A_k x(t - \delta_k(t)), \tag{1}$$

and difference systems with multiple constant delays, governed by continuous-time difference equations

$$x(t) = \sum_{k=1}^{m} A_k x(t - \delta_k). \tag{2}$$

The importance of such systems has been extensively discussed in literature (see e.g. [14, 28]), as they are known to model a large set of systems of practical relevance where the effects of delays can't be neglected without heavily affecting the accuracy of their model. Then, we move to the study of linear coupled differential-difference systems with a single time-varying delay, whose evolution is governed by

$$\begin{aligned} \dot{\xi}(t) &= A\xi(t) + Bx(t - \delta(t)) \\ x(t) &= C\xi(t) + Dx(t - \delta(t)). \end{aligned} \tag{3}$$

This class of systems received a long lasting attention dating back to the first half of the last century, when problems of wide interest as those related to lossless propagation models in electrical and fluid engineering gathered increasing attention from the research community. As discussed in [37] and more extensively in [58], coupled differential-difference equations were used as a key tool to treat models described by PDEs of hyperbolic type, first in the special case of neutral-type equations, and finally in their general formulation [6, 35, 41, 57]. We will discuss the relationship between coupled differential-difference equations and neutral-type equations in some extent, showing how the stability results obtained for coupled systems can be exploited to infer similar results for neutral-type equations.

The main contribution surveyed in this work is the extension of the Internally Positive Representation (IPR) of linear systems to delay systems, in order to exploit results available for positive systems also for systems that are not positive. The IPR technique, introduced in the delay-free discrete-time framework in [10, 24, 25] and in the continuous-time one in [7, 8], aims at constructing internally positive representations of systems whose dynamics is not definite in sign. The method presented in [7], although very easy and straightforward, can produce in some cases an unstable positive system even if the original system is stable. Later works on the IPR focused on this issue, showing how to construct IPRs whose stability properties are equivalent to those of the original system [8]. The IPR technique has also been applied to the state estimation problem to achieve an interval observer for both continuous and discrete-time linear systems [9].

Relying on stability results established for positive delay differential systems [1, 36, 39, 48], positive delay difference systems [19], and positive coupled delay differential-difference systems [61], our work first shows how to construct Internally Positive Representations for those classes of linear systems. Then, we show that the stability of the IPR of a given system is a sufficient condition for the stability of the original system, and this fact can be exploited to obtain easy to check sufficient conditions for the delay-independent stability directly on the original system matrices, which are proved to be less conservative with respect to other similar conditions available in the literature.

This chapter is structured as follows: in Sect. 2 we introduce in some detail the three classes of linear delay systems under investigation, with a special emphasis on the recent results on their positive counterparts. Then, Sect. 3 is devoted to the extension of the Internally Positive Representation to the delayed framework. Section 4 presents the consequences of the IPR method on the stability of arbitrary linear delay systems: novel sufficient conditions for the delay-independent stability of such systems are introduced and compared to similar existing results. Section 5 proposes some references with illustrative numerical examples. Conclusions and ideas for future works conclude the chapter.

Notations. $\mathbb{R}_+$ is the set of nonnegative real numbers. $\mathbb{C}_-$ and $\mathbb{C}_+$ are the open left-half and right-half complex planes, respectively. $\mathbb{R}_+^n$ is the nonnegative orthant of $\mathbb{R}^n$. $\mathbb{R}_+^{m\times n}$ is the cone of nonnegative $m \times n$ matrices. I_n is the $n \times n$ identity matrix. $\mathscr{C}([a, b], \mathbb{R}^n)$ and $\mathscr{PC}([a, b], \mathbb{R}^n)$ respectively denote the Banach spaces of all continuous and bounded functions, and piecewise right-continuous and bounded functions, on $[a, b]$ with values in $\mathbb{R}^n$; both are endowed with the uniform convergence norm $\|\cdot\|_\infty$. $\mathscr{L}_1^p$ and $\mathscr{L}_{1,+}^p$ are the sets of locally integrable functions with values in $\mathbb{R}^p$ and $\mathbb{R}_+^p$, respectively.
$A \in \mathbb{R}^{n\times n}$ is said to be *Metzler* if all its off-diagonal elements are nonnegative. $d(A)$ denotes the diagonal matrix extracted from A, i.e. $d(A)_{ij} = a_{ij}$ if $i = j$ and 0 otherwise. $A^{\mathscr{M}}$ is the Metzler matrix defined as $A^{\mathscr{M}} = d(A) + |A - d(A)|$, where $|\cdot|$ denotes the componentwise absolute value of a matrix. $\sigma(A)$ and $\alpha(A)$ denote the spectrum and the spectral abscissa of A, respectively. A is said to be Hurwitz or *Hurwitz-stable* if $\sigma(A) \subset \mathbb{C}_-$ or, equivalently, if $\alpha(A) < 0$. Similarly, it is said to be Schur or *Schur-stable* if $\sigma(A) \subset \{z \in \mathbb{C} : |z| < 1\}$ or, equivalently, if $\rho(A) < 1$. For a vector $x \in \mathbb{R}^n$, its p-norm, $p \geq 1$, is defined as $\|x\|_p = (\sum_{k=1}^n |x_k|^p)^{1/p}$, with $\|x\|_\infty = \max_k |x_k|$; if no subscript is specified, the Euclidean norm is considered,

i.e. $\|x\| = \|x\|_2$. For a matrix $A \in \mathbb{R}^{m\times n}$, its induced p-norm is defined as $\|A\|_p = \sup_{x\neq 0} {}^{\|Ax\|_p}/_{\|x\|_p}$; moreover, its measure (or logarithmic norm) is defined as $\mu_p(A) = \lim_{h\to 0^+} {}^{(\|I_n + hA\|_p - 1)}/_h$. Finally, the n leading principal minors of A are defined as $\det(A_i)$, where A_i are the matrices obtained removing from A the last $n-i$ rows and columns, for all $i = 1, \dots, n$.

2 Positive Linear Delay Systems

In this section, we introduce in some detail the three classes of linear delay systems under investigation. Moreover, we define positive delay systems of such classes, and examine the necessary and sufficient conditions for positivity. Finally, we first survey the stability of arbitrary linear delay systems, and then highlight the remarkable simplification of the stability analysis provided by the positivity assumption, which guarantees that any positive linear delay system is delay-independent stable if it is stable for the zero-valued delays case.

2.1 *Delay Differential Systems*

The first class of linear delay systems under investigation is that of differential systems with multiple time-varying delays, i.e. systems governed by (1) and described by:

$$\begin{aligned} \dot{x}(t) &= A_0 x(t) + \sum_{k=1}^{m} A_k x(t - \delta_k(t)) + Bu(t), \\ y(t) &= Cx(t) + Du(t), \end{aligned} \qquad t \geq t_0, \qquad (4)$$
$$x(t) = \phi(t - t_0), \qquad t \in [t_0 - \delta, t_0],$$

where u is locally integrable with values in $\mathbb{R}^p$, i.e. $u \in \mathscr{L}^p$, $A_k \in \mathbb{R}^{n\times n}$ for every $k = 0, \dots, m$ and $B \in \mathbb{R}^{n\times p}$, $C \in \mathbb{R}^{q\times n}$, $D \in \mathbb{R}^{q\times p}$, implying $y(t) \in \mathbb{R}^q$. The delays are supposed continuous and bounded, with $0 \leq \delta_k(t) \leq \delta$, $\delta \in \mathbb{R}^+$ for all $k = 1, \dots, m$. Moreover, the initial state is assigned by the function $\phi \in \mathscr{C}([-\delta, 0], \mathbb{R}^n)$. A differential system as in (4) will be denoted by $S = \{\{A_k\}_0^m, B, C, D\}_{n,p,q}$, or briefly S where the class of systems under investigation is unambiguous. The state and output trajectories of S for a given initial condition ϕ and input function u are commonly denoted with $(x(t), y(t)) = \Phi_S(t, t_0, \phi, u)$. If the explicit dependency of the trajectories on the initial condition has to be remarked, then $x(t, \phi)$ and $y(t, \phi)$ are used.

Remark 1 We note that to completely and uniquely describe a delay system and its trajectories both its matrices and the set of admissible delays must be specified. In this sense, denoting a system as in (4) by $S = \{\{A_k\}_0^m, B, C, D\}_{n,p,q}$, and its trajectories by $(x(t), y(t)) = \Phi_S(t, t_0, \phi, u)$, is not fully adequate, since no information on the delays is provided. However, the core results of this work concern delay-independent stability: for this reason, and in order to avoid inessential technicalities, we use the aforementioned notation.

Clearly, much could be said about the mathematical properties of delay differential systems. Since our main focus is on stability via positivity based approach, hereafter we will mainly cover those aspects for (4). For a comprehensive overview, please refer to e.g. [23, 31].

For what concerns positivity, two main definitions exist in the literature. Roughly speaking, one has *external positivity*, if nonnegative input functions produce nonnegative output responses, while *internal positivity* also involves the state trajectory, and implies the external positivity. The formal definition of an internally positive (or briefly *positive*) delay differential system can be stated as follows [12].

Definition 1 A delay differential system $S = \{\{A_k\}_0^m, B, C, D\}_{n,p,q}$ is defined to be internally positive if

$$\left\{ \begin{array}{c} \phi \in \mathscr{C}([-\delta, 0], \mathbb{R}^n_+) \\ u \in \mathscr{L}^p_{1,+} \end{array} \right\} \Rightarrow \left\{ \begin{array}{l} x(t) \in \mathbb{R}^n_+, \\ y(t) \in \mathbb{R}^q_+, \end{array} \quad \forall t \geq t_0 \right\}. \tag{5}$$

Definition 1 means that a delay differential system is internally positive if, when excited by nonnegative inputs and started by nonnegative initial conditions, its state and output trajectories stay nonnegative at any time instant.

As for delay-free systems (see [20]), a necessary and sufficient condition exists for linear delay differential systems to be internally positive. The following result has been presented in [29, 48].

Theorem 1 *A delay differential system $S = \{\{A_k\}_0^m, B, C, D\}_{n,p,q}$ is internally positive if and only if A_0 is Metzler and B, C, D, A_k, for $k = 1, \ldots, m$, are nonnegative.*

Before seeing how the positivity assumption vastly simplifies the stability analysis of delay differential systems, let's briefly recall the main stability definitions for (4). A first major distinction in the concept of stability for delay systems has to be made between delay-dependent and delay-independent stability. We refer to *delay-dependent* stability for a system whose stability can be proved for a set of values of the delays, when constant delays are considered, or for delays taking values in bounded regions, when modelling time-varying delays. Conversely, a system is *delay-independent* stable if the stability holds for all values of the delays. Among delay-independent conditions, some further distinction can be made if constant or time-varying delays are considered (e.g. slowly versus fast varying delays), especially when Lyapunov-based conditions are investigated.

Let's now introduce the main definitions of stability for differential systems with multiple time-varying delays described by (4). First of all, a given stability property is said to hold for a delay system if it holds for its trivial solution $x(t) \equiv 0$.

Definition 2 A differential system with multiple time-varying delays S as in (4) is said to be:

- stable, if for any $\varepsilon > 0$ there exists $\eta_\varepsilon > 0$ such that $\|\phi\|_\infty < \eta_\varepsilon$ implies $\|x(t,\phi)\| < \varepsilon$ for any $t \geq t_0$.
- asymptotically stable, if it is stable and for any $\phi \in \mathscr{C}([-\delta, 0], \mathbb{R}^n)$ it holds that $\lim_{t\to\infty} x(t,\phi) = 0$.
- exponentially stable, if it is asymptotically stable and there exist $k \geq 0, \alpha > 0$ such that $\|x(t,\phi)\| \leq ke^{-\alpha(t-t_0)}\|\phi\|_\infty$ for any $t \geq t_0$.

Our work focuses on the delay-independent stability analysis of linear delay system, thus we will not discuss in detail delay-dependent stability conditions. For what concerns delay-independent stability, necessary and sufficient conditions for a differential system as in (4) to be asymptotically stable can be formulated only for constant delays, since this particular framework allows the definition of the characteristic function associated to the system (which naturally extends the concept of characteristic polynomial to the retarded case). In detail, the stability of (4) is determined by the stability of the differential equation $\dot{x}(t) = A_0x(t) + \sum_{k=1}^{m} A_k x(t-\delta_k)$, which for constant delays admits the characteristic function

$$F(\lambda) = \det\left(\lambda I - A_0 - \sum_{k=1}^{m} A_k e^{-\delta_k \lambda}\right). \tag{6}$$

Then, the location in $\mathbb{C}_-$ of the infinite many roots of the transcendental function contained in (6) determines the stability of the associated system [50]. As a matter of fact, while many interesting properties of (6) have been studied, testing this condition is known to be an NP-hard problem (see [63]). Only in the even more restrictive framework of constant and commensurate delays the necessary and sufficient conditions become numerically tractable, since the multiple delays case reduces, from a complexity viewpoint, to the single delay case [50]. As previously remarked, the focus of our work is on the considerably more complex case of time-varying delays, for which no necessary and sufficient conditions can be formulated with respect to delay-independent stability. This explains why a remarkable amount of research has been done pursuing sufficient stability conditions, both delay-dependent and independent. For the latter, two main approaches are known: matrix-measure conditions, and Lyapunov-based conditions. Since the condition which we will derive by means of a positivity based approach share strong similarities with the former, we will mainly compare our work with this kind of conditions, skipping further details on the Lyapunov-based methods, to which a considerable literature has been devoted (see [23, 50] and references therein).

In [60] and [59] it has been show that a norm-based condition first proposed by [44] and [45] for the case of single and constant delay, and then extended in [65] to the case of multiple commensurate delays, is sufficient for the delay-independent stability of delay differential systems as in (4). The condition is the following.

Theorem 2 *Consider a delay differential system S as in* (4). *If it exists some* $p \in [1, +\infty]$ *such that*

$$\mu_p(A_0) + \sum_{k=1}^{m} \|A_k\|_p < 0 \tag{7}$$

then (4) *is delay-independent asymptotically stable.*

The logarithmic norm of a given matrix A, defined as $\mu_p(A) = \lim_{h \to 0^+} \frac{\|I_n + hA\|_p - 1}{h}$ is actually not a proper norm, since it doesn't satisfy all the defining properties of a norm (e.g. it verifies the triangular inequality but it can assume negative values). The expression of $\mu_p(\cdot)$ can easily be computed for $p = 1, 2, \infty$:

$$\mu_1(A) = \max_{j=1,\dots,n} \left(a_{jj} + \sum_{i=1,\ i \neq j}^{n} |a_{ij}| \right), \quad \mu_2(A) = \frac{1}{2} \lambda_{max}(A^T + A),$$

$$\mu_\infty(A) = \max_{i=1,\dots,n} \left(a_{ii} + \sum_{j=1,\ j \neq i}^{n} |a_{ij}| \right),$$

while computing a matrix p-norm has been shown to be NP-hard for $p \neq 1, 2, \infty$ [33].

Remarkably, the delay-independent stability analysis is a trivial task if the positivity assumption is satisfied, i.e. if (8) satisfies Theorem 1. In this case, the following result has been established [1, 12, 39].

Theorem 3 *An internally positive delay differential system as in* (4) *is delay-independent asymptotically stable if and only if* $\sum_{k=0}^{m} A_k$ *is Hurwitz.*

Hence, the delay-independent stability of a *positive* delay differential system is equivalent to the stability of the zero-valued delays case (i.e. $\delta_k(t) \equiv 0$ for all $k = 1, \dots, m$), a condition which is only necessary if no positivity assumption is made. We will see how the IPR technique allows to export this result to delay difference systems that are not positive, yielding a novel sufficient condition for their delay-independent stability.

2.2 Delay Difference Systems

The second class of linear delay systems we consider is that of continuous-time delay difference systems, i.e. systems whose evolution is governed by (2), which are described by:

$$
\begin{aligned}
x(t) &= \sum_{k=1}^{m} A_k x(t-\delta_k) + Bu(t), && \multirow{2}{*}{$t \geq t_0,$} \\
y(t) &= Cx(t) + Du(t), && \\
x(t) &= \phi(t-t_0), && t \in [t_0-\delta, t_0),
\end{aligned}
\tag{8}
$$

where $u(t) \in \mathbb{R}^p$ is the input, with $u \in \mathscr{PC}([t_0, +\infty), \mathbb{R}^p)$, and $y(t) \in \mathbb{R}^q$ is the output. Moreover, the initial state function ϕ is defined as $\phi \in \mathscr{PC}([-\delta, 0), \mathbb{R}^n)$. It easily follows that $A_k \in \mathbb{R}^{n\times n}$ for any $k = 1, \ldots, m$, and $B \in \mathbb{R}^{n\times p}$, $C \in \mathbb{R}^{q\times n}$, $D \in \mathbb{R}^{q\times p}$. In this case, the delays are considered to be constant and, with no loss of generality, ordered and non vanishing, i.e. $0 < \delta_1 < \delta_2 < \cdots < \delta_m = \delta$, $\delta \in \mathbb{R}^+$. This restriction is due to the fact that results on the stability of positive delay difference systems we use in the remainder of the work are limited to this framework [19]. Also in this case, we will refer to (8) by $S = \{\{A_k\}_1^m, B, C, D\}_{n,p,q}$, or briefly S, when the class of systems under investigation is not ambiguous. The state and output trajectories of S for a given initial condition ϕ and input function u are denoted with $(x(t), y(t)) = \Phi_S(t, t_0, \phi, u)$, and if the explicit dependency of the trajectories with respect to the initial condition is noteworthy, then $x(t, \phi)$ and $y(t, \phi)$ are used.

It has to be noted that while delay differential systems received a huge amount of attention by the scientific community, producing entire monographies, the same is not true for delay difference systems. However, for the basic results on the existence of the solution, its computation, Lyapunov-based approaches to the stability, please refer to [13, 14, 30, 55].

The definition of internal positivity can be easily generalized to this class of systems [17].

Definition 3 A delay difference system $S = \{\{A_k\}_1^m, B, C, D\}_{n,p,q}$ is defined to be internally positive if

$$
\left\{ \begin{array}{c} \phi \in \mathscr{PC}([-\delta, 0), \mathbb{R}^n_+) \\ u \in \mathscr{PC}([t_0, +\infty), \mathbb{R}^p_+) \end{array} \right\} \Rightarrow \left\{ \begin{array}{ll} x(t) \in \mathbb{R}^n_+, & \\ & \forall t \geq t_0 \\ y(t) \in \mathbb{R}^q_+, & \end{array} \right\}. \tag{9}
$$

As in the differential case, a necessary and sufficient condition exists for linear delay difference systems to be internally positive. The following result is a straightforward extension of the work of [19].

Theorem 4 *A delay difference system $S = \{\{A_k\}_1^m, B, C, D\}_{n,p,q}$ is internally positive if and only if A_k is nonnegative for $k = 1, \ldots, m$, and B, C, D are nonnegative.*

For what concerns the delay-independent stability of the trivial solution of (8), similar remarks to those holding in the differential case are in order. First of all, notice that the formal definitions given in Definition 2 hold for this class with trivial modifications (i.e. $\phi \in \mathscr{PC}$). Coming to the stability conditions, a distinction has to be made with respect to the commensurability of the delays. If commensurate delays are considered, then the analysis is trivial, at least conceptually, since (2) admits an augmented state space realization with a single delay $\delta_c : X(t) = AX(t - \delta_c)$, for which the stability analysis is equivalent to that of a sampled system. However, depending on the commensurability ratio of the delays, the dimensions of A could be large enough to pose numerical issues in evaluating the stability of the augmented system. We refer to [13] for further details. In the general case of incommensurate details, the following necessary and sufficient condition for the delay-independent exponential stability was proposed by Silkowski in [62] and further extended by Avellar and Hale in [3]: (8) is delay-independent exponentially stable if and only if

$$\sup_{\theta_1,\ldots,\theta_m \in [0,2\pi]^m} \left\{ \rho\left(\sum_{k=1}^{m} A_k e^{j\theta_k} \right) \right\} < 1. \tag{10}$$

The condition is far from being tractable even from a numerical viewpoint. The only notable case in which it gets readily tested is the scalar one, where it reduces to checking that $\sum_{k=1}^{m} |a_k| < 1$, where a_k stands for the scalar A_k [43].

As for differential systems, this lack of effective necessary and sufficient conditions in the general case of multiple and incommensurate delays, demands that sufficient conditions are investigated. With this respect, the literature has offered a considerable amount of results on the extensions of Lyapunov approaches also to this class of delay systems. We refer to [22, 42, 53], and in particular to [55], where both Lyapunov-Krasovskii and Lyapunov-Razumikhin approaches are discussed. Similar results for the case of time-varying delays are still missing.

For what concerns stability conditions comparable to those obtained in the remainder of our work, a simple and easy to check delay-independent condition is the following (see e.g. [42]).

Theorem 5 *A difference system with multiple constant delays as in* (8) *is delay-independent asymptotically stable if there exists a* $p \in [1, \infty]$ *such that:*

$$\sum_{k=1}^{m} \|A_k\|_p < 1. \tag{11}$$

Remarkably, the positivity assumption on S allows to vastly simplify the stability analysis also for this class of systems. As a matter of fact, the onerous necessary and sufficient delay-independent condition given by (10) becomes the simple and easy to check condition established in [19], and formulated in the following Theorem, which indeed assesses that a positive difference system is stable for all possible delays if it is stable for the zero-valued delays case.

Theorem 6 *An internally positive delay difference system as in* (8) *is delay independent exponentially stable if and only if* $\sum_{k=1}^{m} A_k$ *is Schur.*

We will exploit this result by means of the IPR technique to establish a novel delay-independent sufficient condition for arbitrary delay difference systems.

2.3 Coupled Delay Differential-Difference Systems

When delay differential equations and delay difference equations are coupled, some wider classes of delay systems can be obtained, according to how the coupling is chosen. A relevant literature has been devoted to a particular coupling in which both equations involve delays, but actually only one of them (the difference one) has a proper state delay. This class of equations is described by (3), where the first equation is an ordinary differential equation, and the second one is a functional difference equation. When the investigation is restricted to the linear case with a single delay (as done in the majority of the referenced works), a coupled differential-difference system can be stated, in the general case of time-varying delay, as:

$$\begin{aligned} \dot{\xi}(t) &= A\xi(t) + Bx(t-\delta(t)) + Fu(t) \\ x(t) &= C\xi(t) + Dx(t-\delta(t)) + Gu(t) \quad t \geq t_0, \\ y(t) &= H\xi(t) + Mx(t) + Nu(t), \\ \xi(t_0) &= \psi, \\ x(t) &= \phi(t-t_0), \qquad\qquad t \in [t_0-\delta, t_0). \end{aligned} \tag{12}$$

where $u(t) \in \mathbb{R}^p$ is the input, with $u \in \mathscr{PC}([t_0, +\infty), \mathbb{R}^p)$, and $y(t) \in \mathbb{R}^q$ is the output. The initial state is the pair (ψ, ϕ) defined as $\psi \in \mathbb{R}^n$, $\phi \in \mathscr{PC}([-\delta, 0), \mathbb{R}^\nu)$, and the time-varying delay is continuous and bounded, i.e. $0 \leq \delta(t) \leq \delta$, with $\delta \in \mathbb{R}^+$. It follows that $A \in \mathbb{R}^{n\times n}$, $B \in \mathbb{R}^{n\times \nu}$, $C \in \mathbb{R}^{\nu\times n}$, $D \in \mathbb{R}^{\nu\times \nu}$, $F \in \mathbb{R}^{n\times p}$, $G \in \mathbb{R}^{\nu\times p}$, $H \in \mathbb{R}^{q\times n}$, $M \in \mathbb{R}^{q\times \nu}$, $N \in \mathbb{R}^{q\times p}$. As for the previous two classes of delay systems, we will refer to (12) by $S = \{A, B, C, D, F, G, H, M, N\}_{n,\nu,p,q}$, or briefly S if no ambiguities arise. The state and output trajectories of S for a given initial condition (ψ, ϕ) and input function u are denoted with $(\xi(t), x(t), y(t)) = \Phi_S(t, t_0, \psi, \phi, u)$, and if the explicit dependency of the state trajectories with respect to the initial condition has to be highlighted, then $\xi(t, \psi, \phi)$, $x(t, \psi, \phi)$ are used. We point out that usually no output function is associated or explicitly modelled in the literature for coupled differential-difference system, thus making $y(t)$ inessential in (12). However, we decided to give a complete formulation of a linear coupled system that directly and completely generalizes the system theoretical approach to this class of delay system.

Remarkably, a class of neutral-type equations can be derived as a special case of (12). Indeed, consider the neutral-type equation

$$\frac{d}{dt}(x(t) - Dx(t-\delta(t))) = Ax(t) + Bx(t-\delta(t)), \quad t \geq t_0 \tag{13}$$

for which the initial condition is specified by $x(t) = \phi(t-t_0)$ for $t \in [t_0-\delta, t_0]$, $\phi \in \mathscr{C}^1([-\delta, 0], \mathbb{R}^n)$, $\delta \in \mathbb{R}^+$ being the upper bound for the delay. Then, defining $\xi(t) = x(t) - Dx(t-\delta(t))$, trivial manipulations yield the equivalent representation:

$$\begin{aligned} \dot{\xi}(t) &= A\xi(t) + (AD + B)x(t - \delta(t)) \\ x(t) &= \xi(t) + Dx(t - \delta(t)), \end{aligned} \qquad t \geq t_0, \tag{14}$$

with initial conditions assigned as

$$\begin{aligned} x(t) &= \phi(t - t_0), \quad t \in [t_0 - \delta, t_0], \\ \xi(t_0) &= x(t_0) - Dx(t_0 - \delta(t_0)) = \phi(0) - D\phi(-\delta(t_0)). \end{aligned} \tag{15}$$

Then, results established for coupled differential-difference systems as in (12) can be naturally particularized to neutral-type systems governed by (13) with little effort.

Let's now examine the theory of positive coupled systems. The definition and necessary and sufficient conditions for S to be internally positive have been given in [18, 61] for the single-delay case with no associated output equation. The extension to the case of an explicitly modelled output equation is trivial.

Definition 4 A coupled delay differential-difference system $S = \{A, B, C, D, F, G, H, M, N\}_{n,\nu,p,q}$ is defined to be *internally positive* if

$$\left\{ \begin{array}{c} \psi \in \mathbb{R}^n_+ \\ \phi \in \mathscr{PC}([-\delta, 0), \mathbb{R}^\nu_+) \\ u \in \mathscr{L}^p_{1,+} \end{array} \right\} \Rightarrow \left\{ \begin{array}{l} \xi(t) \in \mathbb{R}^n_+, \\ x(t) \in \mathbb{R}^\nu_+, \quad \forall t \geq t_0 \\ y(t) \in \mathbb{R}^q_+, \end{array} \right\}. \tag{16}$$

Theorem 7 *Assume that D is Schur. Then, a delay system $S = \{A, B, C, D, F, G, H, M, N\}_{n,\nu,p,q}$ is internally positive if and only if A is Metzler and B, C, D, F, G, H, M, N are nonnegative.*

We informally speak of the stability of (12) referring to that of the trivial solution $\xi(t) \equiv 0, x(t) \equiv 0$. In the following definition we use the compact notation $(\xi(t), x(t))$ to denote the column vector obtained by stacking $\xi(t)$ and $x(t)$.

Definition 5 A coupled differential-difference system with delays S as in (12) is said to be:

- stable if for any $\varepsilon > 0$ there exists $\eta_\varepsilon > 0$ such that $\|(\psi, \phi)\|_\infty < \eta_\varepsilon$ implies $\|(\xi(t, \psi, \phi), x(t, \psi, \phi))\| < \varepsilon, \forall t \geq t_0$;
- asymptotically stable if it is stable and for any $(\psi, \phi) \in \mathbb{R}^n \times \mathscr{PC}([-\delta, 0), \mathbb{R}^\nu)$, $\lim_{t\to\infty} \|(\xi(t, \psi, \phi), x(t, \psi, \phi))\| = 0$;
- exponentially stable if it is asymptotically stable and there exist $k \geq 0, \alpha > 0$ such that $\|(\xi(t, \psi, \phi), x(t, \psi, \phi))\| \leq ke^{-\alpha(t-t_0)}\|(\psi, \phi)\|_\infty$ for any $t \geq t_0$.

A necessary and sufficient condition for the delay-independent stability of (12) can be formulated in the special case of a constant delay, as follows [50]: a coupled differential-difference system as in (12), with constant delay $\delta(t) = \delta$, is delay-independent exponentially stable if and only if all the (infinite many) solutions of

$$F(\lambda) = \det \begin{bmatrix} \lambda I_n - A & -Be^{-\lambda\delta} \\ -C & I_\nu - De^{-\lambda\delta} \end{bmatrix} = 0 \tag{17}$$

have negative real part.

Clearly, testing the previous condition is prohibitive. Moreover, it is limited to the special case of constant delays. These reasons explain, once again, why much literature has been devoted to the search of sufficient conditions for both the delay-dependent and delay-independent stability of linear coupled differential-difference systems. To the best of the authors' knowledge, all such conditions deal with the extensions of the Lyapunov method, actually in the Krasovskii approach. Some notable results within this framework can be found in [27, 37, 54, 56]. In the special case of neutral-type systems, widely investigated in the literature, although most works focus on the classical Lyapunov analysis (see e.g. [11, 21, 32, 34, 51, 64]), there are some papers that present explicit (not Lyapunov-based) delay-independent stability criteria ([34, 38, 52]), that to a certain extent are similar to those based on the IPR method, and therefore a comparison is rather natural in terms of computability and conservativeness.

Now that we have summarized the stability properties of arbitrary coupled systems, and recollected and established the main positivity definitions and conditions, we are in a position to examine how the positivity constraint allows to simplify the stability analysis of this class of systems. The next result has been proven in [61].

Theorem 8 *An internally positive delay differential-difference system* $S = \{A, B, C, D, F, G, H, M, N\}_{n,\nu,p,q}$ *as in* (12) *is delay-independent asymptotically stable if and only if* D *is Schur and* $A + B(I_\nu - D)^{-1}C$ *is Hurwitz.*

Note that, also for this class of delay systems, the condition of Theorem 8 states that, under positivity constraint, checking its delay-independent stability is equivalent to check that it is stable for the special case of constant and zero-valued delay.

3 Internally Positive Representation of Delay Systems

The motivations behind the need for positive representations of arbitrary systems arise in technological frameworks where only positive systems can be physically implemented. In detail, in the pioneering works [4, 5] it was shown how such constraints could be overcome in two practical applications: Charge Routing Networks (CRNs) and fiber optic filters. The works tackled the positive representation problem for delay-free systems in the simple single input/single output (SISO) case, with the additional constraint of transfer functions with simple poles, and served as a boost in the study of positive realization (and representations) of arbitrary systems, which indeed continued in later works that, for example, provided similar methods to also face the case of multiple poles transfer functions [46, 47].

The Internally Positive Representation technique offered a general approach to the problem of realizing an arbitrary system through a combination of positive ones, overcoming several limits of [5]. It was introduced, in the delay-free setting, for multiple input/multiple output (MIMO) discrete-time systems in [24, 25] and then for continuous-time systems in [7]. The method provides a simple and systematic approach to construct an internally positive system of larger dimension with respect to the original non positive one, both in space-state and transfer matrices form. Transformations to pass from the original to the positive system are provided, and a stability analysis is also conducted to show how the stability of the original system is necessary to obtain a stable positive representation. Notable advantages of the IPR technique with respect to previous approaches are the following: it can be applied to general MIMO systems with no restriction on the multiplicity of the poles of the original system, it doesn't require to solve any numerical optimization problem to compute the matrices of the positive representation, it applies to systems forced by inputs that can be indefinite in sign. Two notable extensions of the technique followed for both discrete-time systems [10] and continuous-time systems [8], where additional results were presented to show how is always possible to build refined IPRs whose stability properties are equivalent to those of the original systems. As an interesting by-product, the method was also exploited to obtain a new approach to interval observers for both discrete and continuous-time linear systems in [9]. Then, the peculiar stability properties of positive delay systems encouraged the authors to seek for an extension of the IPR method to linear delay systems, which has been derived for the three aforementioned classes of systems. Before introducing the technical details, we present some preliminary concepts related to positive representations of vectors and matrices.

3.1 Positive Representation of Vectors and Matrices

Given a matrix (or vector) $M \in \mathbb{R}^{m \times n}$, the symbols M^+, M^- denote the componentwise *positive* and *negative* parts of M, i.e.:

$$[M^+]_{ij} = \begin{cases} [M]_{ij} & \text{if } [M]_{ij} \geq 0, \\ 0 & \text{if } [M]_{ij} < 0, \end{cases} \tag{18}$$

and $M^- = (-M)^+$.

Moreover, we write $|M|$ for the componentwise absolute value of the matrix M. Hence $M = M^+ - M^-$ and $|M| = M^+ + M^-$.

Now let $\Delta_n = [I_n \ \ -I_n] \in \mathbb{R}^{n \times 2n}$. The following definitions are from [10, 25].

Definition 6 A positive representation of a vector $x \in \mathbb{R}^n$ is any vector $\tilde{x} \in \mathbb{R}_+^{2n}$ s.t.

$$x = \Delta_n \tilde{x}. \tag{19}$$

The min-positive representation of a vector $x \in \mathbb{R}^n$ is the positive vector $\pi(x) \in \mathbb{R}_+^{2n}$ defined as

$$\pi(x) = \begin{bmatrix} x^+ \\ x^- \end{bmatrix}. \tag{20}$$

The min-positive representation of a matrix $M \in \mathbb{R}^{m\times n}$ is the positive matrix $\Pi(M) \in \mathbb{R}_+^{2m\times 2n}$ defined as

$$\Pi(M) = \begin{bmatrix} M^+ & M^- \\ M^- & M^+ \end{bmatrix} \tag{21}$$

while the min-Metzler representation of a matrix $A \in \mathbb{R}^{n\times n}$ is the Metzler matrix $\Gamma(A) \in \mathbb{R}^{2n\times 2n}$ defined as

$$\Gamma(A) = \begin{bmatrix} d(A) + (A - d(A))^+ & (A - d(A))^- \\ (A - d(A))^- & d(A) + (A - d(A))^+ \end{bmatrix}. \tag{22}$$

Note that if $d(A) \in \mathbb{R}_+^{n\times n}$ then $\Gamma(A) = \Pi(A)$. Moreover, for any vector $x \in \mathbb{R}^n$ and matrices $M \in \mathbb{R}^{m\times n}$, $A \in \mathbb{R}^{n\times n}$ the following properties are satisfied:

(a) $x = \Delta_n \pi(x)$;
(b) $\Delta_m \Pi(M) = M\Delta_n$, so that $\Delta_m \Pi(M)\pi(x) = Mx$;
(c) $\Delta_n \Gamma(A) = A\Delta_n$, so that $\Delta_n \Gamma(A)\pi(x) = Ax$.

3.2 *IPR of Delay Differential Systems*

The first class of linear delay system we consider is that of differential systems with multiple time-varying delays. As we have seen in Sect. 2.1, we conduct our analysis for the class of systems described by (4).

We have previously seen that a delay differential system is internally positive if Definition 1 is satisfied, i.e. if nonnegative input and initial conditions produce nonnegative state and output trajectories at any time. Actually, a necessary and sufficient condition exists to fulfil the definition, and has been given in Theorem 1: A_0 has to be a Metzler matrix, and A_k, B, C, D have to be nonnegative for any $k = 1, \ldots, m$.

Now, we see how an Internally Positive Representation can be defined for this class of delay systems [12].

Definition 7 An *Internally Positive Representation* (IPR) of a delay differential system $S = \{\{A_k\}_0^m, B, C, D\}_{n,p,q}$ is an internally positive system $\tilde{S} = \{\{\widetilde{A}_k\}_0^m, \widetilde{B}, \widetilde{C}, \widetilde{D}\}_{\tilde{n},\tilde{p},\tilde{q}}$ together with four continuous transformations $\{T_X^f, T_X^b, T_U, T_Y\}$,

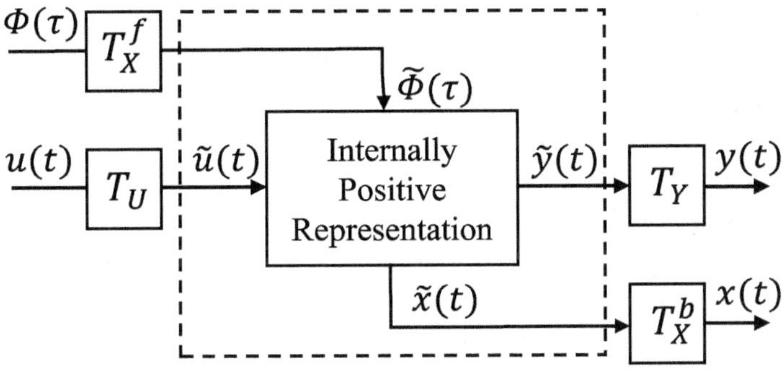

Fig. 1 Block diagram of the IPR in Definition 7

$$\begin{aligned} T_X^f &: \mathbb{R}^n \mapsto \mathbb{R}_+^{\tilde{n}}, & T_X^b &: \mathbb{R}_+^{\tilde{n}} \mapsto \mathbb{R}^n, \\ T_U &: \mathbb{R}^p \mapsto \mathbb{R}_+^{\tilde{p}}, & T_Y &: \mathbb{R}_+^{\tilde{q}} \mapsto \mathbb{R}^q, \end{aligned} \tag{23}$$

such that $\forall t_0 \in \mathbb{R}, \forall(\phi, u) \in \mathscr{C}([-\delta, 0], \mathbb{R}^n) \times \mathscr{L}_1^p$, the following implication holds:

$$\left\{\begin{array}{ll} \tilde{\phi}(\tau) = T_X^f(\phi(\tau)), & \forall \tau \in [-\delta, 0] \\ \tilde{u}(t) = T_U(u(t)), & \forall t \geq t_0 \end{array}\right\} \implies \left\{\begin{array}{ll} x(t) = T_X^b(\tilde{x}(t)), & \\ y(t) = T_Y(\tilde{y}(t)), & \end{array} \forall t \geq t_0 \right\} \tag{24}$$

where

$$\begin{aligned} \big(x(t), y(t)\big) &= \Phi_S\big(t, t_0, \phi, u\big), \\ \big(\tilde{x}(t), \tilde{y}(t)\big) &= \Phi_{\tilde{S}}\big(t, t_0, \tilde{\phi}, \tilde{u}\big). \end{aligned}$$

Definition 7 is a direct extension of that of an IPR for continuous-time (delay free) systems to the delayed case.

T_X^f and T_X^b in (23) are the *forward* and *backward* state transformations of the IPR, respectively, while T_U and T_Y are the input and output transformations. The implication (24) means that **if** the (nonnegative) initial state function $\tilde{\phi}$ of the IPR is computed as the forward state transformation T_X^f of ϕ of the *original* system, **and** the (nonnegative) input $\tilde{u}$ to the IPR is computed as the input transformation T_U of the input u to the *original* system, **then** the state trajectory of the *original* system is given by the backward transformation T_X^b of the (nonnegative) state $\tilde{x}$ of the IPR, **and** the output trajectory y of the *original* system is given by the output transformation T_Y of the (nonnegative) output $\tilde{y}$ of the IPR. For consistency, the *backward* map T_X^b must be a left-inverse of the *forward* map T_X^f, i.e. $x = T_X^b\big(T_X^f(x)\big)$, $\forall x \in \mathbb{R}^n$. The defined IPR is depicted in Fig. 1.

Among the IPRs that could satisfy Definition 7, the simplest and most effective direct generalization of the *normal-form* IPR given for continuous-time delay-free system in [7] (with which it coincides if $A_k = 0$ for all $k = 1, \dots, m$) is constructed as follows (see [12] for the proof).

Theorem 9 *Consider a delay differential system S as in* (4) *with* $S = \{\{A_k\}_0^m, B, C, D\}_{n,p,q}$. *The internally positive system* $\bar{S} = \{\{\bar{A}_k\}_0^m, \bar{B}, \bar{C}, \bar{D}\}_{2n,2p,2q}$, *with*

$$\begin{aligned} \bar{A}_0 &= \Gamma(A_0), \quad \bar{B} = \Pi(B), \quad \bar{C} = \Pi(C), \quad \bar{D} = \Pi(D), \\ \bar{A}_k &= \Pi(A_k), \quad k = 1, \dots, m, \end{aligned} \tag{25}$$

together with the four transformations

$$\bar{x} = T_X^f(x) = \pi(x), \qquad x = T_X^b(\bar{x}) = \Delta_n \bar{x}, \tag{26}$$

$$\bar{u} = T_U(u) = \pi(u), \qquad y = T_Y(\bar{y}) = \Delta_q \bar{y}, \tag{27}$$

is an IPR of S.

For further details, please refer to [12]. We close this section remarking that the IPR method for delay differential systems has recently been extended to the time-varying setting in [16].

In Sect. 4.1 we will exploit the IPR method to export Theorem 3 to arbitrary delay differential systems.

3.3 IPR of Delay Difference Systems

Expectedly, the extension of the IPR method to delay differential systems presented in the previous section has been built on the normal-form IPR defined for differential delay-free systems. Then, if one thinks at discrete-time systems as systems driven by discrete-time difference equations with a single delay, it is not surprising that a direct extension to continuous-time difference equations with multiple delays can be defined in a similar manner. Consider an arbitrary difference system with multiple constant delays described by (8).

In Sect. 2.2 we have seen that necessary and sufficient conditions exist for a system $S = \{\{A_k\}_1^m, B, C, D\}_{n,p,q}$ to fulfil Definition 3: S is internally positive if and only if A_k, B, C, D are nonnegative matrices for any $k = 1, \dots, m$. Before seeing how this condition straightforwardly suggests a particular IPR construction for delay difference systems, first let's see how the IPR definition extends to this class with trivial technical modifications.

Definition 8 An *Internally Positive Representation* of a delay difference system $S = \{\{A_k\}_1^m, B, C, D\}_{n,p,q}$ is an internally positive system $\widetilde{S} = \{\{\widetilde{A}_k\}_1^m, \widetilde{B}, \widetilde{C}, \widetilde{D}\}_{\tilde{n},\tilde{p},\tilde{q}}$ together with four continuous transformations $\{T_X^f, T_X^b, T_U, T_Y\}$,

$$\begin{aligned} T_X^f &: \mathbb{R}^n \mapsto \mathbb{R}_+^{\tilde{n}}, \qquad T_X^b : \mathbb{R}_+^{\tilde{n}} \mapsto \mathbb{R}^n, \\ T_U &: \mathbb{R}^p \mapsto \mathbb{R}_+^{\tilde{p}}, \qquad T_Y : \mathbb{R}_+^{\tilde{q}} \mapsto \mathbb{R}^q, \end{aligned} \tag{28}$$

such that $\forall t_0 \in \mathbb{R}$, $\forall(\phi, u) \in \mathscr{PC}([-\delta, 0), \mathbb{R}^n) \times \mathscr{PC}([t_0, +\infty), \mathbb{R}^p)$, the following implication holds:

$$\left\{\begin{array}{ll} \tilde{\phi}(\tau) = T_X^f(\phi(\tau)), & \forall \tau \in [-\delta, 0) \\ \tilde{u}(t) = T_U(u(t)), & \forall t \geq t_0 \end{array}\right\} \implies \left\{\begin{array}{l} x(t) = T_X^b(\tilde{x}(t)), \\ y(t) = T_Y(\tilde{y}(t)), \end{array} \quad \forall t \geq t_0 \right\} \tag{29}$$

where

$$\begin{aligned} \big(x(t), y(t)\big) &= \Phi_S\big(t, t_0, \phi, u\big), \\ \big(\tilde{x}(t), \tilde{y}(t)\big) &= \Phi_{\tilde{S}}\big(t, t_0, \tilde{\phi}, \tilde{u}\big). \end{aligned}$$

The scheme of Fig. 1 applies also to this Definition.

Among the possible choices, the most simple and interesting IPR satisfying Definition 8 is obtained as follows (see [17] for the proof).

Theorem 10 *Consider a delay difference system S as in* (8) *with* $S = \{\{A_k\}_1^m, B, C, D\}_{n,p,q}$. *The internally positive system* $\bar{S} = \{\{\bar{A}_k\}_1^m, \bar{B}, \bar{C}, \bar{D}\}_{2n,2p,2q}$, *with*

$$\bar{A}_k = \Pi(A_k),\ k = 1, \ldots, m, \quad \bar{B} = \Pi(B), \quad \bar{C} = \Pi(C), \quad \bar{D} = \Pi(D), \tag{30}$$

together with the four transformations

$$\bar{x} = T_X^f(x) = \pi(x), \qquad x = T_X^b(\bar{x}) = \Delta_n \bar{x}, \tag{31}$$

$$\bar{u} = T_U(u) = \pi(u), \qquad y = T_Y(\bar{y}) = \Delta_q \bar{y}, \tag{32}$$

is an IPR of S.

Remark 2 Note that the IPR construction given in this section also holds for the case of bounded time-varying delays. However, the technique has been presented in the constant delays framework since the stability analysis that follows in Sect. 4.2 is restricted to this particular setting.

3.4 IPR of Coupled Delay Differential-Difference Systems

In the previous two sections we have seen how the IPR method can be extended to delay differential systems and delay difference systems. Now we face the more general class of linear delay systems consisting of a coupling of both differential and difference equations, in the special case of a single time-varying delay, described by (12). The IPR for this class of single-delay systems is defined as follows.

Definition 9 An *Internally Positive Representation* (IPR) of a delay differential-difference system $S = \{A, B, C, D, F, G, H, M, N\}_{n,\nu,p,q}$ is an internally positive system $\widetilde{S} = \{\widetilde{A}, \widetilde{B}, \widetilde{C}, \widetilde{D}, \widetilde{F}, \widetilde{G}, \widetilde{H}, \widetilde{M}, \widetilde{N}\}_{\tilde{n},\tilde{\nu},\tilde{p},\tilde{q}}$ together with six continuous transformations $\{T_\Xi^f, T_\Xi^b, T_X^f, T_X^b, T_U, T_Y\}$,

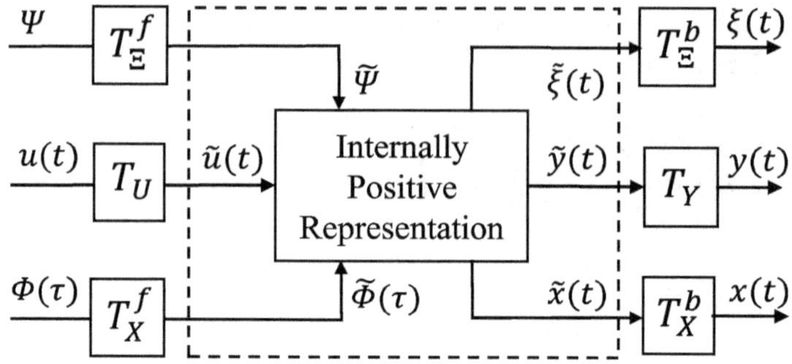

Fig. 2 Block diagram of the IPR in Definition 9

$$
\begin{aligned}
T_\Xi^f &: \mathbb{R}^n \mapsto \mathbb{R}_+^{\tilde{n}}, & T_\Xi^b &: \mathbb{R}_+^{\tilde{n}} \mapsto \mathbb{R}^n,\\
T_X^f &: \mathbb{R}^\nu \mapsto \mathbb{R}_+^{\tilde{\nu}}, & T_X^b &: \mathbb{R}_+^{\tilde{\nu}} \mapsto \mathbb{R}^\nu,\\
T_U &: \mathbb{R}^p \mapsto \mathbb{R}_+^{\tilde{p}}, & T_Y &: \mathbb{R}_+^{\tilde{q}} \mapsto \mathbb{R}^q,
\end{aligned}
\tag{33}
$$

such that $\forall t_0 \in \mathbb{R}, \forall(\psi, \phi, u) \in \mathbb{R}^n \times \mathscr{PC}([-\delta, 0), \mathbb{R}^\nu) \times \mathscr{L}_1^p$, the following implication holds:

$$
\left\{
\begin{aligned}
\tilde{\psi} &= T_\Xi^f(\psi),\\
\tilde{\phi}(\tau) &= T_X^f(\phi(\tau)),\ \forall \tau \in [-\delta, 0)\\
\tilde{u}(t) &= T_U(u(t)),\ \ \forall t \geq t_0
\end{aligned}
\right\}
\implies
\left\{
\begin{aligned}
\xi(t) &= T_\Xi^b(\tilde{\xi}(t)),\\
x(t) &= T_X^b(\tilde{x}(t)),\ \forall t \geq t_0\\
y(t) &= T_Y(\tilde{y}(t)),
\end{aligned}
\right\}
\tag{34}
$$

where

$$
\begin{aligned}
(\xi(t), x(t), y(t)) &= \Phi_S(t, t_0, \psi, \phi, u),\\
(\tilde{\xi}(t), \tilde{x}(t), \tilde{y}(t)) &= \Phi_{\tilde{S}}(t, t_0, \tilde{\psi}, \tilde{\phi}, \tilde{u}).
\end{aligned}
$$

Clearly, also in this joint case, for consistency the T_Ξ^b and T_X^b maps must be left-inverses for T_Ξ^f and T_X^f, respectively. The behaviour of the IPR of Definition 9 is depicted in Fig. 2. The simplest and most effective IPR that fulfil Definition 9 is constructed as follows.

Theorem 11 *Consider a delay system* $S = \{A, B, C, D, F, G, H, M, N\}_{n,\nu,p,q}$ *as in* (12). *An internally positive system* $\bar{S} = \{\bar{A}, \bar{B}, \bar{C}, \bar{D}, \bar{F}, \bar{G}, \bar{H}, \bar{M}, \bar{N}\}_{2n,2\nu,2p,2q}$, *with*

$$
\begin{aligned}
\bar{A} &= \Gamma(A), & \bar{B} &= \Pi(B), & \bar{C} &= \Pi(C),\\
\bar{D} &= \Pi(D), & \bar{F} &= \Pi(F), & \bar{G} &= \Pi(G),\\
\bar{H} &= \Pi(H), & \bar{M} &= \Pi(M), & \bar{N} &= \Pi(N),
\end{aligned}
\tag{35}
$$

together with the six transformations

$$
\begin{aligned}
\bar{\xi} &= T^f_{\Xi}(\xi) = \pi(\xi), & \xi &= T^b_{\Xi}(\bar{\xi}) = \Delta_n \bar{\xi}, \\
\bar{x} &= T^f_X(x) = \pi(x), & x &= T^b_X(\bar{x}) = \Delta_\nu \bar{x}, \\
\bar{u} &= T_U(u) = \pi(u), & y &= T_Y(\bar{y}) = \Delta_q \bar{y},
\end{aligned} \tag{36}
$$

is an IPR of S.

Both the previous Definition and Theorem are a straightforward extension of the results presented in [18] to the case in which an output equation is explicitly associated to a coupled differential-difference system. Refer to the cited work for a detailed proof of Theorem 11.

4 Stability Analysis

In this section we will see how the IPR method can be fruitfully applied to arbitrary linear delay systems of the aforementioned classes, in order to exploit the remarkable stability conditions presented in Sect. 2.

Indeed, the main property of the IPR construction when dealing with the stability analysis is that, given an arbitrary linear system (delayed or not), its IPR is stable only if the original system is stable as well. This is readily illustrated, with no loss of generality, for the case of delay differential systems as in (4). Indeed, consider the state trajectory of the IPR of S (i.e. $\bar{S}$) defined as in Definition 7, and denoted by $\bar{x}(t)$. Then, the continuity of $T^b_X(\cdot)$ yields the implication:

$$
x(t) = T^b_X(\bar{x}(t)) \Rightarrow \|x(t)\| \leq \|T^b_X\| \|\bar{x}(t)\|. \tag{37}
$$

From this, the stability of $\bar{S}$ implies the stability of S. Such property, which is only sufficient (i.e. the IPR method can produce unstable positive representations of stable systems), is readily proved for the IPR of any class of systems and implies that a straightforward way to check the delay-independent stability of a given linear delay system can be pursued as follows:

- Given an arbitrary linear delay system S as in (4),(8) or (12), compute its IPR $\bar{S}$ by means of Theorems 9, 10 or 11, respectively;
- Then, check the stability of the obtained internally positive system $\bar{S}$ by means of the necessary and sufficient conditions of Theorems 3, 6 or 8, respectively;
- If $\bar{S}$ is delay-independent stable, then S is delay-independent stable as well.

While the previous algorithm introduces an interesting way to check the stability of arbitrary linear delay systems (a generally tough task as discussed in Sect. 2), in [12, 15–18] it has been show that the stability conditions obtained via the IPR method can be formulated directly on the matrices of the original (arbitrary) systems, thus avoiding the explicit computation of their positive representations. We will now briefly recall the obtained results.

4.1 Delay Differential Systems

Consider an arbitrary delay differential system S as in (4). Now consider its IPR, computed by means of Theorem 9, i.e. $\bar{S} = \{\{\bar{A}_k\}_0^m, \bar{B}, \bar{C}, \bar{D}\}_{2n,2p,2q}$, where

$$\bar{A}_0 = \Gamma(A_0) = \begin{bmatrix} d(A_0) + (A_0 - d(A_0))^+ & (A_0 - d(A_0))^- \\ (A_0 - d(A_0))^- & d(A_0) + (A_0 - d(A_0))^+ \end{bmatrix} \tag{38}$$

$$\bar{A}_k = \Pi(A_k) = \begin{bmatrix} A_k^+ & A_k^- \\ A_k^- & A_k^+ \end{bmatrix}, \qquad k = 1, \ldots, m. \tag{39}$$

Then, according to Theorem 3 $\bar{S}$ is delay-independent asymptotically stable if and only if $\sum_{k=0}^m \bar{A}_k$ is Hurwitz. As discussed above, the stability of the IPR of S implies the stability of S itself. Moreover, exploiting the properties of block circulant matrices, the Hurwitz stability of $\sum_{k=0}^m \bar{A}_k$ is shown (see [16]) to be equivalent to the Hurwitz-stability of both $\sum_{k=0}^m A_k$ and $A_0^{\mathscr{M}} + \sum_{k=1}^m |A_k|$, where $A_0^{\mathscr{M}} = d(A_0) + |A_0 - d(A_0)|$. The result is formalized in the following Theorem [16].

Theorem 12 *An arbitrary delay differential system S as in* (4) *is delay-independent asymptotically stable if the following conditions are satisfied:*

- $$\sum_{k=0}^m A_k \textit{ is Hurwitz} \tag{40}$$
- $$A_0^{\mathscr{M}} + \sum_{k=1}^m |A_k| \textit{ is Hurwitz.} \tag{41}$$

Note that condition (40) is also necessary for the delay-independent stability of (4), since it is required for the stability of the zero-valued delays case.

Remarkably, condition (41) can been shown [16] less conservative than the comparable norm-based condition (7) for $p = 1, +\infty$, while none of the two conditions is less conservative than the other for $p = 2$. We also remark that checking condition (41) is a straightforward task, since it involves a Metzler matrix, and many easy to check equivalent conditions exist for the Hurwitz-stability of Metzler matrices that do not require the explicit computation of its spectrum nor its characteristic polynomial. For example, the Hurwitz-stability of the Metzler matrix $M = A_0^{\mathscr{M}} + \sum_{k=1}^m |A_k|$ is equivalent to the positivity of all the n leading principal minors of $-M$.

4.2 Delay Difference Systems

Now consider an arbitrary delay difference system S as in (8). Its IPR can be computed according to Theorem 10, and is given by $\bar{S} = \{\{\bar{A}_k\}_1^m, \bar{B}, \bar{C}, \bar{D}\}_{2n,2p,2q}$, where

$$\bar{A}_k = \begin{bmatrix} A_k^+ & A_k^- \\ A_k^- & A_k^+ \end{bmatrix}, \qquad k = 1, \ldots, m. \tag{42}$$

Then, according to Theorem 6 $\bar{S}$ is delay-independent asymptotically stable if and only if $\sum_{k=0}^m \bar{A}_k$ is Schur.

By virtue of the properties of the spectrum of block circulant matrices, the Schur stability of $\sum_{k=0}^m \bar{A}_k$ is shown (see [17]) to be equivalent to the Schur-stability of both $\sum_{k=1}^m A_k$ and $\sum_{k=1}^m |A_k|$. Moreover, $\rho(\sum_{k=1}^m A_k) \leq \rho(\sum_{k=1} |A_k|)$, which yields the following result on the stability of arbitrary difference systems with constant delays.

Theorem 13 *An arbitrary delay difference system S as in* (8) *is delay-independent asymptotically stable if $\sum_{k=1}^m |A_k|$ is Schur.*

Remarkably, the condition presented in the previous Theorem has been shown [17] less conservative than that of Theorem 5 for $p = 1, +\infty$, while none of the two conditions is less conservative than the other for $p = 2$, thus completing an interesting similarity with the delay differential case. Also in this case, we highlight the fact that checking the Schur-stability of $M = \sum_{k=1}^m A_k$ can be easily done due to the fact that M is a nonnegative matrix, and simple equivalent conditions exist for the Schur-stability of this class of matrices. For example, M being Schur can be checked verifying the positivity of all the n leading principal minors of $(I_n - M)$.

4.3 Coupled Delay Differential-Difference Systems

Coming to arbitrary delay-differential systems as in (12), their IPR is computed by means of Theorem 11, which provides the positive system $\bar{S}=\{\bar{A}, \bar{B}, \bar{C}, \bar{D}, \bar{F}, \bar{G}, \bar{H}, \bar{M}, \bar{N}\}_{2n,2\nu,2p,2q}$, where:

$$\begin{aligned} \bar{A} = \Pi(A) &= \begin{bmatrix} A^+ & A^- \\ A^- & A^+ \end{bmatrix}, & \bar{B} = \Pi(B) &= \begin{bmatrix} B^+ & B^- \\ B^- & B^+ \end{bmatrix}, \\ \bar{C} = \Pi(C) &= \begin{bmatrix} C^+ & C^- \\ C^- & C^+ \end{bmatrix}, & \bar{D} = \Pi(D) &= \begin{bmatrix} D^+ & D^- \\ D^- & D^+ \end{bmatrix}. \end{aligned} \tag{43}$$

Then, by virtue of Theorem 8, $\bar{S}$ is delay-independent asymptotically stable if and only if $\bar{D}$ is Schur and $\bar{A} + \bar{B}(I_{2\nu} - \bar{D})^{-1}\bar{C}$ is Hurwitz. Interestingly, the conditions (which imply the stability of S), can be further reduced in dimensionality, as discussed in [18] and summarized in the following theorem.

Theorem 14 *An arbitrary coupled differential-difference system S as in (12) is delay-independent asymptotically stable if the following conditions are satisfied:*

- $|D|$ *is Schur* (44)
- $A + B(I_\nu - D)^{-1}C$ *is Hurwitz* (45)
- $A^{\mathscr{M}} + |B|(I_\nu - |D|)^{-1}|C|$ *is Hurwitz.* (46)

As discussed in Sect. 2.3, a class of neutral-type equation can be represented by means of the coupled differential-difference system (12). Then, Theorem 14 readily applies as follows.

Theorem 15 *A neutral-type equation with time-varying delay descibed by*

$$\frac{d}{dt}(x(t) - Dx(t - \delta(t))) = Ax(t) + Bx(t - \delta(t)), \quad t \geq t_0 \tag{47}$$

is delay-independent asymptotically stable if the following conditions are satisfied:

- $|D|$ *is Schur* (48)
- $A + (AD + B)(I_n - D)^{-1}$ *is Hurwitz* (49)
- $A^{\mathscr{M}} + |AD + B|(I_n - |D|)^{-1}$ *is Hurwitz.* (50)

For what concerns similar results presented in the literature, to the best of our knowledge no comparable condition exists for time-varying delays. If the comparison is restricted to the case of constant delays, similar conditions have been discussed [34, 38, 52]. In detail, in [38] it has been shown that (47), in the special case of constant delay ($\delta(t) = \delta \in \mathbb{R}^+$) is delay-independent asymptotically stable if both the following conditions are satisfied for some $p \in [1, \infty]$:

$$\|D\|_p < 1 \quad \text{and} \quad \mu_p(A) + \frac{\|A\|_p\|D\|_p + \|B\|_p}{1 - \|D\|_p} < 0. \tag{51}$$

In [18] it is shown that the conditions of Theorem 15 are less conservative than those in (51) for $p = 1, \infty$, and numerical examples in which Theorem 15 is less conservative than the results of [34, 52] are also reported. A result similar to Theorem 15 has been obtained in [2], Theorem 5.1, for the class of switching neutral-type systems, in the special case of a single constant delay. Moreover, the result of Theorem 15 is analogous to the one given in [49], Corollary III.4, if a single common constant delay is considered and no distributed delay is modelled. Yet again, we stress that our condition holds for possibly time-varying delays, while the cited other conditions, while holding for wider classes of systems, are limited to constant delays.

5 Examples

Due to space limitations, we refer to [12, 16–18] for various examples for each class of system addressed in this chapter. In the cited works, the proposed stability conditions are illustrated by means of numerical examples, and favourably compared to the existing similar results available in the literature.

6 Conclusions and Future Work

This chapter has presented an overview of the Internally Positive Representation (IPR) method for linear time-delay systems. After a brief survey of the available results for positive linear delay systems, the IPR technique for delay differential, delay difference and coupled delay differential-difference systems has been established. Finally, novel sufficient conditions for the delay-independent stability of such classes of linear systems have been discussed and favourably compared to similar results available in the literature. Future works may concern the extension of IPR method to other classes of delay systems, possibly with similar consequences on the stability of arbitrary systems of such classes. Moreover, as done in [9] for delay-free systems, the IPR method could be exploited to obtain a novel approach to designing interval observers for linear delay systems.

References

1. M. Ait Rami, Stability analysis and synthesis for linear positive systems with time-varying delays, in *Positive Systems* (Springer, 2009), pp. 205–2015
2. A.Y. Aleksandrov, O. Mason, Absolute stability and Lyapunov-Krasovskii functionals for switched nonlinear systems with time-delay. J. Franklin Inst. **351**(8), 4381–4394 (2014)
3. C.E. Avellar, J.K. Hale, On the zeros of exponential polynomials. J. Math. Anal. Appl. **73**(2), 434–452 (1980)
4. L. Benvenuti, L. Farina, The design of fiber-optic filters. J. Lightwave Technol. **19**(9), 1366–1375 (2001)
5. L. Benvenuti, L. Farina, B.D. Anderson, Filtering through combination of positive filters. IEEE Trans. Circuits Syst. I: Fund. Theory Appl. **46**(12), 1431–1440 (1999)
6. R.K. Brayton, Nonlinear oscillations in a distributed network. Q. Appl. Math. **24**(4), 289–301 (1967)
7. F. Cacace, L. Farina, A. Germani, C. Manes, Internally positive representation of a class of continuous time systems. IEEE Trans. Autom. Control **57**(12), 3158–3163 (2012)
8. F. Cacace, A. Germani, C. Manes, Stable internally positive representations of continuous time systems. IEEE Trans. Autom. Control **59**(4), 1048–1053 (2014)
9. F. Cacace, A. Germani, C. Manes, A new approach to design interval observers for linear systems. IEEE Trans. Autom. Control **60**(6), 1665–1670 (2015)
10. F. Cacace, A. Germani, C. Manes, R. Setola, A new approach to the internally positive representation of linear MIMO systems. IEEE Trans. Autom. Control **57**(1), 119–134 (2012)
11. J. Chen, C. Lien, K. Fan, J. Cheng, Delay-dependent stability criterion for neutral time-delay systems. Electron. Lett. **36**(22), 1897–1898 (2000)

12. F. Conte, V. De Iuliis, C. Manes, Internally positive representations and stability analysis of linear delay systems with multiple time-varying delays, in *Positive Systems: Theory and Applications* (POSTA 2016) Rome, Italy, 14–16 Sept. 2016 (Springer, 2017), pp. 81–93
13. S. Damak, M. Di Loreto, W. Lombardi, V. Andrieu, Stability analysis for a class of linear systems governed by difference equations (2013). arXiv:1312.7210 (2013)
14. S. Damak, M. Di Loreto, W. Lombardi, V. Andrieu, A bridge between Lyapunov-Krasovskii and spectral approaches for difference equations, in *Recent Results on Time-Delay Systems* (Springer, 2016), pp. 107–124
15. V. De Iuliis, Internally Positive Representations and Stability Analysis of Linear Delay Systems. Ph.D. thesis (2018)
16. V. De Iuliis, A. D'Innocenzo, A. Germani, C. Manes, Internally positive representations and stability analysis of linear differential systems with multiple time-varying delays. IET Control Theory Appl. **13**(7), 920–927 (2019). https://doi.org/10.1049/iet-cta.2018.5280
17. V. De Iuliis, A. Germani, C. Manes, Internally positive representations and stability analysis of linear difference systems with multiple delays. IFAC-PapersOnLine **50**(1), 3099 – 3104 (2017). 20th IFAC World Congress
18. V. De Iuliis, A. Germani, C. Manes, Internally Positive Representations and stability analysis of coupled differential-difference systems with time-varying delays. IEEE Trans. Autom. Control **64**(6), 2514–2521 (2019). https://doi.org/10.1109/TAC.2018.2866467
19. M. Di Loreto, J.J. Loiseau, On the stability of positive difference equations, in *Time Delay Systems: Methods, Applications and New Trends* (Springer, 2012), pp. 19–33
20. L. Farina, S. Rinaldi. *Positive Linear Systems: Theory and Applications*, vol. 50 (Wiley, 2011)
21. E. Fridman, New Lyapunov-Krasovskii functionals for stability of linear retarded and neutral type systems. Syst. Control Lett. **43**(4), 309–319 (2001)
22. E. Fridman, Stability of linear descriptor systems with delay: a Lyapunov-based approach. J. Math. Anal. Appl. **273**(1), 24–44 (2002)
23. E. Fridman, *Introduction to Time-Delay Systems* (Birkhäuser Basel, 2014)
24. A. Germani, C. Manes, P. Palumbo, State space representation of a class of MIMO systems via positive systems, in *2007 46th IEEE Conference on Decision and Control* (IEEE, 2007), pp. 476–481
25. A. Germani, C. Manes, P. Palumbo, Representation of a class of MIMO systems via internally positive realization. Eur. J. Control **16**(3), 291–304 (2010)
26. K. Gu, V.L. Kharitonov, J. Chen, *Stability of Time-Delay Systems* (Birkhäuser Basel, 2003)
27. K. Gu, Y. Liu, Lyapunov-Krasovskii functional for uniform stability of coupled differential-functional equations. Automatica **45**(3), 798–804 (2009)
28. K. Gu, S.I. Niculescu, Survey on recent results in the stability and control of time-delay systems. J. Dyn. Syst. Meas. Control **125**(2), 158–165 (2003)
29. W.M. Haddad, V. Chellaboina, Stability theory for nonnegative and compartmental dynamical systems with time delay. Syst. Control Lett. **51**(5), 355–361 (2004)
30. J.K. Hale, M.A. Cruz, Existence, uniqueness and continuous dependence for hereditary systems. Annali di matematica pura ed applicata **85**(1), 63–81 (1970)
31. J.K. Hale, S.M.V. Lunel, *Introduction to Functional Differential Equations* (Springer, New York, 1993)
32. Y. He, M. Wu, J.H. She, G.P. Liu, Delay-dependent robust stability criteria for uncertain neutral systems with mixed delays. Syst. Control Lett. **51**(1), 57–65 (2004)
33. J.M. Hendrickx, A. Olshevsky, Matrix p-norms are NP-hard to approximate if $p \neq 1, 2, \infty$. SIAM J. Matrix Anal. Appl. **31**(5), 2802–2812 (2010)
34. G.D. Hu, G.D. Hu, Some simple criteria for stability of neutral delay-differential systems. Appl. Math. Comput. **80**(2–3), 257–271 (1996)
35. I. Kabakov, On steam pressure control. Inzh. shornik **2**, 27–60 (1946)
36. T. Kaczorek, Stability of positive continuous-time linear systems with delays, in *Control Conference (ECC), 2009 European* (IEEE, 2009), pp. 1610–1613
37. H. Li, K. Gu, Discretized Lyapunov-Krasovskii functional for coupled differential-difference equations with multiple delay channels. Automatica **46**(5), 902–909 (2010)

38. L.M. Li, Stability of linear neutral delay-differential systems. Bull. Austral. Math. Soc. **38**, 339–344 (1988)
39. X. Liu, W. Yu, L. Wang, Stability analysis for continuous-time positive systems with time-varying delays. IEEE Trans. Autom. Control **55**(4), 1024–1028 (2010)
40. D. Luenberger, *Introduction to Dynamic Systems: Theory, Models, and Applications* (Wiley, 1979)
41. P. Martinez-Amores, Periodic solutions of coupled systems of differential and difference equations. Annali di Matematica Pura ed Applicata **121**(1), 171–186 (1979)
42. D. Melchor-Aguilar, Exponential stability of linear continuous time difference systems with multiple delays. Syst. Control Lett. **62**(10), 811–818 (2013)
43. W.R. Melvin, Stability properties of functional difference equations. J. Math. Anal. Appl. **48**(3), 749–763 (1974)
44. T. Mori, N. Fukuma, M. Kuwahara, Simple stability criteria for single and composite linear systems with time delays. Int. J. Control **34**(6), 1175–1184 (1981)
45. T. Mori, N. Fukuma, M. Kuwahara, On an estimate of the decay rate for stable linear delay systems. Int. J. Control **36**(1), 95–97 (1982)
46. B. Nagy, M. Matolcsi, Minimal positive realizations of transfer functions with nonnegative multiple poles. IEEE Trans. Autom. Control **50**(9), 1447–1450 (2005)
47. B. Nagy, M. Matolcsi, M. Szilvasi, Order bound for the realization of a combination of positive filters. IEEE Trans. Autom. Control **52**(4), 724–729 (2007)
48. P.H.A. Ngoc, Stability of positive differential systems with delay. IEEE Trans. Autom. Control **58**(1), 203–209 (2013)
49. P.H.A. Ngoc, H. Trinh, Novel criteria for exponential stability of linear neutral time-varying differential systems. IEEE Trans. Autom. Control **61**(6), 1590–1594 (2016)
50. S.I. Niculescu, *Delay Effects on Stability: A Robust Control Approach*, vol. 269 (Springer, 2001)
51. S.I. Niculescu, On delay-dependent stability under model transformations of some neutral linear systems. Int. J. Control **74**(6), 609–617 (2001)
52. J.H. Park, S. Won, A note on stability of neutral delay-differential systems. J. Franklin Inst. **336**(3), 543–548 (1999)
53. P. Pepe, The Liapunov's second method for continuous time difference equations. Int. J. Robust Nonlinear Control **13**(15), 1389–1405 (2003)
54. P. Pepe, On the asymptotic stability of coupled delay differential and continuous time difference equations. Automatica **41**(1), 107–112 (2005)
55. P. Pepe, Direct and converse Lyapunov theorems for functional difference systems. Automatica **50**(12), 3054–3066 (2014)
56. P. Pepe, E.I. Verriest, On the stability of coupled delay differential and continuous time difference equations. IEEE Trans. Autom. Control **48**(8), 1422–1427 (2003)
57. V. Rasvan, *Absolute Stability of a Class of Control Processes Described by Functional Differential Equations of Neutral Type* (Hermann, Equations Differentielles et Fonctionelles Non-lineaires, Paris, 1973)
58. V. Rasvan, Functional differential equations of lossless propagation and almost linear behavior, in *Plenary Lecture at 6th IFAC Workshop on Time-Delay Systems*, L'Aquila, Italy (2006)
59. G.M. Schoen, Stability and stabilization of time-delay systems. Ph.D. thesis, Swiss Federal Institute of Technology, Zurich, Switzerland (1995). http://e-collection.library.ethz.ch
60. G.M. Schoen, H.P. Geering, On stability of time delay systems, in *Proceedings of the 31st Annual Allerton Conference on Communication, Control and Computer*, Monticello, Il, pp. 1058–1060 (1993)
61. J. Shen, W.X. Zheng, Positivity and stability of coupled differential-difference equations with time-varying delays. Automatica **57**, 123–127 (2015)
62. R.A. Silkowski, Star-shaped regions of stability in hereditary systems. Ph.D. thesis (1976)
63. O. Toker, H. Özbay, Complexity issues in robust stability of linear delay-differential systems. Math. Control Signals Syst. (MCSS) **9**(4), 386–400 (1996)

64. E.I. Verriest, S.I. Niculescu, Delay-independent stability of linear neutral systems: a Riccati equation approach, in *Stability and control of time-delay systems* (Springer, 1998), pp. 92–100
65. S.S. Wang, C.H. Lee, T.H. Hung, New stability analysis of system with multiple time delays, in *American Control Conference* (ACC 1991), pp. 1703 – 1704 (1991)

Control by Set Invariance of a Class of Convolution Systems

Berna Bou Farraa, Rosa Abbou, and Jean Jacques Loiseau

Abstract We consider the control of a class of convolution systems subject to constraints on the state and the control, using the search of invariant sets. Our motivations come from production management. The papers provides an explicit characterization of the $\mathscr{D}$-invariance property of polytope sets for continuous-time systems. An interesting application of the concept of $\mathscr{D}$-invariance is to investigate the possibility of controlling a class of convolution system. The considered system corresponds to a supply chain. The controller design problem for supply chain, with constrained orders and inventories, and unknown customer demands variations, has been investigated. The problem is stated in terms of controlled input time delay system, with positivity constraints, subject to bounded disturbances.

1 Introduction

The papers provides an explicit characterization of the $\mathscr{D}$-invariance property of polytope sets for continuous-time systems. An interesting application of the concept of $\mathscr{D}$-invariance is to investigate the possibility of controlling a constrained system. Forrester [8] has observed that a distribution system of cascaded inventories and ordering procedures amplifies small disturbances that occur at the retail level. The same effect is observed in shops with many workstations. The flows between work stations, and the levels of buffer stores, are subject to variations, that are amplified along the production chain. This phenomenon is called Bullwhip Effect, concept first noticed, as said previously, by Forrester and thus it is also known as the Forrester

B. Bou Farraa (✉) · R. Abbou · J. Jacques Loiseau
Université Bretagne Loire, Laboratory of Digital Science of Nantes-LS2N,
1 rue de la Noë, 44321 Nantes cedex 03, France
e-mail: Berna.Boufarraa@ls2n.fr

R. Abbou
e-mail: Rosa.Abbou@ls2n.fr

J. Jacques Loiseau
e-mail: Jean-Jacques.Loiseau@ls2n.fr

G. Valmorbida et al. (eds.), *Accounting for Constraints in Delay Systems*,
Advances in Delays and Dynamics 12,
https://doi.org/10.1007/978-3-030-89014-8_3

effect. The bullwhip effect was named for the way the amplitude of a whip increases down its length. The further from the originating signal, the greater the distortion of the wave pattern. It has generated a continuous flow of research work for the last fifty years. Plenty of clever strategies already exist to avoid the Bullwhip Effect as well as minimizing the costs. We can cite stochastic optimization, echelon policies (MRP), lot sizing, Kanban and its variants, CONWIP, and also control approaches. A classical study of the Bullwhip effect is provided by Lee et al. [12], and a recent survey by Wang and Disney [22]. Simon (Nobel prize winner, 1978) introduced in 1952 [19] a fundamental model, and proposed to use the theory of servomechanisms to analyze the supply chain stability. This seminal paper was the first to formulate the management of inventories as a control problem. Forrester, (another Nobel prize, 1961) introduced the system dynamics [8], to model and study the logistic systems and more generally the economical systems. Towill used in 1982 [21] the Laplace transform to study inventory and order based production control system. More recently, some authors like Blanchini [2], Blanchini et al. [4], Ignaciuk and Bartoszewicz [11], and Hennet [9], have used the invariance principle for the design of control laws in this context.

In this chapter, we carry out another study on this problem. We are particularly interested into the description of a supply chain as an interconnected system. From the point of view of control, each node of a supply chain is a system, for which the inventory level is the output, the production order is the control input and the demand is a perturbation. Indeed, the production orders (and corresponding supplies orders) are calculated online so as to satisfactorily answer the demand, regulating the inventory level. The instantaneous production order at a given place is directly related to the consumption of goods that are ordered to the upstream nodes of the network, so that it plays the role of the demand for these ones. The characteristic features of such a system is the presence of delays and constraints. At a first level, the production rate, inventory levels, and demand are non-negative and bounded. At a second level, their range are intervals, which size depends on the demand variations. Our aim is to propose a control law that takes into account both delays and constraints, and permits to regulate the inventory level and limit the Forrester effect.

The chapter is organized as follows. We introduce the basic concepts of convex sets, polytopes, and invariance in Sect. 2. New conditions are introduced for the so-called $\mathscr{D}$-invariance of a polytope for a system without delay subject to a disturbance evolving in a given polytope $\mathscr{D}$. In Sect. 3, we consider a general model for logistic systems subject to product losses and variable demand, with multiple delays. This model is transformed into an equivalent system without delay using Artstein reduction [1]. The question of regulating the inventory levels of the nodes of this logistic system is then reduced to a pair of sub-problems, that are the regulation of the equivalent system without delay, and the relationship between the state of the system and that of the reduced system. A pair of conditions is therefore obtained, for the existence of a control law that permits to satisfy the external demand, while meeting the constraints that are imposed to the system. In Sect. 4, we show that the conditions lead to an explicit solution, in the case of a supply chain with two elements. Some conclusions are pointed out in Sect. 5.

2 Background Concepts

2.1 Polytopes

2.1.1 Definition

A polytope is a bounded polyhedron. The celebrated Minkowski-Weyl theorem (see e.g. [18]) states that every polytope can be represented in two different ways, either as a polyhedron, defined by a finite set of linear inequalities, or as a convex set, that is the set of convex combinations of a finite number of generators.

Theorem 1 (Minkowski-Weyl theorem) *A set $\mathscr{P} \subset \mathbb{R}^n$ being given, the following statements are equivalent.*

(i) $\mathscr{P}$ is bounded and there exists an integer q, a real matrix $P \in \mathbb{R}^{q\times n}$ and a real vector $\pi \in \mathbb{R}^q$ such that the following equality holds true:

$$\mathscr{P} = \mathrm{Pol}\,(P, \pi) := \{z \in \mathbb{R}^n \mid Pz \leq \pi\} \ . \tag{1}$$

(ii) There exists an integer p and a matrix $M \in \mathbb{R}^{n\times p}$ such that $\mathscr{P}$ is the convex set defined by

$$\mathscr{P} = \mathrm{Conv}\, M := \{z \in \mathbb{R}^n \mid \exists v \in \Gamma(p), z = Mv\} \ , \tag{2}$$

where $\Gamma(p)$ is the set of p-dimensional convex combinations, defined by:

$$\Gamma(p) = \left\{ v \in \mathbb{R}^p \mid v \geq 0, \sum_{i=1}^{p} v_i = 1 \right\} \ .$$

Furthermore, notice the following result, that is well-known (see e.g. [14]) and will be useful in the sequel.

Lemma 1 *With the notation introduced in Theorem 1, and for every vector $w \in \mathbb{R}^p$, we have the equality:*

$$\max\nolimits_{v\in\Gamma(p)} w^T v = \max\nolimits_i w_i \ . \tag{3}$$

As a consequence, the following holds true:

$$\max\nolimits_{z\in\mathrm{Conv}\, M} w^T z = \max\nolimits_i (w^T M)_i \ . \tag{4}$$

This permits to easily calculate the maximal value taken by a linear functional over a polytope. Following [14], we shall use the result in Sect. 2.3 to obtain bounds for input-output operators.

2.1.2 Multivariable Intervals

Definition 1 A multivariable interval is a set of the form $\mathscr{P} = \{z \in \mathbb{R}^n \mid z^{\min} \leq z \leq z^{\max}\}$, for some vectors $z^{\min}, z^{\max} \in \mathbb{R}^n$ satisfying $z^{\min} \leq z^{\max}$. It is denoted by $\mathscr{P} = [z^{\min}, z^{\max}]$.

The faces of a multivariable interval are in number $2 \times n$. They correspond to the subsets of those vectors that satisfy either $z_i = z_i^{\min}$ or $z_i = z_i^{\max}$, for some index i. The vertices of the multivariable interval are the vectors that, for each i, satisfy either $z_i = z_i^{\min}$ or $z_i = z_i^{\max}$. This set of vertices form a matrix M that has 2^n elements. These remarks are used to get polyhedral and convex representations of a multivariable interval.

Proposition 1 *The multivariable interval defined by $z^{\min}$ and $z^{\max}$ satisfies*

$$[z^{\min}, z^{\max}] = \text{Conv}\, M = \text{Pol}\,(P, \pi)\ ,$$

with the matrices M, P, and the vector π defined as follows:

$$M_{ij} = \begin{cases} z_i^{\min}, \text{ if } \beta_i(j-1) = 0\ , \\ z_i^{\max}, \text{ if } \beta_i(j-1) = 1\ , \end{cases}$$

for $j = 1$ to 2^n, where $\beta_i(k)$ denotes the ith *bit of a binary representation of any positive integer k,*

$$P = \begin{pmatrix} I_n \\ -I_n \end{pmatrix}\ ,$$

where I_n is the identity matrix, and

$$\pi = \begin{pmatrix} z^{max} \\ -z^{min} \end{pmatrix}\ .$$

Proof The interval $[z^{\min}, z^{\max}]$ is the set of vectors $z \in \mathbb{R}^n$ that satisfies both conditions $z \leq z^{\max}$ and $z^{\min} \leq z$. The definition of the matrix P and the vector π as in Proposition 1 corresponds to these inequalities. The columns of the matrix M are the vertices of the interval $[z^{\min}, z^{\max}]$. They form the family $\{z \in \mathbb{R}^n \mid \text{for } i = 1 \text{ to } n,\ z_i = z_i^{\min} \text{ or } z_i = z_i^{\max}\}$. The matrix M is obtained by ordering this family as indicated in Proposition 1. ■

2.2 Controlled Invariance and $\mathscr{D}$-invariance of Polytopes

2.2.1 Definitions

We first introduce the following definitions.

Definition 2 A set $\mathscr{P}$ is said to be invariant for the system $\dot{z} = -Az$, if for every initial condition $z(0) \in \mathscr{P}$, the trajectory $z(t)$ remains in $\mathscr{P}$ for $t \geq 0$.

Definition 3 A set $\mathscr{P}$ is said to be controlled invariant for the system $\dot{z} = -Az + Bu$, with admissible control in a set $\mathscr{U}$, if for every initial condition $z(0) \in \mathscr{P}$, there exists a control $u(t) \in \mathscr{U}$ such that the trajectory $z(t)$ remains in $\mathscr{P}$ for $t \geq 0$.

Definition 4 A set $\mathscr{P}$ is said to be $\mathscr{D}$-invariant for the system $\dot{z} = -Az + Bu - Ed$, with admissible control in a set $\mathscr{U}$, if for every initial condition $z(0) \in \mathscr{P}$, there exists a control $u(t)$ such that the trajectory $z(t)$ remains in $\mathscr{P}$ for $t \geq 0$, whatever be $d(t) \in \mathscr{D}$.

2.2.2 Characterization of the Invariant Polytopes

Dorea and Hennet have obtained in [7] an explicit characterization of the controlled invariance property of polyhedral sets with respect to linear continuous-time systems. Necessary and sufficient conditions for a polyhedron to be (A, B)-invariant are established in the form of linear matrix relations, and presented in the Theorem 2. The study of controlled invariance is then extended in their work, to control the constrained disturbed systems. So Theorem 3 present the conditions of $\mathscr{D}$-invariance for general convex polyhedra. In the following, we shall denote P_{r_i} the ith row of a matrix P, and M_{c_j} the jth column of a matrix M.

Theorem 2 *A polytope $\mathscr{P} = \text{Pol}\,(P, \pi)$ is controlled invariant if and only if, for every point z on the face corresponding to the ith row of P, such that $P_{r_i} z = \pi_i$, and $P_{r_j} z \leq \pi_j$ for $j \neq i$, there exists a vector $u \in \mathscr{U}$ such that $P_{r_i}(-Az + Bu) \leq 0$.*

Theorem 3 *A polytope $\mathscr{P} = \text{Pol}\,(P, \pi)$ is $\mathscr{D}$-invariant if and only if, for every point z on the face corresponding to the ith row of P, such that $P_{r_i} z = \pi_i$, and $P_{r_j} z \leq \pi_j$ for $j \neq i$, there exists a vector $u \in \mathscr{U}$ satisfying $P_{r_i}(-Az + Bu - Ed) \leq 0$, for every $d \in \mathscr{D}$.*

Dorea and Hennet [7] actually reformulated the conditions of the controlled invariance of general convex polyhedral sets in terms of the existence of constrained matrices, that can be checked using linear programming and differential analysis. Blanchini and Miani [3] provided another characterization, in terms of a convex representation of a polytope and using the dual analysis for the (A, B)-invariance of polyhedral region. We provide here another slightly different characterization, using a double description of the polytope that is considered. This new formulation of the conditions for the controlled invariance has the advantage to construct explicitly the matrix U which allows the calculation of an admissible control law. This approach reduces the algorithmic complexity with respect to the approach of Blanchini and Miani. On the contrary, the formulation of Dorea and Hennet provides a simpler test to verify the existence of admissible solutions. But when it comes to verify the sufficiency and to build the control law u, they come back to our characterization defined in Corollary 1, so they used our characterization for practical implementation of the control.

Corollary 1 *Let D be any matrix such that If $\mathscr{D} = \text{Conv } D$, and denote*

$$\delta_i = \min_j \, (P_{r_i} E D)_j \; for \;\; i = 1 \; to \; q.$$

The polytope $\mathscr{P} = \text{Conv } M = \text{Pol}\,(P, \pi)$ is $\mathscr{D}$-invariant if and only if for each vertex of $\mathscr{P}$, say M_{c_j}, there exists a vector, say $U_{c_j} \in \mathscr{U}$, such that

$$P_{r_i}(-AM_{c_j} + BU_{c_j}) \leq \delta_i \;, \tag{5}$$

for each face containing the vertex M_{c_j}, such that $P_{r_i} M_{c_j} = \pi_i$. In addition, if $\mathscr{U}$ is a convex set, then an admissible control law is given in the form

$$u(t) = Uv(t) \;, \tag{6}$$

where for $t \geq 0$, $v(t) \in \Gamma(p)$ is any convex combination vector satisfying

$$z(t) = Mv(t) \;. \tag{7}$$

Proof According to Theorem 3, the conditions (5) are necessarily satisfied by each vertex if the polytope is $\mathscr{D}$-invariant. The vectors U_{c_j} that are successively determined for each vertex form a matrix U that permits to define the control law (6) for every vector $z(t)$ in $\mathscr{P}$. Notice that, by definition of the matrix M, a vector $v(t) \in \Gamma(p)$ satisfying (7) can be defined for each vector $z(t) \in \mathscr{P}$. The sufficiency of the conditions comes from the fact that this control law can be used to check the conditions of Theorem 3. Indeed, let us consider a vector z that is on the face of $\mathscr{P}$ that corresponds to the i-th row of matrix P. This vector is a convex combination of the vertices that satisfy $P_{r_i} M_{c_j} = \pi_i$. If J_i denotes the set of indexes of these vertices, we have $z = \sum_{j \in J_i} \alpha_j M_{c_j}$, with $\alpha_j \geq 0$, for $j \in J_i$, and $\sum_{j \in J_i} \alpha_j = 1$. Defining v by $v_j = \alpha_j$, for $j \in J_i$, and $v_j = 0$, if $j \notin J_i$, one can verify that $u = Uv$ is an admissible control that verifies $P_{r_i}(-Az + Bu) \leq 0$. Indeed, u is admissible since it is a convex combination of admissible control vectors, and $\mathscr{U}$ is convex, and the condition of Theorem 3 holds true since, for every $d \in \mathscr{D}$, we have $\sum_{j \in J_i} \alpha_j \times d = d$. ∎

2.2.3 Algorithmic Remarks

2.2.4 Parameterization of the Control Laws

The vector v is not uniquely defined in general. This lets a first degree of freedom in the definition of the matrix U. Another degree of freedom comes from the fact that (5) is an inequality. Therefore, we have the following parameterization for each column U_{c_j} of the matrix $U \in \mathbb{R}^{m*p}$:

$$U_{c_j} \in \left\{ u \in \mathbb{R}^p \mid u \in \mathscr{U} \text{ and } P_{r_i} Bu \leq P_{r_i} AM_{c_j} + \delta_i \text{ if } i \text{ satisfies } P_{r_i} M_{c_j} = \pi_i \right\} .$$

This set is a polyhedron if the set $\mathscr{U}$ of control vectors is itself a polyhedron. This parameterization permits to easily check the existence of a solution.

2.2.5 Complexity of the Verification

It is important to notice that the number of vertices of a polyhedron, say p, increases more rapidly than its number of faces, that is q. For instance, in the case of a multivariable interval, we already noticed that $q = 2n$ and $p = 2^n$. Table 1 gives the number of faces and vertices of an interval for $n = 3$, 4, 10 and 20.

For this reason, the initial formulation of Dorea and Hennet [7] provides a verification of the existence of a solution that is cheaper from the point of view of the algorithmic complexity. As far as the construction of the control law is concerned, Dorea and Hennet also use in their paper the determination of the vertices of the polyhedron that is considered. The method of Corollary 1 actually follows the same line of though (see the proof of the theorem in [3]). One can estimate that the limit number of vertices that can be handled in practice is of the order of 10^6. In our study we must calculate a matrix $U \in \mathbb{R}^{m \times p}$, which means that our formulation is limited to $n = 20$. Whereas in the approach of Blanchini and Miani, we must calculate in addition of U a non-negative matrix $P \in \mathbb{R}^{p \times p}$, whose columns are elements of the convex set $\Gamma(p)$. So this calculation leads to a great increase in complexity. As a consequence, the formulation in [3] is limited to the treatment of intervals in dimension $n = 10$, in practice.

Table 1 Number of faces and vertices of an interval for $n = 3$, 4, 10 and 20

n	q	p
3	6	8
4	8	16
10	20	1024
20	40	1 048 576

2.3 Bounds for Input-Output Convolution Systems

Let us consider the input-output system defined by

$$y(t) = \int_0^t H(\tau)u(t-\tau)d\tau \ , \tag{8}$$

together with two matrices M, P, and a vector π, of convenient dimensions.

Moussaoui et al. [14] and Loiseau [13] formulated the following result.

Theorem 4 *The output $y(t)$ belongs to* Pol (P, π), *for $t \geq 0$, for every input $u(t)$ evolving in* Conv M, *for $t \geq 0$, if and only if the following inequalities hold true:*

$$\int_0^t \max_j (PH(\tau)M)_{ij}\, d\tau \leq \pi_i \ .$$

for $i = 1$ to q.

Proof We proceed by equivalences, successively obtained using the definition (1) of Pol (P, π), the definition (8) of $y(t)$, the definition (2) of Conv M, and Lemma 1:

$$\begin{aligned} y(t) \in \text{Pol}\,(P,\pi) &\Leftrightarrow\ Py \leq \pi \ , \\ &\Leftrightarrow \int_0^t PH(\tau)u(t-\tau)d\tau \leq \pi \ , \\ &\Leftrightarrow \int_0^t PH(\tau)Mv(t-\tau)d\tau \leq \pi \ , \\ &\Leftrightarrow \int_0^t \max_{j=1}^{q}(PH(\tau)M)_{ij}\, d\tau \leq \pi_i \ , \text{for } i = 1 \text{ to } q \ . \end{aligned}$$

This ends the proof. ■

3 Control of Time Delay System by Set Invariance

3.1 Problem Formulation

3.1.1 Elementary Logistic System Subject to Constraints

We consider an elementary logistic system composed of a supplying unit and a storage unit. Simon [19] explains that the open-loop model for the inventory dynamic level $x(t)$ is described by the first order delayed equation

$$\dot{x}(t) = u(t-\theta) - d(t), \quad \text{for } t \geq \theta.$$

$u(t)$ denotes the production order and $d(t)$ the external demand. The three variables $x(t)$, $u(t)$, and $d(t)$ are real variable, well defined for $t \geq 0$. The delay θ corresponds to the time needed to complete the task. This model can actually be used in many situations, to model an elementary production system, a buffer, or a simple transportation device, for instance. The delay is a production time in the first case, a propagation delay in the second case, a transportation delay in the latter one. We consider in the sequel a production system.

In addition to the delay, the considered system is subject to different constraints. Our problem consists in calculating $u(t)$ online so that

$$u^{\min} \leq u(t) \leq u^{\max} ,$$

$$x^{\min} \leq x(t) \leq x^{\max} ,$$

where $u^{\min}$, $u^{\max}$, $x^{\min}$, and $x^{\max}$ are given real numbers, for every demand satisfying

$$d^{\min} \leq d(t) \leq d^{\max} ,$$

where $d^{\min}$ and $d^{\max}$ are given real numbers. This type of constraints is frequently found in practical applications, being generally associated to physical limitations. For instance, in the case of a production system, $u^{\min}$ and $u^{\max}$ define the capacity of production, and $x^{\min}$, $x^{\max}$ the storage capacity of the system. The bounds $d^{\min}$ and $d^{\max}$ for the demand are determined by the market and specified to the considered production system. At a first level, one may consider that the variables are positive and bounded. At a second level, the range of variation of the demand is precisely estimated, and one wants to contain the variations of the instantaneous production orders and inventory level, which leads to the actual formulation.

Since $u(t)$ is defined for $t \geq 0$, we will assume that for $t \in [0, \theta]$, we have

$$\dot{x}(t) = \phi(t) - d(t) ,$$

where $\phi(t)$ is a function that depends on the work-in-progress at initial time.

3.1.2 Multivariable Model of Delayed Logistic System

The interconnection of n elementary logistic systems leads to a continuous-time system described by the differential equation:

$$\dot{x}(t) = -Ax(t) + \sum_{k=1}^{h} B_k u(t-\theta_k) - B_0 u(t) - Ed(t) , \tag{9}$$

where $x(t) \in \mathbb{R}^n, u(t) \in \mathbb{R}^m, d(t) \in \mathbb{R}^r$ are, respectively, the state vector, the control input vector and the disturbance vector, while A, B_k, for $k = 0$ to h, and E are constant matrices of appropriate dimensions. The components of the vectors $x(t)$, $u(t)$ and $d(t)$ denote, respectively, the inventory level of the different nodes, the instantaneous production order of the different edges, and the external demand levels. The goods are used to fulfill customer demands at each node, and they are acquired with delays θ_k from multiple production stages, while $B_0 u(t)$ presents the external production orders that can feed the different nodes of production. The inventory level is subject to a continuous decrease due to the losses on the goods modeled by the matrix A.

They are basically subject to constraints of the form:

$$x^{\min} \leq x(t) \leq x^{\max} , \tag{10}$$

$$u^{\min} \leq u(t) \leq u^{\max} , \tag{11}$$

$$d^{\min} \leq d(t) \leq d^{\max} , \tag{12}$$

where the inequalities are meant component by component, and $x^{\min}, x^{\max} \in \mathbb{R}^n$, $u^{\min}, u^{\max} \in \mathbb{R}^m$, and $d^{\min}, d^{\max} \in \mathbb{R}^r$. These constraints define three interval polytopes, that are denoted by $\mathscr{P}_x$, $\mathscr{P}_u$, and $\mathscr{P}_d$ in the sequel.

3.1.3 Control Approach

In terms of control theory, the fundamental problem we face is to design online an admissible control $u(t)$ which will stabilize the delayed system while ensuring that the constraints on the state $x(t)$ are met for any bounded perturbation $d(t)$. This control problem can be formulated in terms of $\mathscr{D}$-invariance.

Our aim is to formulate conditions for the existence of an admissible control law, and describe the control design procedure. The method starts with the elimination of the delays using Artstein's reduction. The problem is then twofold. A first subproblem is to determine conditions for the resulting reduced model, while the second subproblem is to express the relationship between the variables of the initial model and the reduced one, in order to obtain finally the conditions for the original delayed system.

3.2 Artstein's Reduction

3.2.1 Reduced Model

Using Artstein reduction [1], delayed systems are transformed into systems without delay. The new system is an ordinary differential control equation, so that various control problems can be analyzed using the reduced model.

Given first the model of a logistic system, say (9), we define a reduced variable $z(t)$ by the following equality:

$$z(t) = x(t) + \sum_{k=1}^{h} \int_{t-\theta_k}^{t} \mathrm{e}^{-A(t-\tau)} \mathrm{e}^{A\theta_k} B_k u(\tau)\, d\tau \ . \tag{13}$$

Proposition 2 *Using the definition (13) of the variable $z(t)$, and verifying the system model (9), we can deduce the following differential equation, called the reduced system:*

$$\dot{z}(t) = -Az(t) + Bu(t) - Ed(t) \ , \tag{14}$$

with $B = \sum_{k=1}^{h} \mathrm{e}^{A\theta_k} B_k - B_0$.

Proof One first remark that the interval that appears in the definition (13) can be rewritten as $\mathrm{e}^{-At} \int_{t-\theta_k}^{t} \mathrm{e}^{A\tau}\, \mathrm{e}^{A\theta_k} B_k u(\tau)\, d\tau$. Its derivative is then calculated using the usual formulas for the product rule of derivation, and of an integral which integrator does not depend on the integration bounds. We obtain the following result

$$\frac{d}{dt} \int_{t-\theta_k}^{t} \mathrm{e}^{-A(t-\tau)} \mathrm{e}^{A\theta_k} B_k u(\tau)\, d\tau \ =$$

$$-A \int_{t-\theta_k}^{t} \mathrm{e}^{-A(t-\tau)} \mathrm{e}^{A\theta_k} B_k u(\tau)\, d\tau \ + \ \mathrm{e}^{A\theta_k} B_k u(t) \ - \ B_k u(t-\theta_k).$$

The conclusion is then obtained by the time derivation of the identity (13), and using the definition (9) of the system. ■

In the following section, we address the control of the reduced model (14).

3.2.2 $\mathscr{D}$-invariance Conditions for the Reduced System

Let us denote $\mathscr{D}$ the polytope $\mathscr{P}_d = [d^{\min}, d^{\max}]$, and let D be a matrix such that $\mathscr{D} = \mathrm{Conv}\, D$. Then, according to Corollary 1, a given polyhedron of $\mathbb{R}^n$, say $\mathscr{P} = \mathrm{Pol}\,(P, \pi) = \mathrm{Conv}\, M$, is $\mathscr{D}$-invariant for the reduced system if and only if there exists a matrix $U \in \mathbb{R}^{m \times p}$ satisfying the conditions (5) of Corollary 1.

The matrix U is not uniquely defined. According to Sect. 3.2, the jth column of this matrix can be chosen in the polyhedron

$$\mathscr{P}_u \cap \{u \in \mathbb{R}^m \mid P_{r_i} Bu \leq \delta_i + P_{r_i} AM_{c_j} \text{ if } P_{r_i} M_{c_j} = \pi_i\} \ . \tag{15}$$

This characterization permits to effectively check the existence of a solution. The fact that the polyhedron corresponding to the p vertices of $\mathscr{P}$ are non-empty is a necessary and sufficient condition for the $\mathscr{D}$-invariance of the polytope $\mathscr{P}$, and for the existence of an admissible matrix U.

In addition, the computation of the matrix U gives a way to implement a control law. We use this feature in the following section. An admissible control law is given by $u(t) = Uv(t)$, for every vector $v(t)$ satisfying $z(t) = Mv(t)$, which necessarily exists, provided that $z(t) \in \mathscr{P}$. This control law makes the closed-loop system meet the constraints $z(t) \in \mathscr{P}$, $u(t) \in \mathscr{P}_u$, for every demand verifying $d(t) \in \mathscr{P}_d$.

Remark that we hence have the equality:

$$\begin{pmatrix} z(t) \\ u(t) \end{pmatrix} = \begin{pmatrix} M \\ U \end{pmatrix} v(t) . \tag{16}$$

3.3 $\mathscr{D}$-invariance of the Time Delay System

3.3.1 Conditions of $\mathscr{D}$-invariance

From the definition of $z(t)$ by (13), and using the control characterization expressed by (16), we obtain

$$x(t) = Mv(t) - \sum_{k=1}^{h} \int_{t-\theta_k}^{t} \mathrm{e}^{-A(t-\tau)} \mathrm{e}^{A\theta_k} B_k U v(\tau)\, d\tau , \tag{17}$$

where $v(t)$ is at each instant a vector of convex combination, that lies in $\Gamma(q)$ at each instant t. One can remark that (17) is the definition of a convolution system, which kernel is equal to:

$$H(t) = M\delta(t) - \sum_{k=1}^{h} H_k(t) ,$$

where the terms $H_k(t)$ are $n \times q$ matrices, function of the time, and defined by:

$$H_k(t) = \begin{cases} \mathrm{e}^{-At} \mathrm{e}^{A\theta_k} B_k U & \text{if } t \in [0, \theta_k] , \\ 0 & \text{if } t \geq \theta_k . \end{cases}$$

This convolution system has $v(t)$ for input, and $x(t)$ as output, and the input is constrained by the condition $v(t) \in \Gamma(q)$. Remark that $\Gamma(q) = \mathrm{Conv} I_q$. These preliminary remarks permit to apply Theorem 4 to obtain conditions under which the variable $x(t)$ meets a given set of constraints.

Proposition 3 *The pair* $(G, \gamma) \in \mathbb{R}^{s\times n} \times \mathbb{R}^n$ *being given, the variable* $x(t)$ *of system (9) looped by the control law (16) evolves in the polyhedron* $\mathrm{Pol}(G, \gamma)$ *if the variable* $z(t)$ *evolves in the polyhedron* $\mathscr{P}$ *if, for* $i = 1$ *to* s, *we have:*

$$\max_{j=1}^{q}(GM)_{ij} - \sum_{k=1}^{h}\int_0^{\theta_k} \min_{j=1}^{q}(G\mathrm{e}^{-A\tau}B_kU)_{ij}d\tau \leq \gamma_i \ . \tag{18}$$

The result is directly deduced from Theorem 4. Finally, we obtain the following conclusion.

Theorem 5 *The variables* $x(t)$ *and* $u(t)$ *resulting from the interconnection of system (9) looped by the control law (16) evolves in the polyhedra* $[x^{\min}, x^{\max}]$ *and* $[u^{\min}, u^{\max}]$, *for every demand* $d(t)$ *in the range* $[d^{\min}, d^{\max}]$ *if there exists a polytope* $\mathscr{P} = \mathrm{Pol}\,(P, \pi) = \mathrm{Conv}\, M$ *and a matrix* U *satisfying the condition (5) of Corollary 1, such that for* $i = 1$ *to* n, *the following inequalities hold true:*

$$\max_{j=1}^{q} M_{ij} - \sum_{k=1}^{h}\int_0^{\theta_k} \min_{j=1}^{q}(\mathrm{e}^{-A\tau}B_kU)_{ij}d\tau \leq x_i^{\max} \ ,$$

$$x_i^{\min} \leq \min_{j=1}^{q} M_{ij} - \sum_{k=1}^{h}\int_0^{\theta_k} \max_{j=1}^{q}(\mathrm{e}^{-A\tau}B_kU)_{ij}d\tau \ .$$

Proof The interval $[x^{\min}, x^{\max}]$ is the set of vectors $x \in \mathbb{R}^n$ that satisfy both inequalities $x \leq x^{\max}$ and $-x \leq -x^{\min}$. As a consequence, the inequalities of the theorem are obtained using twice Proposition 3, the first time taking $G = I_n$ and $\gamma = x^{\max}$, and the second time taking $G = -I_n$ and $\gamma = -x^{\min}$. ∎

This result permits to check whether or not a given polytope $\mathscr{P}$ provides a solution to our control problem.

3.3.2 Study Analysis

Our study is presented by a production network composed of n elementary systems expressed by (9), where the intrinsic parameters are the delays θ_k for $k = 1$ to h, and the loss factors τ on the storage level of each node. The system is modeled by the state vector $x(t)$, the control vector $u(t)$, and the disturbance vector $d(t)$ that are constrained and evolve respectively in the polytopes $\mathscr{P}_x = [x^{\min}, x^{\max}]$, $\mathscr{P}_u = [u^{\min}, u^{\max}]$ and $\mathscr{P}_d = [d^{\min}, d^{\max}]$.

We use the Artstein reduction in order to eliminate the delays and to obtain a free-delay dynamic system expressed by (14). After defining a polyhedron for the variation of $z(t)$, we have determined the $\mathscr{D}$-invariance conditions for the reduced system using Corollary 1 and condition (15). Then a way to implement a control law is chosen and expressed by (16), in such a way that the closed-loop system meets the constraints

$z(t) \in \mathscr{P}$, $u(t) \in \mathscr{P}_u$, for every demand verifying $d(t) \in \mathscr{P}_d$. Furthermore, using the expression (17) that describes the relation between the variables of the reduced model and the delayed model, we have found the $\mathscr{D}$-invariance conditions for the delayed system using Proposition 3 and Theorem 5.

From a theoretical point of view, the obtained sufficient conditions are used to define properly the polytope $\mathscr{P}$. This latter permits to check the admissibility of the proposed control $u(t)$ which stabilizes the delayed system while ensuring that the constraints on the state $x(t)$ are met for any bounded perturbation $d(t)$. But in fact, the research of a convenient polytope $\mathscr{P}$ is a difficult problem, of high complexity in general. The computation is anyway possible in many practical situations, as illustrated in the Sect. 4 for the case of a supply chain with two nodes and equal delays.

In the following application, we will study the case of a supply chain with two nodes, equal delays and equal loss factors. In this example, we observe that B is square and invertible. This property is actually verified for any supply chain with an arbitrary number of nodes, that is an important class of system in production management. For this reason, this example is particularly significant. The transformation $w(t) = B^{-1}z(t)$ is then performed on the reduced system (14), in order to obtain a new decoupled system that reads

$$\dot{w}(t) = -B^{-1}AB\, w(t) + u(t) - B^{-1}E\, d(t)\ . \tag{19}$$

Since this system is decoupled, it is natural to reach the region where evolves $w(t)$ in an interval of the form

$$w(t) \in [w^{\min}, w^{\max}]\ . \tag{20}$$

Accordingly, our control variable will evolve in a polyhedron $[u^1, u^2]$ that is included into the polyhedron $\mathscr{P}_u = [u^{\min}, u^{\max}]$:

$$u^1,\ u^2 \in [u^{\min}, u^{\max}]\ . \tag{21}$$

As consequence, we obtain the polyhedron $\mathscr{P}$ in the form

$$\mathscr{P} = B[w^{\min}, w^{\max}]\ . \tag{22}$$

Therefore, we obtain a set of inequalities describing the admissible region in the space of parameters, taking into account the invariance conditions for the Theorem 5, together with the constraints (10), (11), (12) and our conditions of the control, that are (15) completed by (21 and (22). These conditions are affine in the parameters $d^{\min}$, $d^{\max}$, $u^{\min}$, $u^{\max}$, $x^{\min}$, $x^{\max}$, and $w^{\min}$, $w^{\max}$, u^1, and u^2. By eliminating the control parameters $w^{\min}$, $w^{\max}$, u^1, and u^2 by projection, we finally obtain a final set of conditions, that only depend from the system parameters, and that are necessary and sufficient conditions for the existence of a control law of form (16) parametrized through (21 and (22), that solves our constrained control problem for a supply chain.

3.3.3 Remark on the Initialization

We have remarked in Sect. 3.1.1 that the evolution of a production system during the first instants may depend from an initial fonction, that can be interpreted in terms of initial work-in-progress. The same actually occurs in the case of a multivariable logistic system. As a consequence, the Eqs. (13) and (14) may be satisfied only for t greater than or equal to the maximal element of the set of delays θ_k, where $k = 1$ to h. We can always assume that the sequence θ_k is arranged in increasing order, so that the maximal element is θ_h. During the initialization period, $t \in [0, \theta_h]$, the reduced variable $z(t)$ also depends on some initial functions $\phi_k(t)$, that correspond to the initial works in progress along the edges of the logistic network that is considered. As a consequence, the invariance of $\mathscr{P}$ is actually obtained for $t \geq \theta_h$, if the conditions of Corollary 1 are satisfied.

The interpretation of the work-in progress depends on the kind of systems that is modelized, and on the interpretation of the initial time. In the case of a production system, the initial time may be the start-up of the process, after a long stop for instance, or the time where a new controller starts to manage the system, after a release of the management decision process. In the first case, the work-in-progress is initially zero, while in the second case, it could be nonzero, for instance if the new management procedures are launched online, without interruption of the production. The former case requires a specific analysis of the initialization period, and the second one requires another specific analysis, to study the commutation and the adaptation period of the new controller. For this reason, and for the sake of the available space, we shall not discuss further this point. Notice anyway that such a study was done in the monovariable case [15].

4 Application on a Supply Chain with Two Elements

We consider a supply chain with two nodes, the first one feeding the second, that in turn satisfies an external demand. We suppose that the production delays of both elementary systems are identical, say $\theta = 1$, as well as their loss factor τ also equal to 1. The interconnection of the two elementary logistic systems is illustrated in the following schema (Fig. 1).

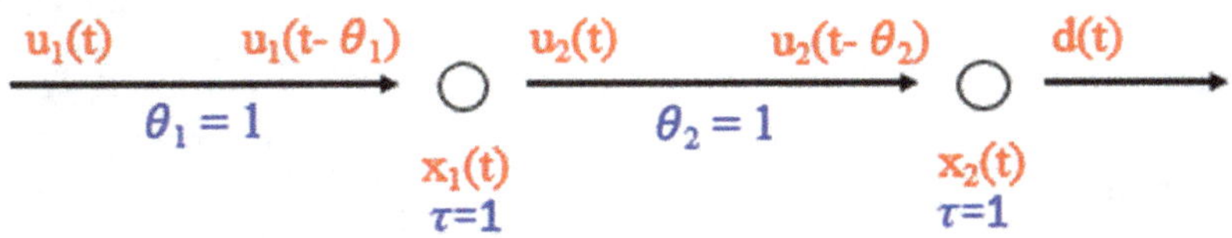

Fig. 1 Application of supply chain with dimension 2

The corresponding matrices defining the system and the reduced system are:

$$A = \begin{pmatrix} 1 & 0 \\ 0 & 1 \end{pmatrix}, \quad B_1 = \begin{pmatrix} 1 & 0 \\ 0 & 1 \end{pmatrix}, \quad B_0 = \begin{pmatrix} 0 & 1 \\ 0 & 0 \end{pmatrix}, \quad E = \begin{pmatrix} 0 \\ 1 \end{pmatrix}, \quad B = \begin{pmatrix} \mathrm{e} & -1 \\ 0 & e \end{pmatrix}.$$

As suggested in Sect. 3.3.2, we construct the invariant polytope $\mathscr{P}$ in the form (22). The variable $w(t) = B^{-1}z(t)$ is the solution of the system (19), with

$$B^{-1} = \begin{pmatrix} \mathrm{e}^{-1} & \mathrm{e}^{-2} \\ 0 & \mathrm{e}^{-1} \end{pmatrix}, \text{ and } B^{-1}E = \begin{pmatrix} \mathrm{e}^{-2} \\ \mathrm{e}^{-1} \end{pmatrix}.$$

Using (15) and Corollary 1, we see that an interval $[w^{\min}, w^{\max}]$ is invariant for (19), and the polytope $\mathscr{P} = B[w^{\min}, w^{\max}]$ is invariant for (14), using the control law of the form (16). As explained in Sect. 2.1.2, $[w^{\min}, w^{\max}]$ can be represented as

$$[w^{\min}, w^{\max}] = \mathrm{Conv}\begin{pmatrix} w_1^{\min} & w_1^{\max} & w_1^{\min} & w_1^{\max} \\ w_2^{\min} & w_2^{\min} & w_2^{\max} & w_2^{\max} \end{pmatrix}.$$

Because the system (19) is decoupled, the matrix U is chosen in such a way that $\mathrm{Conv}\, U = [u^1, u^2]$, say:

$$U = \begin{pmatrix} u_1^1 & u_1^2 & u_1^1 & u_1^2 \\ u_2^1 & u_2^1 & u_2^2 & u_2^2 \end{pmatrix}.$$

Accordingly to (15), the vectors u^1 and u^2 should be in the interval $[u^{\min}, u^{\max}]$ and the conditions of Corollary 1, say

$$u^1 \geq u^{\min}, \quad u^1 \leq u^{\max}, \tag{23}$$

$$u^2 \geq u^{\min}, \quad u^1 \leq u^{\max}, \tag{24}$$

$$u^1 \leq w^{\max} + B^{-1}Ed^{\min}, \tag{25}$$

$$u^2 \geq w^{\min} + B^{-1}Ed^{\max}. \tag{26}$$

In addition, $w^{\min}$ and $w^{\max}$ must verify

$$w^{\min} \leq w^{\max}. \tag{27}$$

The matrix M is expressed by

$$M = \begin{pmatrix} ew_1^{\min} - w_2^{\min} & ew_1^{\max} - w_2^{\min} & ew_1^{\min} - w_2^{\max} & ew_1^{\max} - w_2^{\max} \\ ew_2^{\min} & ew_2^{\min} & ew_2^{\max} & ew_2^{\max} \end{pmatrix} . \tag{28}$$

Using Theorem 5, we obtain the following additional conditions:

$$x^{\min} \leq \begin{pmatrix} ew_1^{\min} - w_2^{\max} \\ ew_2^{\min} \end{pmatrix} - (e-1)\max(u^1, u^2) , \tag{29}$$

$$\begin{pmatrix} ew_1^{\max} - w_2^{\min} \\ ew_2^{\max} \end{pmatrix} - (e-1)\min(u^1, u^2) \leq x^{\max} . \tag{30}$$

We have obtained 9 inequalities, (23), (24), (25), (26), (27), (29) and (30). These inequalities are necessary and sufficient conditions that satisfy the constraints (10) and (11) for every demand $d(t)$ in the range (12), when the system (9) is looped by the control law defined by (13), (16) and (22). One can remark that these conditions are affine. They define an admissible region in the space of parameters $d^{\min}$, $d^{\max}$, $u^{\min}$, $u^{\max}$, $x^{\min}$, $x^{\max}$, $w^{\min}$, $w^{\max}$, u^1, and u^2. This region is polyhedral. Finally by eliminating successively the variables u^1, u^2, $w^{\min}$ and $w^{\max}$, we obtain the following conditions,

$$(e-1)u_1^{\max} + u_2^{\max} + x_1^{\max} \geq e^{-1}d^{\max} , \tag{31}$$

$$u_1^{\max} \geq (1 - e^{-1} + e^{-2})d^{\max} , \tag{32}$$

$$u_2^{\max} \geq d^{\max} , \tag{33}$$

$$x_2^{\max} \geq (1 - e^{-1})d^{\max} , \tag{34}$$

$$x_2^{\max} + (e-1)u_2^{\max} \geq (e-1)d^{\max} . \tag{35}$$

that are independent from the control parameters, and are necessary ans sufficient for the existence of an admissible control law defined by (13), (16) and (22).

A fortiori, these conditions are just sufficient for the existence of an admissible control law solving the constrained control problem that is considered. The conditions are not conservative, since the definitions (16) and (22) cover a large class of control laws, parametrized in terms of the vectors $w^{\min}$, $w^{\max}$, u^1, and u^2.

5 Conclusion

The papers provides a new characterization of the $\mathscr{D}$-invariance property of polytope sets for continuous-time systems. An interesting application of the concept of $\mathscr{D}$-invariance is to investigate the possibility of controlling a class of constrained convolution systems. The considered system is the model of an interconnected logistic network, and our motivation is to design management procedures that permit to answer a demand, that evolves in a specified range, without exceeding the system sizing. The controller design problem for supply chain, with constrained orders and inventories, and unknown customer demands variations, has been investigated. The problem is stated in terms of controlled input time delay system, with positivity constraints, subject to bounded disturbances. Sufficient conditions for the existence of an admissible control were determined on a two dimensional example.

References

1. Z. Artstein, Linear systems with delayed control: a reduction. IEEE Trans. Autom. Control **27**(4), 869–879 (1982)
2. F. Blanchini, Feedback control for linear time-invariant systems with state and control bounds in the presence of disturbances. IEEE Trans. Autom. Control **35**(11), 1231–1234 (1990)
3. F. Blanchini, S. Miani, C.E.T. Dorea, J.-C. Hennet, Discussion on '(A, B)-invariance conditions of polyhedral domains for continuous-time systems' by CET. Dorea and JC. Hennet. Eur. J. Control 5(1), 82–86 (1999)
4. F. Blanchini, R. Pesenti, F. Rinaldi, W. Ukovich, Feedback control of production-distribution systems with unknown demand and delays. IEEE Trans. Robot. Autom. **16**(3), 313–317 (2000)
5. F.M. Callier, C.A. Desoer, An algebra of transfer functions for distributed linear time-invariant systems. IEEE Trans. Circuits Syst. **25**(9), 651–663 (1978)
6. I.I. Delice, R. Sipahi, Inventory dynamics models of supply chains with delays; system-level connection & stability, in *Topics in Time Delay Systems*, ed. by J.J. Loiseau et al. Lecture Notes in Control and Information Sciences, vol. 388. (Springer, Berlin, Heidelberg, 2009), pp. 349-358
7. C.E.T. Dorea, J.-C. Hennet, (A, B)-invariance conditions of polyhedral domains for continuous-time systems. Eur. J. Control **5**(1), 70–81 (1999)
8. J.W. Forrester, *Industrial Dynamics* (MIT Press, Wiley, Cambridge, Mass, New York, 1961)
9. J.-C. Hennet, A bimodal scheme for multi-stage production and inventory control. Automatica **39**(3), 793–805 (2003)
10. J.-C. Hennet, S. Tarbouriech, Stability and stabilization of delay differential systems. Automatica **33**(3), 347–354 (1997)
11. P. Ignaciuk, A. Bartoszewicz, *Congestion Control in Data Transmission Networks: Sliding Modes and Other Designs* (Springer, London, 2013)
12. H.L. Lee, V. Padmanabhan, A.S. Whang, Information distorsion in a supply chain: the bullwhip effect. Manag. Sci. **43**(4), 546–558 (1997)
13. J.J. Loiseau, Polyhedral invariance for convolution systems over the Callier-Desoer class, in *Positive Systems*, ed. by F. Cacace, L. Farina, R. Setola, A. Germani. POSTA 2016. Lecture Notes in Control and Information Sciences, vol. 471 (Springer, Cham, 2017), pp. 151–170
14. C. Moussaoui, R. Abbou, J.J. Loiseau, On bounds of input-output systems. Reachability set determination and polyhedral constraints verification, in *Proceedings of the IFAC World Congress 2014, IFAC Proceedings*, vol. 47(3), pp. 11012–11017 (2014)

15. C. Moussaoui, R. Abbou, J.J. Loiseau, Robust inventory control of production systems subject to uncertainties on demand and lead times. Int. J. Prod. Res. **55**(8), 2177–2196 (2017)
16. M. Nagumo: Über die lage der integralkurven gewhönlicher differential-gleichungen, in *Proceedings of the Physico Mathematical Society of Japan*, pp. 551–559 (1942)
17. C.E. Riddals, S. Bennett, The stability of supply chains. Int. J. Prod. Res. **40**(2), 459–475 (2002)
18. A. Schrijver, *Theory of Linear and Integer Programming* (Wiley, 1986)
19. H. Simon, On the application of servomechanism theory in the study of production control. Econometrica **69**(1), 247–268 (1952)
20. S. Tarbouriech, J.-C. Hennet, Conditions of stability via positive invariance for delay systems. J. Européen des Systèmes Automatisés **31**(6), 983–999 (1997)
21. D.R. Towill, Dynamic analysis of an inventory and order based production control systems with transportation delay. Eur. J. Prod. Econ. **92**, 3–12 (1982)
22. X. Wang, S.M. Disney, The bullwhip effect: progress, trends and directions. Eur. J. Oper. Res. **250**(3), 691–701 (2016)

Sub-platooning via Agent Separation for Improved Traffic Flow Metrics in a Car-Following Model

Min Koh, Adrián Ramírez, and Rifat Sipahi

Abstract An agent separation paradigm is presented over the well-known Optimal Velocity Model (OVM) with the aim to reduce jam formations. The main idea is to pose a constraint on the connectivity of a vehicle with the preceding one, and whenever this constraint is violated, the particular vehicle gets disconnected and acts as a leader of the vehicles following it. This separation from the downstream of traffic can then yield improved traffic flow metrics including reduced jam formations in the upstream of the traffic.

1 Introduction

Traffic jam formations are the source of many undesirable effects, including human safety, emissions, and time loss during commute. Traffic accidents in 2002 for example were the leading cause of death for individuals of every age from 3 to 33 [28]; emissions impact negatively the environment with significant carbon footprint, and time lost in a poorly operating traffic has considerable impact on workforce productivity [5, 16, 33–35]. Due to these reasons, a large body of literature has been

This work was supported in part by the US National Science Foundation Award 1536397.

M. Koh · A. Ramírez · R. Sipahi (✉)
Department of Mechanical and Industrial Engineering, Northeastern University, Boston, MA 02115, USA
e-mail: rifat@coe.neu.edu; r.sipahi@northeastern.edu

M. Koh
e-mail: koh.m@northeastern.edu

A. Ramírez
e-mail: adrian.ramirez@ipicyt.edu.mx

A. Ramírez
CONACYT - Division of Control and Dynamical Systems, IPICYT, San Luis Potosí, SLP 78216, México

G. Valmorbida et al. (eds.), *Accounting for Constraints in Delay Systems*,
Advances in Delays and Dynamics 12,
https://doi.org/10.1007/978-3-030-89014-8_4

devoted to the study of traffic flow, primarily based on the analysis of mathematical models developed to represent the flow dynamics [9, 11, 24, 25, 29].

Two main approaches exist in the literature when it comes to modeling traffic flow dynamics, see reviews in [9, 26]; one of them is based on a macroscopic representation of traffic in terms of density and flow rates, and the other takes a microscopic viewpoint by treating each vehicle as a particle interacting with other vehicles in its immediate vicinity. While each approach has its own advantages, one key advantage of microscopic traffic flow models is their ability to represent the granularity of traffic flow at the resolution of vehicles, to thereby model particular and distinct characteristics of vehicles as well as drivers.

Given that jam formations can be in part attributed to human imperfections including human reaction delays [3, 8, 30], microscopic models which explicitly model human driving offer opportunities to explain in what ways humans could inadvertently promote jam formations. To this end, a large body of literature focuses on "car following" models with the aim to understand how human reaction delays and human sensitivity in making acceleration/deceleration decisions could influence traffic flow, see also [6, 10, 21, 22, 25]. Other focus points include receiving stimuli from multiple vehicles [18], human anticipation and perception capabilities [30], and the use of feedback control to modulate vehicle behavior [15].

Reducing jam formations in traffic is challenging. This is because in a scenario where all vehicles are operated by human drivers, one does not have too many parameters to control to prevent jam formations. Self-driving vehicles can remedy this issue as their behavior can be strategically modulated in a way to influence the fate of traffic flow dynamics. To this end, many results indicate that these vehicles can be controlled in certain ways to render improved traffic flow metrics [12, 20, 32].

Overall, published work is focused on considering that each vehicle in a platoon receives stimulus from at least one preceding vehicle. In the case of self-driving vehicles, more information can be available through wireless communication between vehicles and/or through highway infrastructure that could inform these vehicles. On the other hand, we argue that more connectivity between vehicles is not necessarily the most ideal for several reasons. First of all, we know that jam formations become more apparent in larger size platoons mainly because velocity perturbations from downstream could amplify upstream of traffic causing jam formations [15, 30]. Secondly, on an analogous problem of multi-agent system (MAS) with inter-agent time delays, our results show that when few edges in the agent network are removed, this can help agents reach consensus faster [13]. In other words, in the presence of delays, a network of agents can dissipate perturbations much more effectively if not all the agents are connected to each other. Along these lines, we recently showed that when edges in a MAS can exist *subject to* proximity constraint, an unstable agent network can divide into multiple sub-networks within each of which stability can be recovered [14].

In view of the above observations, we suggest that a proximity constraint when implemented on a car-following model has the potential to split a platoon of vehicles into several sub-platoons with their respective leaders, and with each leader acting as a stop valve not allowing perturbations to leak from a downstream platoon to an upstream one. We propose that, if the separation of a platoon into sub-platoons can be designed properly, then each sub-platoon will achieve jam-free flow, and hence jam formations can be reduced from an otherwise jam-prone large platoon of vehicles. We explore these concepts in this chapter over a well-known car following model, Optimal Velocity Model (OVM) [3].

2 A Self-Regulation Mechanism via Agent Separation

Here, a self-regulation mechanism is summarized from [14]. Before presenting the details of this mechanism, we provide the model of the network dynamics under consideration, given by

$$\dot{x}_i(t) = \sum_{j=1, j\neq i}^{n} a_{ij}[x_j(t-\tau) - x_i(t-\tau)],\ i = 1, \ldots, n, \tag{1}$$

subject to

$$a_{ij} = \begin{cases} a > 0 \text{ when } |x_{ji}^{\tau}| < T \\ 0 \text{ otherwise,} \end{cases} \tag{2}$$

where $|x_{ji}^{\tau}| = |x_j(t-\tau) - x_i(t-\tau)|$ is the distance between nodes j and i at time $t-\tau$. The constant threshold T determines the connectivity between the agents, and thereby the graph structure $\mathscr{G}$ of the network.

Notice that (1) can be written in matrix form

$$\dot{\mathbf{x}}(t) = \mathbf{A}\mathbf{x}(t-\tau), \tag{3}$$

where $\mathbf{A} = \{a_{ij}\}$ is the negative Laplacian matrix. Types of models in Eq. (3) have been broadly studied in the literature including the consideration of threshold, see [14] and the references therein. The main motivation behind the threshold condition is that an agent does not necessarily remain within the communication coverage of all the remaining agents at all times.

Without considering the threshold condition, we know that the *stability of consensus* in (3) holds for delays $\tau \in [0, \tau^*)$, where τ^* is defined as the *delay margin*. In (3), τ^* is closely related to the eigenvalues of $\mathbf{A}$ [23]. For homogeneous agents, it is given by

Lemma 1 *The delay margin τ^* of (3), where $\mathbf{A}$ is subject to the complete graph with homogeneous agents $a_{ij} = a$, is given by $\pi/(2an)$ [13].*

Notice that delay margin is inversely proportional to both the coupling strength a and the size of the network n. For a given coupling strength, the multi-agent system of smaller network size can maintain its consensus stability for larger delays.

When considering the threshold condition, *if* the agents in the network lose connectivity due to threshold in such a way that the graph $\mathcal{G}$ of the network of size n is split into two separate subgraphs $\mathcal{G}_1$ and $\mathcal{G}_2$ respectively of size n_1 and n_2, then one expects, as per Lemma 1, that the delay margins τ_1^* and τ_2^* of the subgraphs $\mathcal{G}_1$ and $\mathcal{G}_2$ are larger than that of $\mathcal{G}$.

The idea of subgrouping of agents can be valuable when the system with $\mathcal{G}$ is unstable, i.e., its delay margin is smaller than the delay in the network $\tau^* < \tau$. Under the threshold condition (2), if the graph splits into smaller size subgraphs $\mathcal{G}_1, \mathcal{G}_2, \ldots, \mathcal{G}_k$, then each subnetwork is guaranteed to have a larger delay margin $\tau_k^* > \tau^*$ as per Lemma 1. If the subgraphs are of sufficiently small size, that is, their delay margins are sufficiently large, then within each subnetwork, stability can be recovered since $\tau_k^* > \tau$ will also hold.

Let us demonstrate the above described scenario over a simulation with $n = 6$ agents. To properly study the effects of the threshold in (2), we select the threshold value T to be larger than the largest difference between agents' initial states. That is, none of the agents at time $t = 0$ are subject to the threshold condition, and hence $\mathcal{G}$ at

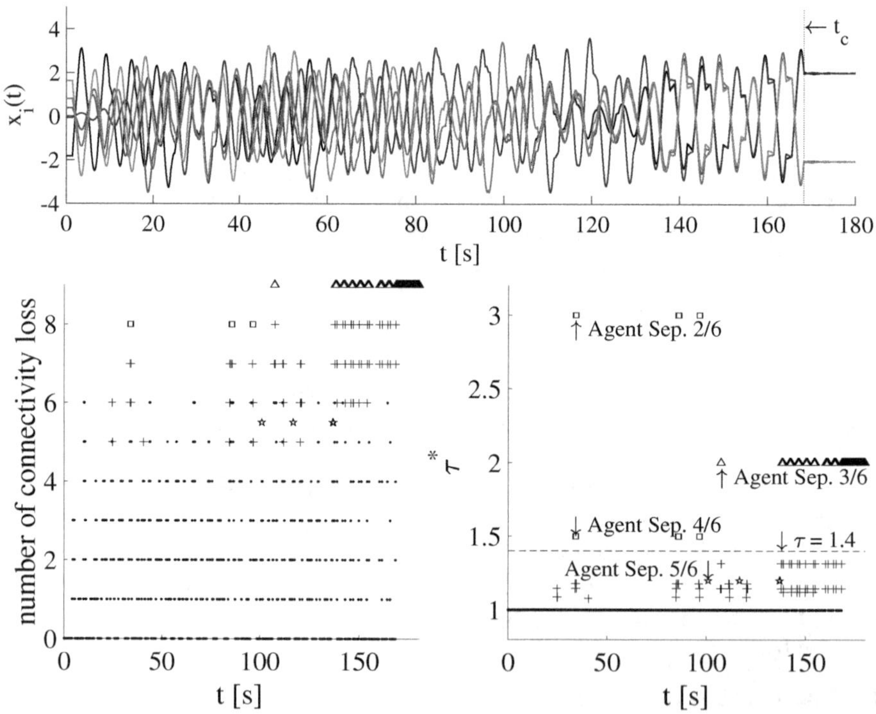

Fig. 1 Time simulation of $n = 6$ agents in MATLAB, with fixed time step $\Delta t = 0.01$. Initial condition $x_i(0) = [-1.8380, -0.8963, -0.0749, 0.3718, 0.8144, -1.6230]^T$, threshold value $T = 3.8071$, inter-agent delay $\tau = 1.4$. Top: Agents' states subject to threshold condition. The complete graph $\mathcal{G}$ separates at $t_c = 168.25[s]$. Bottom: Due to connectivity losses, delay margin of the network changes

time $t = 0$ is a complete graph. For simplicity, we select agents' coupling strengths as $a = \pi/(2n)$ such that the delay margin of $\mathcal{G}$ is normalized to $\tau^* = 1$. In this example, the delay in the network is set to $\tau = 1.4$. That is, the consensus dynamics is unstable ($\tau > \tau^*$) at time $t = 0$. Next, time simulations are performed in MATLAB as shown in Fig. 1 (top).

One interesting result in Fig. 1 is that agent regrouping through graph splitting could regulate toward stability. This can be understood from Fig. 1 (bottom) where through graph separation different delay margins arise for different subgraphs. Graph separations and which agents separate are marked with arrows in the figure panel. In addition, even if there is no graph separation, delay margin can increase due to connectivity losses [13], see Fig. 1 (bottom). In terms of graph separation, there are three different regimes:

- For the graph $\mathcal{G}$, whenever the agents are unstable, e.g., at time $t = 0$, they will eventually be subjected to the threshold condition (2) and hence some agents will lose their connectivity with some other agents.
- At several time instants, graph separation is observed however at least one of the subgraphs despite being of smaller size does not have sufficiently large delay margin against the network delay τ. In this case, agents' states in this unstable subnetwork grow in amplitude and *re-connect* with the stable subgraph. With this re-connection, the graphs become connected and the delay margin drops due to increased network size, see for example time $t = 137.2$ in Fig. 1.
- If and whenever all the subnetworks achieve sufficiently large delay margins, then they will all seek stabilization and this can then lead to a separated graph, see, e.g., $t \geq 168.25$ in Fig. 1.

Notice that the above graph separation mechanism is triggered solely by the physics of the problem under the threshold condition, and it promotes stability through reduced size of subgraphs. In some sense, this is an inherent self-regulation mechanism that will be triggered whenever agents' states grow and reach the threshold level. In such cases, one expects a combination of the three regimes described above. We now wonder whether a similar type of self-regulation mechanism could exist in a car following model, and whether or not one can impose graph separation in order to improve certain traffic flow metrics, e.g., related to accelerations, travel time, stop time, and fuel consumption.

3 A Car Following Model and Linear Stability

We start with a fundamental car-following model from [3], called the Optimal Velocity Model (OVM) and given by

$$\ddot{x}_i(t) = a\Big[V\big(h_i(t-\tau)\big) - \dot{x}_i(t-\tau)\Big], \tag{4}$$

$$h_i(t) = x_{i-1}(t) - x_i(t), \tag{5}$$

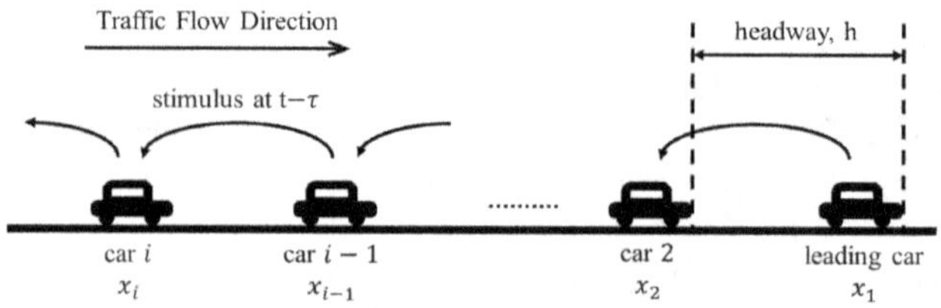

Fig. 2 A typical car-following model with human reaction delay τ

where the constants $a \geq 0$ and $\tau \geq 0$ are respectively the driver's sensitivity and reaction time, $h_i(t)$ is the headway between the $(i-1)$-st and i-th vehicle, and $x_i(t)$ is the position of the i-th vehicle at time t, see Fig. 2. Here, the Optimal Velocity Function (OVF) denoted by $V(\cdot)$, which depends on the headway, can be considered as the "desired speed" and is given by $V(h) = 16.8\big[\tanh\big(0.0860(h-25)\big) + 0.913\big]$, see [4]. This model which was shown to explain both free and congested flow in a unified framework, has attracted broad interest in the physics and engineering communities; see also [2]. We acknowledge that this model and its respective graph structure are different from the MAS considered in the previous section. However, a qualitatively similar self-regulation mechanism could as well exist on this dynamics.

Following next standard approaches, we linearize the above described dynamics around its equilibrium of fixed headway h^* and velocity $V(h^*) > 0$. It reads

$$\ddot{\epsilon}(t) + a\dot{\epsilon}(t-\tau) + aV(h^*)\epsilon(t-\tau) = aV(h^*)\delta(t-\tau), \tag{6}$$

where $\delta(t)$ and $\epsilon(t)$, both continuously differentiable, represent respectively the perturbations in the $(i-1)$-st and i-th vehicle from the headway equilibrium. Notice in (6) that the driver's reaction time $\tau \neq 0$ introduces a time shift and hence the linear dynamics has infinitely many eigenvalues.

We now investigate the interplay between driver's sensitivity a, delay τ, and stability of the perturbation dynamics $\epsilon(t)$. Since each vehicle follows only one preceding vehicle, stability can be studied based on the dynamics of a generic pair of vehicles. In this setting, δ in (6) is the heterogeneous part and therefore does not contribute to linear stability. Hence, stability of (6) is studied through the roots of its characteristic equation (CE)

$$f(s,\tau) = s^2 + ase^{-s\tau} + aV(h^*)e^{-s\tau} = 0. \tag{7}$$

Notice that the presence of the exponential term $e^{-\tau s}$ induces infinitely many characteristic roots hence it is impossible to compute all of the roots. This is the main reason why the stability analysis of (6) through (7) is not a trivial task [17].

The main idea behind analyzing the stability of (6) is to recognize that the real parts of the roots of CE change continuously with respect to system parameters, including delay [7]. For this reason, the only way the system can switch from a stable to unstable behavior (or, vice versa) for delay $\tau = \tau_c > 0$, is that these roots

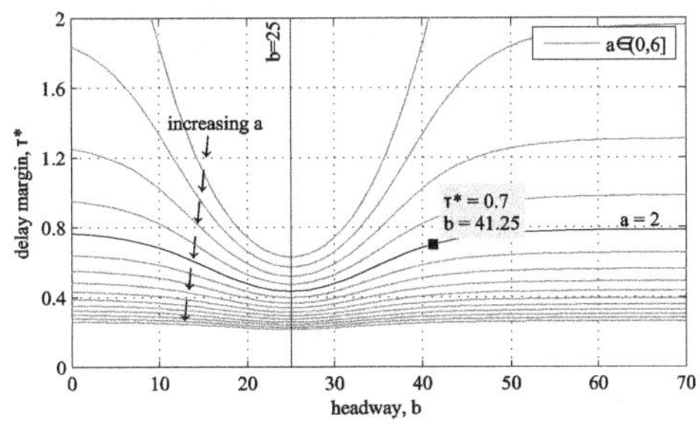

Fig. 3 Delay margin τ^* for the linearized dynamics (6), with respect to driver's sensitivity a and headway equilibrium $h^* b$

touch the imaginary axis at $s = j\omega_c$, where $\omega_c \geq 0$ without loss of generality. Since for the particular problem at hand, the delay-free ($\tau = 0$) system is stable for $a > 0$ and $V(h^*) > 0$, we have that the system will remain stable up to a delay margin τ^*, which is the smallest positive of all $\tau = \tau_c$ satisfying $f(j\omega, \tau) = 0$ for some $\omega = \omega_c$.

Next, we find the set of delay values τ_c that creates a characteristic root on the imaginary axis $j\omega$, see also [17, 19, 27]. To this end, following [31], let s be a pure imaginary root for some $\tau > 0$ then, as per the symmetricity of characteristic roots with respect to the real axis of the complex plane, we have that $-s$ is also a pure imaginary root for the same delay value. That is, $f(s, \tau) = 0$ and $f(-s, \tau) = 0$ share a common root. It then follows from (7), $s^2 = -a\big[V(h^*) + s\big]e^{-s\tau}$ and $(-s)^2 = -a\big[V(h^*) - s\big]e^{s\tau}$. After the elimination of $e^{-s\tau}$ from these two equations, we obtain $s^4 = a^2\big[V(h^*)^2 - s^2\big]$. Substituting $s = j\sqrt{\lambda} = j\omega$ and since $\lambda = \omega^2$, one obtains a single critical frequency given by $\omega_c = \sqrt{[(a^2 + \sqrt{a^4 + 4a^2V(h^*)^2}]/2}$.

We next proceed with the computation of the associated critical delays $\tau = \tau_c$. Using Euler's formula $e^{-j\omega\tau} = \cos(\omega\tau) - j\sin(\omega\tau)$ in CE, and solving the real and the imaginary parts, we have the expression of countably infinite, periodically spaced, critical delays $\tau_{c,l} = \{\tan^{-1}[-\omega_c/-V(h^*)] \pm 2\pi l\}/(-\omega_c)$, $l = 0, \pm 1, \pm 2$, where l is adjusted such that $\tau_{c,l} > 0$. In light of the above developments, given (6) with $a > 0$ and $V(h^*) > 0$, we formally express its delay margin as follows $\tau^* = \min_l \tau_{c,l}$. That is, the linearized leader-follower OVM in (6) remains stable for $\tau \in [0, \tau^*)$ and that the system due to its imaginary roots $s = \mp j\omega_c$ for $\tau = \tau^*$ oscillates in steady state at the frequency ω_c. Finally, Fig. 3 shows the delay margin with respect to variations of the headway h^* for different values of drivers' sensitivities $a \in (0, 6]$. Interestingly, τ^* exhibits a minimum at $h^* = 25$ independent of this sensitivity range. In addition, we highlight the setting $(h, \tau^*) = (41.25, 0.7)$, which is used in the next section. When $\tau = 0.7$ in the system, then the initial headway larger/less than 41.25 indicates that the linear dynamics is stable/unstable. This information can be used in future work to select the threshold condition, T.

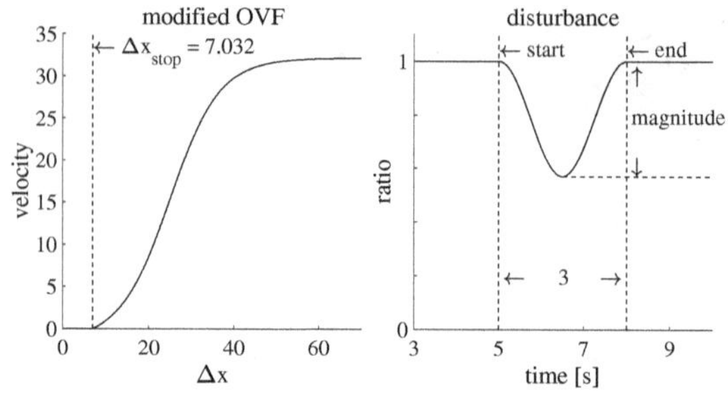

Fig. 4 Modified OVF and disturbance generated by the leader of the platoon. Left: non-negative "desired" velocity determined by modified OVF. Right: Velocity profile of the leader car, ratio$\times V(h^*)$ where $h^* = 50$, obtained based on a modified cosine function with magnitude $= 0.43$ and frequency $1/3$ [Hz]

4 Case Study

In this section, we utilize the parameter settings from the previous section, obtained on the linearized OVM dynamics to perform several simulations. Primarily, the interest is to investigate how human reaction delay τ could influence the nonlinear regime of the dynamics in (4)–(5), and how this could lead to undesirable traffic flow metrics. We shall then propose a network separation idea to improve those metrics. To this end, consider first the modified OVF adjusted to enforce zero "desired" vehicle speed for headway values less than Δx_{stop}, see Fig. 4 (left). This modification however does not affect the linear stability analysis performed in the previous section, for equilibrium headway greater than Δx_{stop}. The modified OVF is given by

$$V(\Delta x) = \begin{cases} 16.8\big[\tanh\big(0.0860(\Delta x - 25)\big) + 0.913\big] & \text{if } \Delta x_{stop} < \Delta x \\ 0 & \text{otherwise} \end{cases} \tag{8}$$

where $\Delta x_{stop} = \tanh^{-1}(-0.913)\big/0.086 + 25 = 7.0319$. Next, to disturb the equilibrium dynamics, we propose to create a perturbation on the traffic flow, induced by the leader of the platoon, whose velocity profile is designed as ratio$\times V(h^*)$, see Fig. 4 (right). For terminology purposes, this leader car is not numbered, and car number 1 follows the leader car.

We now perform numerical simulations under the above settings and those described in Fig. 4. Simulations are performed in MATLAB/SIMULINK with a time resolution of $\Delta t = 0.001$ using ODE4 solver based on Runge-Kutta method. Here, we consider a platoon of 11 cars in a linear path with identical driver sensitivities $a = 2$, reaction time delays $\tau = 0.7$, initial headway $\Delta x = h^* = 50$, and initial velocity $V(50) = 31.6886$. These values are selected to maintain linear stability in view of Fig. 3. To perturb the equilibrium, the disturbance in Fig. 4 (right) is provided next as an exogenous input from the leader car to the first car, whose influence then propagates upstream of traffic.[1] One important observation in the simulations is that vehicles $1 \sim 6$ travel without a stop and go action although their velocities fluctuate, but cars $7 \sim 11$ exhibit stop and go behavior, see Fig. 5. Whenever the velocity reaches zero, or drops below a certain threshold, this reflects as traffic jams.

4.1 Distance Threshold Condition on Car 7 and Platoon Separation

We now introduce a threshold condition on car 7, which is the car that exhibits first the stop and go behavior in Fig. 5 when no threshold condition is considered. The threshold condition is determined based on the headway as follows,

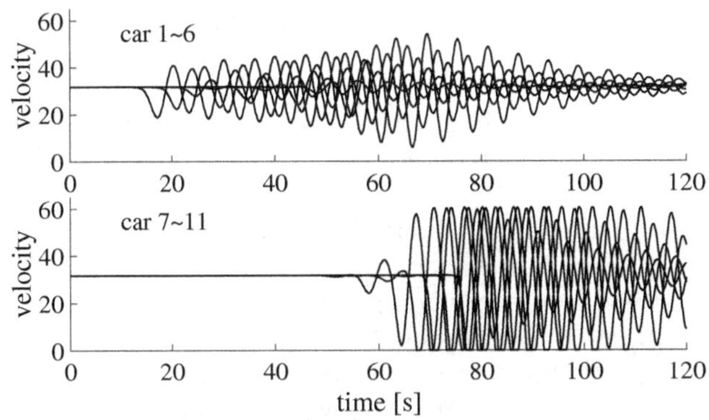

Fig. 5 Traffic jam in car $7 \sim 11$ based on the perturbation at the equilibrium

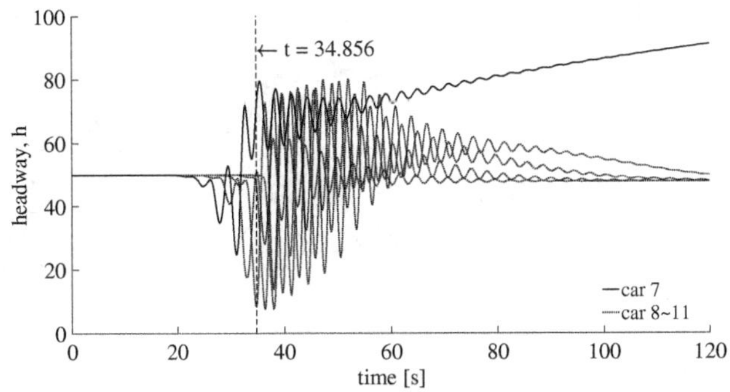

Fig. 6 Platoon separation at $t = 34.856$ under threshold condition $T = 56.4$ on car 7

[1] It is important to remind that the mathematical model considered herein does not take into account the effects of collision. Therefore, only the simulations that render non-negative headway readings are admissible.

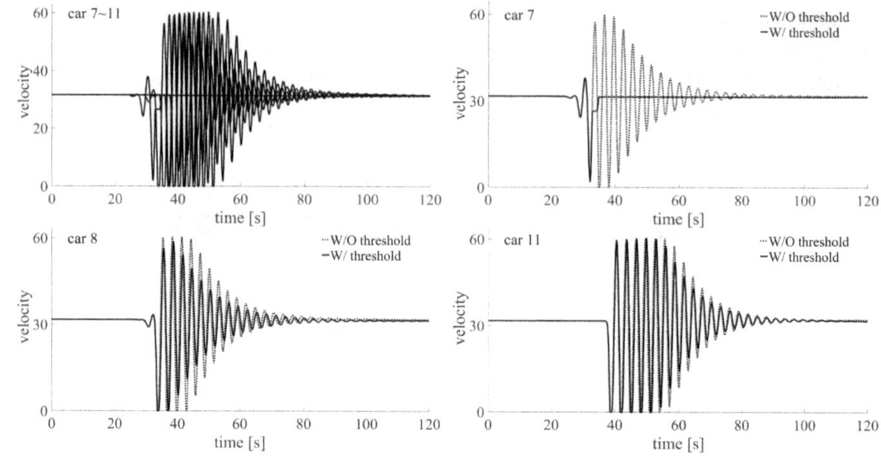

Fig. 7 Velocity profiles on car 7 $\sim$ 11 under threshold condition (9) on car 7

$$a = \begin{cases} a & \text{if } x_{i-1}(t-\tau) - x_i(t-\tau) < T \\ 0 & \text{otherwise.} \end{cases} \tag{9}$$

With the proposed threshold condition, if the headway at time $t - \tau$ is less than T, car 7 operates under OVM. Otherwise, car 7 drives with constant velocity. With $T = 56.4$ and $t > 34.856$, the platoon is separated into two groups, car 1 $\sim$ 6 and car 7 $\sim$ 11, see Fig. 6. This separation behavior then reflects itself on the velocity profiles, Fig. 7 where we find that car 7 no longer exhibits stop and go behavior, and the magnitude of velocity oscillations in car 8 $\sim$ 11 is reduced. Similar qualitative conclusions also hold for cars 9 and 10 (suppressed). Moreover, with the threshold, the duration of exhibiting zero velocity in cars 8 $\sim$ 11 is reduced from 1.401 to 0.725, 1.953 to 1.571, 2.394 to 2.126, and 2.924 to 2.658, respectively.

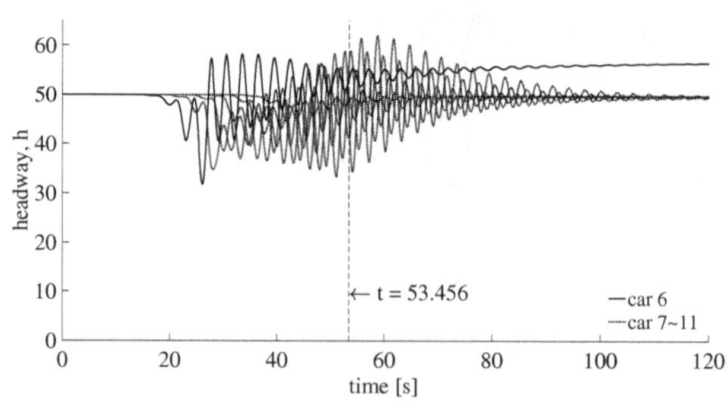

Fig. 8 Platoon separation at $t = 53.456$ under threshold condition $T = 51$ on car 6

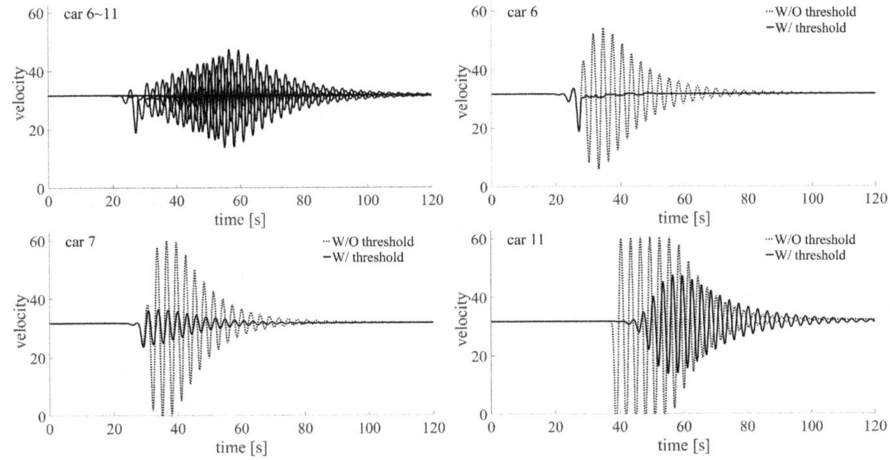

Fig. 9 No stop and go behavior is observed in the velocity profiles of car 6 $\sim$ 11 under the threshold condition on car 6

4.2 Threshold Condition on Car 6 and Sub-platooning

We next explore adding the threshold condition (9) on car 6. Figure 8 shows the headway plural, headways of car 6 $\sim$ 11. At $t = 53.456$, complete group separation occurs under the threshold $T = 51$. Even if this threshold value is close to the vehicle's initial headway, the car 6 subjected to the threshold condition repeatedly connects and disconnects from car 5 until it completely disconnects. Car 6 then becomes the leader of the vehicles following it and in this case, those followers drive without exhibiting stop and go behavior, see Fig. 9.

Considering the above two group-separation examples, we notice that the velocity profile of the separated car at the time of separation can be considered as disturbance to its follower. The waveform as well as its magnitude contribute to the follower's dynamics and ultimately influence whether or not the subsequent followers could be free of stop and go behavior.

4.3 Threshold Condition Modification Based on Linear Analysis

Here, we slightly modify the threshold condition on car 6 as follows

$$a = \begin{cases} 0 & \text{if } x_{i-1}(t-\tau) - x_i(t-\tau) < T \\ a & \text{otherwise.} \end{cases} \tag{10}$$

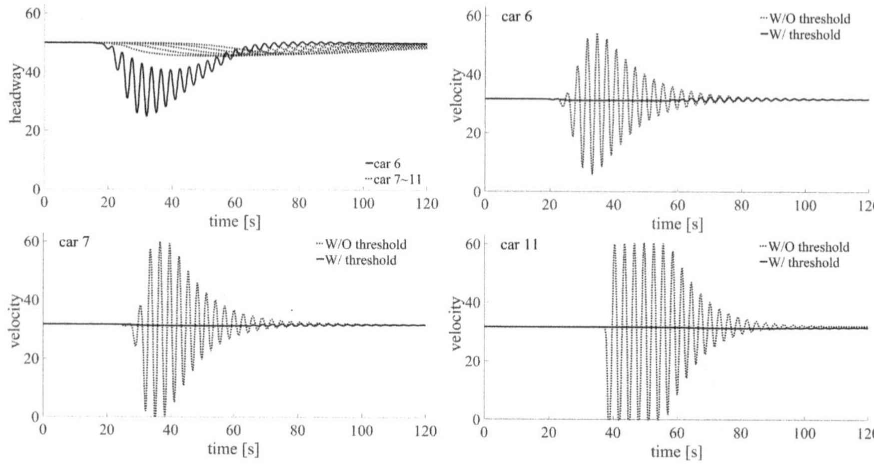

Fig. 10 No traffic jam with threshold condition (10) where $T = 45$

This threshold condition produces actions in the opposite sense of (9). It forces the vehicle to travel at a constant speed whenever it is within the headway range, else the vehicle follows the preceding vehicle based on OVM. As seen in Fig. 3, for a given a, if the headway is larger than 25, then the delay margin is larger. The proposed threshold condition prevents narrowing the headway in the linear regime, to avoid reduced delay margin. In this simulation scenario, we take $a = 2$ which corresponds to a delay margin of 0.74 – larger than the considered human reaction delay $\tau = 0.7$. Moreover, we let $T = 45$. If the threshold condition (10) can prevent the headway to becomes less than $T = 45$ with minimal deviations from the equilibrium, we then expect the dynamics of perturbations around the equilibrium to remain stable. If this can be accomplished, then the nonlinear regime could be avoided and traffic flow characteristics can potentially be improved based on linear theory. In Fig. 10, we present a representative example, where we indeed observe that there are no longer traffic jams.

4.4 Metrics

We next describe two different metrics as figure of merit toward assessing the influence of threshold on the traffic flow dynamics. The first metric collects the time that each car remains stopped in traffic for $t \in [0, 120]$. Then, for the i-th car, we denote its time stopped in traffic by $t_{\text{stop},i}$. Furthermore, the total stopped time $\sum_{i=1}^{n} t_{\text{stop},i}$ indicates the accumulation of all $t_{\text{stop},i}$. Note that cars 1 through 6 travel without interruptions and hence their $t_{\text{stop},i} = 0$. The second metric is related to fluctuations in speed v_i due to variations on the acceleration a_i, which are assumed to be proportional to the rate of fuel consumption. In order to quantify the expenditure of fuel

Table 1 Total stopped time and integral of acceleration as proxy of energy consumption

Car i	**7**	**8**	**9**	**10**	**11**	**Total**
Without threshold condition						
$t_{\text{stop},i}$ $[s]$	0.364	1.396	1.947	2.388	2.917	**9.012**
IA_i	3.374	3.691	4.057	4.461	4.864	**20.446**
Threshold condition (9) on car 7						
$t_{\text{stop},i}$	0	0.722	1.567	2.121	2.652	**7.062**
IA_i	0.369	2.802	3.606	3.984	4.375	**15.137**
Threshold condition (9) on car 6						
$t_{\text{stop},i}$	0	0	0	0	0	**0**
IA_i	0.700	1.369	1.893	2.255	2.533	**8.749**
Threshold condition (10) on car 6						
$t_{\text{stop},i}$	0	0	0	0	0	**0**
IA_i	0.083	0.053	0.035	0.023	0.014	**0.208**

of the i-th vehicle, let us define the integral of acceleration $\text{IA}_i = \frac{1}{\Delta t}\int_0^{\Delta t} a_i(t)dt$, which is related to energy consumed by the vehicle due to stop and go waves. Then, $\sum_{i=1}^{n} \text{IA}_i$ measures the energy consumption. These metrics are computed as summarized in Table 1 for the simulation scenarios at hand. We find indeed that the threshold condition under the considered parametric settings yields more favorable readings on the three metrics.

5 Conclusions

The well-known Optimal Velocity Model is taken as a benchmark to investigate how the idea of sub-platooning through a headway-based threshold condition could be beneficial in reducing traffic flow metrics related to jam formations and energy consumption. Preliminary results reported herein indicate that the idea of strategically creating separation of platoons into sub-platoons presents opportunities in creating improved traffic flow characteristics. Our recent results on the delay margin computation of *large scale* linear time-invariant systems with single delay [1] will support future studies on sub-platooning in large scale traffic flow scenarios.

References

1. A. Ramírez, M. Koh, R. Sipahi, An approach to compute and design the delay margin of a large-scale matrix delay equation. Int. J. Rob. Nonlinear Control **29**(4), 1101–1121 (2019)
2. M. Bando, K. Hasebe, A. Nakayama, A. Shibata, Y. Sugiyama, Dynamical model of traffic congestion and numerical simulation. Phys. Rev. E **51**(2), 1035–042 (1995)

3. M. Bando, K. Hasebe, K. Nakanishi, A. Nakayama A, Analysis of optimal velocity model with explicit delay. Phys. Rev. E **58**(5), 5429–5435 (1998)
4. M. Bando, K. Hasebe, K. Nakanishi, A. Nakayama, Delay of vehicle motion in traffic dynamics. Jpn. J. Ind. Appl. Math. **17**(2), 275–294 (2000)
5. M. Barth, K. Boriboonsomsin, Real-World carbon dioxide impacts of traffic congestion. Trans. Res. Rec.: J. TRB **2058**, 163–171 (2008)
6. A. Bose, P.A. Ioannou, Analysis of traffic flow with mixed manual and semi automated vehicles. IEEE Trans. Intell. Transp. Syst. **4**(4), 173–188 (2003)
7. R. Datko, A procedure for determination of the exponential stability of certain differential-difference equations. Q. Appl. Math. **36**(3), 279–292 (1978)
8. M. Green, "How Long Does It Take to Stop?" Methodological analysis of driver perception-brake times. Transp. Hum. Fact. **2**(3), 195–216 (2000)
9. D. Helbing, Traffic and related self-driven many-particle systems. Rev. Mod. Phys. **73**(4), 1067–1141 (2001)
10. R. Jiang, M.B. Hu, H.M. Zhang, Z.Y. Gao, B. Jia, Q.S. Wu, B. Wang, M. Yang, Traffic experiment reveals the nature of car-following. PLoS One **9**(4), e94351 (2014)
11. B.S. Kerner, *Introduction to Modern Traffic Flow Theory and Control: The Long Road to Modern Three-Phase Traffic Theory* (Springer, Heidelberg, 2009)
12. A. Kesting, M. Treiber, M. Schönhof, D. Helbing, Adaptive cruise control design for active congestion avoidance. Trans. Res. Part C: Emerg. Technol. **16**(6), 668–683 (2008)
13. M. Koh, R. Sipahi, Effects of edge elimination on the delay margin of a class of LTI consensus dynamics. IEEE Trans. Autom. Control **63**(12), 4397–4404 (2018)
14. M. Koh, R. Sipahi, A consensus dynamics with delay-induced instability can self-regulate for stability via agent regrouping. Chaos **26**(11), 116313 (2016)
15. K. Konishi, H. Kokame, K. Hirata, Coupled map car-following model and its delayed-feedback control. Phys. Rev. E **60**(4), 4000 (1999)
16. S. Matthias, Traffic congestion's economic impact: evidence from US metropolitan regions. Urban Stud. **51**(10), 2088–2110 (2014)
17. W. Michiels, S.I. Niculescu, *Stability and Stabilization of Time-Delay Systems: An Eigenvalue Based Approach* (SIAM, Philadelphia, 2007)
18. A. Nakayama, Y. Sugiyama, K. Hasebe, Effect of looking at the car that follows in an optimal velocity model of traffic flow. Phys. Rev. E **65**(1), 016112 (2001)
19. N. Olgac, R. Sipahi, An exact method for the stability analysis of time-delayed linear time-invariant (LTI) systems. IEEE Trans. Autom. Control **47**(5), 793–797 (2002)
20. G. Orosz, Connected cruise control: modelling, delay effects, and nonlinear behavior. Veh. Syst. Dyn. **54**(8), 1147–1176 (2016)
21. G. Orosz, R.E. Wilson, G. Stépán, Traffic jams: dynamics and control. Philos. Trans. R. Soc. A **368**(1928), 4455–4479 (2010)
22. G. Orosz, G. Stépán, Subcritical Hopf bifurcations in a car-following model with reaction-time delay. Proc. R. Soc. London A **462**(2073), 2643–2670 (2006)
23. R. Sipahi, W. Qiao, Responsible eigenvalue concept for the stability of a class of single-delay consensus dynamics with fixed topology. IET Control Theory Appl. **5**(4), 622–629 (2011)
24. R.W. Rothery, Transportation Research Board Special Report 165. Traffic Flow Theory (1998)
25. R.W. Rothery, Car following models, in *Traffic Flow Theory* (1992)
26. R. Sipahi, S.-I. Niculescu, Mathematical models of traffic flow dynamics with time delays, in *Complex Time-Delay Systems* (Springer, 2010)
27. G. Stépán, Retarded dynamical systems: stability and characteristic functions. *Pitman Research Notes in Mathematics Series*, vol. 210. (Longman Scientific & Technical, New York, 1989)
28. R. Subramanian, Motor Vehicle Traffic Crashes as a Leading Cause of Death in the United States, 2002. Report no. DOT HS 809 843, NHTSA (2005)
29. M. Treiber, A. Kesting, *Traffic Flow Dynamics: Data* Models and Simulation. (Springer, Heidelberg, 2013)
30. M. Treiber, A. Kesting, D. Helbing, Delays, inaccuracies and anticipation in microscopic traffic models. Physica A **360**(1), 71–88 (2006)

31. K. Walton, J.E. Marshall, Direct method for TDS stability analysis, in *IEE Proceedings D-Control Theory and Applications*, vol 134. no 2. IET: 101–107 (1987)
32. L. Zhang, G. Orosz, Consensus and disturbance attenuation in multi-agent chains with nonlinear control and time delays. Int. J. R. Nonlinear Control **27**(5), 781–803 (2017)
33. The Cost of Traffic Jams, The Economist explains (2014). http://www.economist.com/blogs/economist-explains/2014/11/economist-explains-1. Accessed 5 Feb 2019
34. Traffic Congestion Costs Americans $124 Billion a Year. Forbes (2014). http://www.forbes.com/sites/federicoguerrini/2014/10/14/traffic-congestion-costs-americans-124-billion-a-year-report-says/#6b54bf616252. Accessed 5 Feb 2019
35. Traffic Congestion and Greenhouse Gases, Access magazine (2009). http://www.accessmagazine.org/articles/fall-2009/traffic-congestion-greenhouse-gases/. Accessed 5 Feb 2019

On the Ore Extension Ring of Differential Time-Varying Delay Operators

Alban Quadrat and Rosane Ushirobira

Abstract In this work, we propose an algebraic method to study linear differential time-varying delay (DTVD) systems. Our goal is to give an effective construction of the ring of DTVD operators as an Ore extension, thanks to the concept of skew polynomial rings developed by Ore in the 30s. Some algebraic properties of the DTVD operators ring are analyzed, such as its Noetherianity, its homological and Krull dimensions, and the existence of Gröbner bases, all given in terms of the time-varying delay function. The algebraic analysis framework for linear systems theory allows us to study linear DTVD systems and essential properties such as the existence of autonomous elements, controllability, parametrizability, flatness, etc., through methods coming from module theory, homological algebra, and constructive algebra.

1 Introduction

In control theory, differential time-delay systems are a vastly studied subject in the literature, most commonly considering constant or distributed time-delays. Numerous applications have been investigated in this domain. Nevertheless, the class of *differential time-varying delay (DTVD) systems* has been studied mainly from a stability analysis viewpoint, such as in incompressible fluid flows in pipes [17], material or vehicular flows [26], metal-rolling processes [10], communication networks [29]. More details on linear DTVD systems can be found in [2, 8, 13, 16, 19, 28] and in their references.

In the 90s, Oberst, Fliess, and Pommaret introduced the *algebraic analysis method* [1, 11, 12] for linear systems theory, see [7, 20, 22, 30]. In this context, an intrinsic

A. Quadrat (✉)
Inria Paris, Ouragan project-team, IMJ—PRG, Sorbonne University, 75252 Paris, France
e-mail: alban.quadrat@inria.fr

R. Ushirobira
UMR, Inria, University Lille, CNRS, 9189—CRIStAL, 59000 Lille, France
e-mail: rosane.ushirobira@inria.fr

G. Valmorbida et al. (eds.), *Accounting for Constraints in Delay Systems*,
Advances in Delays and Dynamics 12,
https://doi.org/10.1007/978-3-030-89014-8_5

study of linear systems can be elaborated based on module theory, homological algebra, and functional analysis. In particular, built-in properties of linear systems can be characterized through module properties providing a representation independent of the system representation itself. Furthermore, *Willems' behavioral approach* [21] can also be realized and developed within this framework [20].

An *effective algebraic analysis method* to linear functional systems was initiated in [3, 6], based on properties of *skew polynomials* [18] and *Ore algebras* [5], computer algebra methods (e.g., *Gröbner* or *Janet bases* [3, 5]) and a constructive approach to both module theory and homological algebra [25]. This effective method allows the study of large classes of linear functional systems (e.g., differential systems, constant time-delay systems, discrete systems). Moreover, this approach yields the development of the OREMODULES and OREMORPHISMS packages [4, 6], allowing a complete study of control concepts such as controllability, observability, parametrizability, flatness, and equivalences.

In this work, our goal is to develop an algebraic analysis method for linear DTVD systems. First, the ring of DTVD operators is shown to be an *Ore extension* $\mathscr{D}$ [18]. Important algebraic and homological properties of $\mathscr{D}$ are then characterized. Since $\mathscr{D}$ is also a *bijective skew Poincaré-Birkhoff-Witt (PBW) extension* [9], Gröbner basis techniques can be applied over $\mathscr{D}$ (using [14]), allowing to test effectively standard module properties and thus system properties. The corresponding Gröbner basis algorithms will be implemented in the future in a computer algebra system.

The outline of the paper is the following. In Sect. 2, we give basic definitions of skew polynomial rings and Ore extensions and some examples. An explicit construction of the ring $\mathscr{D}$ of DTVD operators as an Ore extension is provided in Sect. 3. Algebraic properties of this Ore extension are analyzed in Sect. 4, in particular within a homological framework. In Sect. 5, system properties are characterized in terms of module theory and homological algebra. Finally, in Sect. 6, we discuss some open questions concerning $\mathscr{D}$.

2 Skew Polynomial Rings and Ore Extensions

In the sequel, basic definitions and properties of skew polynomials rings and Ore structures are recalled.

Definition 1 ([18]) Let A be a ring.

(i) We denote by $\mathrm{End}(A)$ the ring of endomorphisms of A, where a map $\sigma : A \longrightarrow A$ is an *endomorphism* of A if it satisfies the following conditions:

$$\forall\, a,\, b \in A, \quad \begin{cases} \sigma(1) = 1, \\ \sigma(a+b) = \sigma(a) + \sigma(b), \\ \sigma(a\, b) = \sigma(a)\, \sigma(b). \end{cases}$$

(ii) For $\sigma \in \text{End}(A)$, a *σ-derivation* δ of A is a map $\delta : A \longrightarrow A$ satisfying the following conditions:

$$\forall\, a,\, b \in A, \quad \begin{cases} \delta(a+b) = \delta(a) + \delta(b), \\ \delta(a\, b) = \delta(a)\, b + \sigma(a)\, \delta(b). \end{cases}$$

Definition 2 Let A be a ring.

(i) A *derivation* $\delta : A \to A$ is a linear map satisfying Leibniz's rule, that is δ is a id_A-derivation (where id_A is the identity map of A).
(ii) A *differential ring* (A, δ) is a ring equipped with a derivation δ.

Let us introduce the definition of an *Ore extension* which will play a fundamental role in what follows.

Definition 3 ([18]) Let A be a ring, $\sigma \in \text{End}(A)$ and δ be a σ-derivation of A. An *Ore extension* $A[\partial; \sigma, \delta]$ *of* A is the noncommutative polynomial ring in ∂ with coefficients in A, formed by the elements of the form $\sum_{i=0}^{n} a_i\, \partial^i$ with $a_i \in A$, and equipped with a new product given by:

$$\forall\, a \in A, \quad \partial\, a = \sigma(a)\, \partial + \delta(a). \tag{1}$$

The Ore extension $A[\partial; \sigma, \delta]$ of A is also called a *skew polynomial ring* over A.

Example 1 Consider a differential ring (A, δ) such as, e.g., $A = \mathbb{K}[t]$ (resp., $\mathbb{K}(t)$) the commutative ring of polynomials (resp., rational functions) in t with coefficients in a field $\mathbb{K}$ (e.g., $\mathbb{K} = \mathbb{Q}, \mathbb{R}$) or A the ring of smooth/analytic/meromorphic functions in t, and $\delta = \frac{d}{dt}$. The Ore extension $\mathcal{O} = A\left[\partial;\, \text{id}_A,\, \frac{d}{dt}\right]$ of A is the ring of differential operators in ∂ with coefficients in A. An element $a \in A$ acts as an operator of A by multiplication $b \in A \mapsto a\, b \in A$ and an element of $\mathcal{O}$ can be written as $\sum_{i=0}^{n} a_i\, \partial^i$, where the concatenation denotes the composition of the operators acting on A:

$$\begin{aligned} a : A &\longrightarrow A \\ b &\longmapsto a\, b, \end{aligned} \qquad \begin{aligned} \partial : A &\longrightarrow A \\ a &\longmapsto \delta(a). \end{aligned}$$

For all $a \in A$, we have

$$(\partial\, a)(b) = \partial(a\, b) = \delta\,(a\, b) = a\, \delta(b) + \delta(a)\, b = (a\, \partial + \delta(a))\, (b),$$

showing that (1) is satisfied for $\sigma = \text{id}_A$.

Example 2 Let $h \in \mathbb{R}_{>0} = \mathbb{R}_{\geq 0} \setminus \{0\}$, A be a ring of real-valued functions equipped with $\sigma \in \text{End}(A)$ defined by $\sigma(a(t)) = a(t-h)$ for all $a \in A$. As in Example 1, given $a \in A$, we can associate an operator $a : A \to A$ defined by the multiplication $b \in A \mapsto a\, b \in A$ of functions. The Ore extension $\mathcal{O} = A\,[S;\, \sigma, 0]$ of A is the *ring of time-delay (TD) operators with coefficients in* A. An element of $\mathcal{O}$ can be written as $\sum_{i=0}^{n} a_i\, S^i$, where the concatenation means the composition of two operators and:

$$\begin{aligned} S : A &\longrightarrow A \\ a(t) &\longmapsto a(t-h). \end{aligned}$$

Example 3 In a discrete-time version, if A is a ring of real-valued sequences on $\mathbb{Z}$, i.e. $A = \mathbb{R}^{\mathbb{Z}}$, then an element $a \in A$ can be written as $a = (a_i)_{i \in \mathbb{Z}}$. Let $\sigma \in \mathrm{End}(A)$ be defined by forward (resp., backward) shift $\sigma\,(a_i)_{i \in \mathbb{Z}} = (a_{i+1})_{i \in \mathbb{Z}}$ (resp., $\sigma\,(a_i)_{i \in \mathbb{Z}} = (a_{i-1})_{i \in \mathbb{Z}}$). Given $a \in A$, there is an operator $a : A \to A$ defined by the multiplication of functions: $a\,b = (a_i\,b_i)_{i \in \mathbb{Z}}$ for all $b = (b_i)_{i \in \mathbb{Z}} \in A$. Then, the Ore extension $A\,[S; \sigma, 0]$ is the *skew polynomial ring of forward (resp., backward) shift operators*. An element of $\mathscr{O}$ can be written as $\sum_{k=0}^{n} a_k\,S^k$, where the concatenation means the composition of two operators and:

$$\begin{aligned} S : A &\longrightarrow A \\ (a_i)_{i \in \mathbb{Z}} &\longmapsto (a_{i+1})_{i \in \mathbb{Z}}\,. \end{aligned}$$

Skew polynomial rings satisfy the following important *universal property*.

Proposition 1 ([18], 1.2.5) *Let A and E be two rings and $f : A \longrightarrow E$ a ring homomorphism (i.e., $f(a+b) = f(a) + f(b)$, $f(a\,b) = f(a)\,f(b)$, for all $a,\,b \in A$, and $f(1) = 1$). Let $\sigma \in \mathrm{End}(A)$, δ be a σ-derivation of A, and $A[\partial; \sigma, \delta]$ the corresponding Ore extension of A. Moreover, let $e \in E$ be such that:*

$$\forall\, a \in A, \quad e\,f(a) = f(\sigma(a))\,e + f(\delta(a)). \tag{2}$$

Then, there exists a unique ring homomorphism $g : A[\partial; \sigma, \delta] \longrightarrow E$ extending f, satisfying:

$$\begin{cases} g_{|A} = f, \\ g(\partial) = e. \end{cases}$$

In Proposition 1, considering $E = A[\partial; \sigma, \delta]$, $\sigma' \in \mathrm{End}(A)$, and

$$\begin{aligned} f : A &\longrightarrow E \\ a &\longmapsto \sigma'(a), \end{aligned}$$

we obtain the following corollary.

Corollary 1 *Let A be a ring, σ, $\sigma' \in \mathrm{End}(A)$, and δ a σ-derivation of A. Moreover, let $e \in A[\partial; \sigma, \delta]$ be such that:*

$$\forall\, a \in A, \quad e\,\sigma'(a) = \sigma'(\sigma(a))\,e + \sigma'(\delta(a)). \tag{3}$$

Then, there exists a unique ring homomorphism $\overline{\sigma} \in \mathrm{End}(A[\partial; \sigma, \delta])$ satisfying:

$$\overline{\sigma}_{|A} = \sigma', \quad \overline{\sigma}(\partial) = e.$$

Proposition 1 and Corollary 1 will play an important role in what follows.

In the case of multiple σ-derivations and ring endomorphisms, for instance, for partial differential equations, or differential-difference equations, or differential time-delay (DTD) equations, an iteration of Ore extensions can be constructed as follows. If $\mathcal{O} = A[\partial; \sigma, \delta]$ is an Ore extension of A, $\sigma_2 \in \mathrm{End}(\mathcal{O})$ and δ_2 is a σ_2-derivation of $\mathcal{O}$, then we define the following Ore extension of $\mathcal{O}$:

$$B = \mathcal{O}\,[\partial_2; \sigma_2, \delta_2] = A\,[\partial; \sigma, \delta]\,[\partial_2; \sigma_2, \delta_2]\,.$$

By definition, the following equation holds:

$$\forall\, o \in \mathcal{O}, \quad \partial_2\, o = \sigma_2(o)\, \partial_2 + \delta_2(o).$$

Hence, we have $\partial_2\, \partial = \sigma_2(\partial)\, \partial_2 + \delta_2(\partial)$, therefore usually $\partial_2\, \partial \neq \partial\, \partial_2$ so ∂ and ∂_2 do not commute. Nevertheless, in many cases, the operators might commute, leading to the definition of an *Ore algebra*.

Definition 4 ([5]) If A is a $\mathbb{K}$-algebra, then an Ore extension $A\,[\partial_1; \sigma_1, \delta_1] \cdots [\partial_n; \sigma_n, \delta_n]$ of A is called an *Ore algebra* if the following conditions are satisfied

$$\begin{cases} \sigma_j(A) \subseteq A, & \delta_j(A) \subseteq A, \quad 1 \leq j \leq n, \\ \sigma_j(\partial_i) = \partial_i, & \delta_j(\partial_i) = 0, \quad 1 \leq i < j \leq n, \end{cases}$$

and the restrictions $\sigma_i|_A$ and $\delta_j\big|_A$ commute two by two, for all $1 \leq i, j \leq n$.

Example 4 Consider the polynomial ring $A = \mathbb{R}\,[x_1, x_2]$. Notice that $\beta_1 := \frac{\partial}{\partial x_1}$ is a derivation of A. The Ore extension $\mathcal{O}$ can be defined, $\mathcal{O} = A\,[\partial_1; \mathrm{id}_A, \beta_1]$ of A consists of the differential operators in ∂_1 with coefficients in A. Now consider the following derivation of $\mathcal{O}$:

$$\begin{aligned} \beta_2 := \frac{\partial}{\partial x_2} &: \mathcal{O} \longrightarrow \mathcal{O} \\ \textstyle\sum_{i=0}^r a_i(x_1, x_2)\, \partial_1^i &\longmapsto \textstyle\sum_{i=0}^r \frac{\partial a_i(x_1, x_2)}{\partial x_2}\, \partial_1^i. \end{aligned}$$

Then we can construct the Ore extension $B = \mathcal{O}\,[\partial_2; \mathrm{id}_{\mathcal{O}}, \beta_2]$ of $\mathcal{O}$. Since $\beta_2(\partial_1) = 0$, the operators ∂_1 and ∂_2 commute and B is an Ore algebra.

Example 5 Let A be a differential $\mathbb{R}$-algebra of real-valued functions of t with support on $\mathbb{R}_{\geq 0}$ or on $\mathbb{R}$, and $\delta = \frac{d}{dt}$. The Ore extension $\mathcal{O} = A\,[\partial; \mathrm{id}_A, \delta]$ of A is the ring of ordinary differential operators with coefficients in A. Let $h \in \mathbb{R}_{>0}$ and consider $\sigma \in \mathrm{End}(A)$ defined by $\sigma(a(t)) = a(t-h)$ for all $a \in A$. Using (1), we obtain:

$$\forall\, a \in A, \quad \partial\, \sigma(a) = \partial\, a(\cdot - h) = a(\cdot - h)\, \partial + \dot{a}(\cdot - h) = \sigma(a)\, \partial + \sigma(\delta(a)).$$

By Corollary 1, there exists a unique $\overline{\sigma} \in \mathrm{End}(\mathcal{O})$ such that $\overline{\sigma}_{|A} = \sigma$ and $\overline{\sigma}(\partial) = \partial$. To simplify the notations, we write σ for $\overline{\sigma}$. Hence, $\sigma \in \mathrm{End}(\mathcal{O})$ is defined by:

$$\begin{aligned} \sigma : \mathcal{O} &\longrightarrow \mathcal{O} \\ \sum_{i=0}^{r} a_i(t)\, \partial^i &\longmapsto \sum_{i=0}^{r} a_i(t-h)\, \partial^i. \end{aligned}$$

Then, by Proposition 1, we can construct the Ore extension $B = \mathcal{O}[S; \sigma, 0]$ of $\mathcal{O}$. For $a \in A$, we have

$$(\delta \circ \sigma)\,(a(t)) = \frac{d}{dt}\, a(t-h) = \dot{a}(t-h) = (\sigma \circ \delta)\,(a(t)),$$

which shows that $(\delta \circ \sigma)|_A = (\sigma \circ \delta)|_A$. By construction, it is clear that the remaining conditions for B to be an Ore algebra hold. In particular, we have:

$$S\,\partial = \sigma(\partial)\, S = \partial\, S.$$

Hence, $B = A\,[\partial; \mathrm{id}_A, \delta]\,[S; \sigma, 0]$ is the commutative polynomial ring of differential constant time-delay operators with coefficients in A.

Now, we consider the ring of *differential time-varying delay* (DTVD) operators. This ring is not an Ore algebra, as we now briefly explain.

In the remainder of the paper, we shall suppose that the function $h : \mathbb{R} \to \mathbb{R}$ is smooth or analytic function, and, since we only consider time-delay operators, that:

$$\forall\, t \geq 0, \quad h(t) \geq 0. \tag{4}$$

Remark 1 It is usually assumed that h satisfies $\dot{h}(t) < 1$ for all $t \in \mathbb{R}_{\geq 0}$. That can be interpreted as the fact that the delay cannot be faster, or as fast as, than time itself.

Let us consider a differential ring of real-valued functions $\left(A, \delta = \frac{d}{dt}\right)$ equipped with $\sigma \in \mathrm{End}(A)$ defined by:

$$\forall\, a \in A, \quad \sigma\,(a(t)) = a\,(t - h(t))\,. \tag{5}$$

For instance, we can consider $A = C^\infty(\mathbb{R})$. Then, we have:

$$\begin{aligned} (\delta \circ \sigma)(a(t)) &= \delta(a(t - h(t))) = (1 - \dot{h}(t))\, \dot{a}(t - h(t)) \\ &= (1 - \dot{h}(t))\, (\sigma \circ \delta)(a(t)). \end{aligned}$$

That implies:

$$(\delta \circ \sigma)|_A = \left(1 - \dot{h}\right)\, (\sigma \circ \delta)|_A\,. \tag{6}$$

If h is not a constant function (i.e., $\dot{h} \neq 0$), δ and σ do not commute, and it is then not possible to construct an Ore algebra of DTVD operators (see Definition 4).

3 An Ore Extension Construction

In the above Section, we have seen that the ring of DTVD operators cannot give rise to an Ore algebra. Nevertheless, it is possible to construct an Ore *extension* for this operator ring, and that is the goal of the present Section.

Let $\left(A, \delta = \frac{d}{dt}\right)$ be a differential ring of real-valued functions (see Definition 2) equipped with $\sigma \in \mathrm{End}(A)$ defined by (5). Generalizing the construction made in Example 2, we can consider the skew polynomial ring $\mathcal{O} = A[S; \sigma, 0]$ of time-varying delay operators. If we note $e = (1 - \dot{h}(t))\, S \in \mathcal{O}$, then we have:

$$\forall\, a \in A, \quad e\, a = (1 - \dot{h}(t))\, S\, a = (1 - \dot{h}(t))\, a(t - h(t))\, S = \sigma(a)\, e.$$

By Corollary 1, there exists a unique $\sigma' \in \mathrm{End}(\mathcal{O})$ such that:

$$\forall\, a \in A, \quad \sigma'(a) = a, \quad \sigma'(S) = e = (1 - \dot{h}(t))\, S.$$

Let us consider the following map:

$$\begin{aligned} \delta' : A[S; \sigma, 0] &\longrightarrow A[S; \sigma, 0] \\ o = \textstyle\sum_{i=0}^{r} a_i(t)\, S^i &\longmapsto \textstyle\sum_{i=0}^{r} \dot{a}_i(t)\, S^i = \frac{do}{dt}. \end{aligned}$$

In particular, we get $\delta'(a) = \dot{a}$ and $\delta'(S) = 0$. Moreover, we have

$$\begin{cases} \delta'(S\, a) = \delta'(\sigma(a)\, S) = \delta'(a(t - h(t))\, S) = (1 - \dot{h}(t))\, \dot{a}(t - h(t))\, S, \\ \delta'(S)\, a + \sigma'(S)\, \delta'(a) = e\, \delta'(a) = (1 - \dot{h}(t))\, \dot{a}(t - h(t))\, S, \end{cases}$$

for all $a \in A$, which shows that:

$$\forall\, a \in A, \quad \delta'(S\, a) = \delta'(S)\, a + \sigma'(S)\, \delta'(a) = e\, \delta'(a) = \sigma(\delta'(a))\, e.$$

More generally, we can check that δ' is a σ'-derivation of $\mathcal{O}$.
Hence, the Ore extension $\mathcal{D} = A[S; \sigma, 0]\left[\partial; \sigma', \delta'\right]$ of $\mathcal{O}$ can be defined. We have:

$$\forall\, o \in \mathcal{O}, \quad \partial\, o = \sigma'(o)\, \partial + \delta'(o).$$

In particular, we obtain:

$$\begin{cases} \partial\, a(t) = \sigma'(a(t))\, \partial + \delta'(a(t)) = a(t)\, \partial + \dot{a}(t), \\ \partial\, S = \sigma'(S)\, \partial + \delta'(S) = (1 - \dot{h}(t))\, S\, \partial. \end{cases}$$

This last identity encodes the commutation rule of ∂ and S in terms of h (see Sect. 6).

An element of $\mathscr{D}$ is of the form $d = \sum_{i=0}^{r} o_i\, \partial^i = \sum_{i=0}^{r} \sum_{j=0}^{s} a_{ij}\, S^j\, \partial^i$ with $a_{ij} \in A$. In the remaining of the paper, this ring $\mathscr{D}$ will be called a *ring of differential time-varying delay (DTVD) operators over A*.

Remark 2 Under the hypothesis that the function h satisfies the condition

$$\forall\, t, \quad \dot{h}(t) \neq 1, \tag{7}$$

a second construction of $\mathscr{D}$ can be given. Note that (7) is satisfied if the hypothesis of Remark 1 is assumed. Let $\mathscr{O}' = A[\partial;\, \mathrm{id}_A, \delta]$ be the Ore extension of A defined in Example 1, i.e., the ring of OD operators with coefficients in A. Moreover, let $\sigma \in \mathrm{End}(A)$ be defined by (5) and note $e = (1 - \dot{h})^{-1}\, \partial \in \mathscr{O}'$. Then, we have

$$\begin{aligned} e\, \sigma(a) &= \frac{1}{1 - \dot{h}(t)}\, \partial\, a(t - h(t)) = a(t - h(t))\, \frac{1}{1 - \dot{h}(t)}\, \partial + \dot{a}(t - h(t)) \\ &= a(t - h(t))\, e + \dot{a}(t - h(t)) = \sigma(a)\, e + \sigma(\dot{a}), \end{aligned}$$

for all $a \in A$. By Corollary 1, there exists a unique endomorphism of $\mathscr{O}'$, simply denoted by σ, which satisfies $\sigma(\partial) = e = (1 - \dot{h})^{-1}\, \partial$, i.e. $\sigma \in \mathrm{End}(\mathscr{O}')$ is given by:

$$\begin{aligned} \sigma : \mathscr{O}' &\longrightarrow \mathscr{O}' \\ \textstyle\sum_{i=0}^{r} a_i(t)\, \partial^i &\longmapsto \textstyle\sum_{i=0}^{r} a_i(t - h(t)) \left(\frac{1}{1 - \dot{h}}\, \partial\right)^i. \end{aligned} \tag{8}$$

If $h(t) = h \in \mathbb{R}_{\geq 0}$, then we note that $\sigma(\partial) = \partial$ and we find again Example 5. Let us interpret (8). First denote $x(t) := t - h(t)$. Then, we have

$$\frac{d}{dt} = \frac{dx}{dt}\, \frac{d}{dx} = \left(1 - \dot{h}(t)\right) \frac{d}{dx},$$

and from that, it results:

$$\frac{d}{dx} = \frac{1}{1 - \dot{h}(t)}\, \frac{d}{dt}.$$

Set $\partial_t := \partial = \dfrac{d}{dt}$ and $\partial_x := \dfrac{d}{dx}$. Then, the map σ satisfies:

$$\sigma(a(t)) = a(x), \quad \sigma(\partial_t) = \partial_x.$$

Hence, σ is a *change of time-scale*, from time t to time $x(t) := t - h(t)$, and:

$$\sigma\left(\sum_{i=0}^{r} a_i(t)\, \partial_t^i\right) = \sum_{i=0}^{r} a_i(x)\, \partial_x^i. \tag{9}$$

Now, thanks to $\sigma \in \mathrm{End}(\mathscr{O}')$, an Ore extension of $\mathscr{O}'$ can be constructed by:

$$\mathscr{D}' = \mathscr{O}'\,[S; \sigma, 0] = A\left[\partial; \mathrm{id}_A, \frac{d}{dt}\right][S; \sigma, 0]\,.$$

An element $d' \in \mathscr{D}'$ can be written as $d' = \sum_{j=0}^{s} o'_j\, S^j = \sum_{j=0}^{s} \sum_{i=0}^{r} a_{ij}\, \partial^i\, S^j$ with $a_{ij} \in A$. By definition of an Ore extension, we have $S\, o = \sigma(o)\, S$ for all $o \in \mathscr{O}'$. In particular, setting $o = \partial$, we get $S\, \partial = (1 - \dot{h})^{-1}\, \partial\, S$, i.e. $\partial\, S = (1 - \dot{h})\, S\, \partial$ (see (6)).

We can check that $\mathscr{D}' = \mathscr{D}$, which gives a second construction of the ring of DTVD operators over A.

4 Algebraic Properties of the Ore Extension *D*

In this section, general properties of the ring $\mathscr{D}$ of DTVD operators defined in Sect. 3 are studied. For general concepts and structural definitions, we refer to [3, 18, 25]. Let us start with standard results on Ore extensions.

Theorem 1 ([18]) *Let A be a noncommutative ring and $D = A[\partial; \sigma, \delta]$ a skew polynomial ring.*

1. *If A is a domain*[1] *and σ is injective, then D is a domain.*
2. *If A is a left Ore domain*[2] *and σ is injective, then D is a left Ore domain.*
3. *If A is a left (right) Noetherian ring*[3] *and σ is bijective, then D is a left (right) Noetherian ring. Moreover, if A is a domain, then D is a left (right) Ore domain.*

A very useful corollary of Theorem 1 can then be proven:

Corollary 2 *Let $\mathscr{D} = A\left[\partial; \mathrm{id}_A, \frac{d}{dt}\right][S; \sigma, 0]$ be the ring of DTVD operators defined in Sect. 3, where $\dot{h} \neq 1$. Then, we have:*

1. *If A is a domain, then so is $\mathscr{D}$.*
2. *If A is a left Ore domain, then so is $\mathscr{D}$.*
3. *If A is a left (right) Noetherian ring and if $\sigma\,|_A$ is an automorphism of A, then $\mathscr{D}$ is a left (right) Noetherian ring. Moreover, $\sigma\,|_A$ is an automorphism of A if and only if the function ℓ defined by*

$$\begin{aligned} \ell : \mathbb{R}_{>0} &\longrightarrow \mathbb{R} \\ t &\longmapsto t - h(t), \end{aligned} \tag{10}$$

admits an inverse $\ell^{-1} \in A$ such that $a \circ \ell^{-1} \in A$, for all $a \in A$. Finally, if A is a domain, then $\mathscr{D}$ is a left (right) Ore domain.

---

[1] A ring A is a *domain* if A has no non-zero zero divisors.

[2] A ring A is a *left Ore domain* if A is a domain such that for $a_1, a_2 \in A \setminus \{0\}$, the intersection of the left ideals $A\,a_1 \cap A\,a_2 \neq 0$.

[3] A ring A is a *left (right) Noetherian ring* if every left (right) ideal of A is finitely generated.

Proof Since id_A and σ (as defined in (9), Sect. 3) are injective endomorphisms of A, then (1) and (2) are direct consequences of (1) and (2) of Theorem 1.

To prove (3), consider first that σ is an automorphism of A, so for every $a \in A$, there exists a unique $b \in A$ such that $a(t) = \sigma(b(t)) = b(\ell(t))$. In particular, for $t \mapsto t \in A$, we obtain $t = (b \circ \ell)(t)$. That is $b \circ \ell = \mathrm{id}_{\mathbb{R}_{>0}}$, so ℓ is injective and thus bijective on its image, and $b = \ell^{-1} \in A$. In addition, given $a \in A$, then $\sigma^{-1}(a) \in A$, where $\sigma^{-1}(a(t)) = a(\sigma^{-1}(t)) = a(\ell^{-1}(t))$ and that implies $a \circ \ell^{-1} \in A$. Conversely, if (10) admits an inverse $\ell^{-1} \in A$ and $a \circ \ell^{-1} \in A$ for all $a \in A$, then we can define $\sigma' \in \mathrm{End}(A)$ by:

$$\begin{aligned} \sigma' : A &\longrightarrow A \\ a(t) &\longmapsto a(\ell^{-1}(t)). \end{aligned}$$

We can check that $\sigma' \circ \sigma = \mathrm{id}_A = \sigma \circ \sigma'$, i.e., $\sigma = \sigma'^{-1}$ is an automorphism of A.

Next, we show that the automorphism σ of A extends to an automorphism of the ring $\mathcal{O} = A\left[\partial; \mathrm{id}_A, \frac{d}{dt}\right]$ of differential operators. For that, we consider

$$\begin{aligned} \sigma' : \mathcal{O} &\longrightarrow \mathcal{O} \\ a(t) &\longmapsto a(y) \\ \partial_t &\longmapsto \partial_y, \end{aligned}$$

where $y := \ell^{-1}$, hence $\partial_y := \frac{1}{\dot{y}(t)}\,\partial_t$. Similarly as it was proven for (8), it results that $\sigma' \in \mathrm{End}\,(\mathcal{O})$:

$$\begin{aligned} \sigma' : \mathcal{O} &\longrightarrow \mathcal{O} \\ \sum_{i=0}^{r} a_i(t)\,\partial_t^i &\longmapsto \sum_{i=0}^{r} a(y)\,\partial_y^i, \end{aligned}$$

Using $y \circ \ell = \mathrm{id} = \ell \circ y$, we have:

$$\begin{aligned} \ell(y(t)) = t \quad &\Rightarrow \quad \dot{\ell}(y(t))\,\dot{y}(t) = 1, \\ y(\ell(t)) = t \quad &\Rightarrow \quad \dot{y}(\ell(t))\,\dot{\ell}(t) = 1. \end{aligned}$$

Since $\dot{\ell} = 1 - \dot{h} \in A$, then $\frac{1}{\dot{y}} = \dot{\ell}(y) \in A$. Furthermore,

$$\begin{aligned} (\sigma' \circ \sigma)\left(\sum_{i=0}^{r} a_i(t)\,\partial_t^i\right) &= \sigma'\left(\sum_{i=0}^{r} a_i(\ell(t))\left(\frac{1}{\dot{\ell}(t)}\,\partial_t\right)^i\right) = \sum_{i=0}^{r} a_i(t)\left(\frac{1}{\dot{\ell}(y(t))\,\dot{y}(t)}\,\partial_t\right)^i \\ &= \sum_{i=0}^{r} a_i(t)\,\partial_t^i, \end{aligned}$$

$$(\sigma\circ\sigma')\left(\sum_{i=0}^r a_i(t)\,\partial_t^i\right)=\sigma\left(\sum_{i=0}^r a_i(y(t))\left(\tfrac{1}{\dot y(t)}\,\partial_t\right)^i\right)=\sum_{i=0}^r a_i(t)\left(\tfrac{1}{\dot y(\ell(t))\,\dot\ell(t)}\,\partial_t\right)^i$$
$$=\sum_{i=0}^r a_i(t)\,\partial_t^i$$

and it follows that σ is an automorphism of $\mathscr{O}$ and $\sigma'=\sigma^{-1}$. Finally, the results are consequences of (3) of Theorem 1. ■

An example of a function h as in (10) is given by $t\mapsto q\,t+\alpha\in A$ with $\alpha\in\mathbb{R}_{>0}$ and $0<q<1$.

Remark 3 In the proof of Corollary 2, we used the second construction of the ring of DTVD operators given in Remark 2. But the first construction can also be used. Note that the condition $\dot h\neq 1$ is also required so that $\sigma'(S)=(1-\dot h)\,S$ is injective.

Example 6 Consider two functions $h_1: t\mapsto \frac{1}{1+t^2}$ and $h_2: t\mapsto 1-h_1(t)$. Then, we have $h_1, h_2:\mathbb{R}_{>0}\to [0,\,1[$. Moreover, the equations $1-\dot h_1(t)=1+\frac{2t}{(1+t^2)^2}=0$ and $1-\dot h_2(t)=1-\frac{2t}{(1+t^2)^2}=0$ have no real (positive) solutions. In addition, the functions $\ell_i: t\in\mathbb{R}_{>0}\longmapsto t-h_i(t)\in[0,\,1[$ are bijective for $i=1,\,2$. Given $x\geq 0$, the equation $\ell_1(t)=t-h_1(t)=x$ admits a unique real positive solution defined by:

$$\ell_1^{-1}(x):=\frac{x}{3}+\frac{\alpha_1^{\frac{1}{3}}}{6}-\frac{1}{3\,\alpha_1^{\frac{1}{3}}}\,(3-x^2),$$

where $\alpha_1=8\,x^3+72\,x+108+12\,\sqrt{3\,(4\,x^4+4\,x^3+8\,x^2+36\,x+31)}$. In a similar way, given $x\geq 0$, the equation $\ell_2(t)=t-h_2(t)=x$ admits a unique real solution defined by:

$$\ell_2^{-1}(x)=\frac{1}{3}\,(x+1)+\frac{1}{6}\,\alpha_2^{\frac{1}{3}}+\frac{(x^2+2\,x-2)}{3\,\alpha_2^{\frac{1}{3}}},$$

where $\alpha_2=8\,x^3+24\,x^2+96\,x+188+\sqrt{3\,(4\,x^4+20\,x^3+44\,x^2+80\,x+83)}$. Hence, for $i=1,\,2$, if A_i is a differential field such that $a\circ\ell_i\in A$ and $a\circ\ell_i^{-1}\in A_i$ for all $a\in A_i$, then using Corollary 2, $\mathscr{D}_i=A_i\left[\partial;\,\mathrm{id}_{A_i},\,\frac{d}{dt}\right][S_i;\,\sigma_{h_i},\,0]$ are Noetherian domains.

In what follows, for a given ring $\mathbf{A}$, we denote by $\mathbf{A}^{p\times q}$ the $\mathbf{A}$-module of $p\times q$-matrices with entries in $\mathbf{A}$. The following Theorem provides a helpful result on the effective study of linear systems over the ring $\mathscr{D}$ of DTVD operators.

Theorem 2 *Every left ideal of the Ore extension* $\mathscr{D}=A\left[\partial;\,\mathrm{id}_A,\,\frac{d}{dt}\right][S;\,\sigma,\,0]$, *where* A *is a field and* σ *an automorphism, admits a Gröbner basis for an admissible monomial order [15] which can be computed by means of Buchberger's algorithm*

[15]. More generally, every left $\mathscr{D}$-submodule of $\mathscr{D}^{1\times p}$ admits a Gröbner basis for $p \in \mathbb{Z}_{\geq 0}$.

Proof If A is a field, then the Ore extension $\mathscr{D} = A\left[\partial; \mathrm{id}_A, \frac{d}{dt}\right][S; \sigma, 0]$ of DTVD operators is a bijective skew PBW extension (see [9] for the definitions). The authors in [14] proved that Gröbner techniques [15] hold over a bijective skew PBW extension, so the Theorem follows. ■

4.1 Algebraic Analysis Framework

This subsection contains some generalities on the algebraic analysis framework for linear systems theory. Here, D will denote an arbitrary ring.

Definition 5 Let D be a ring (of functional operators), $R \in D^{q\times p}$ and $\mathscr{F}$ a left D-module. A *linear functional system* or a *behavior* is defined by the Abelian group:

$$\ker_{\mathscr{F}}(R.) := \{\eta \in \mathscr{F}^{p\times 1} \mid R\,\eta = 0\}.$$

Let us consider the left D-submodule $D^{1\times q}\,R = \{\mu\, R \mid \mu \in D^{1\times q}\}$ of $D^{1\times p}$ consisting of all the left D-linear combinations of the columns of R. Let us denote by $M = D^{1\times p}/(D^{1\times q}\,R)$ the factor left D-module. The Abelian group of D-homomorphisms (i.e., left D-linear maps) from M to $\mathscr{F}$ [25] is denoted by $\hom_D(M, \mathscr{F})$. An important note concerns the D-module isomorphism,[4] following a standard homological algebra result:

$$\ker_{\mathscr{F}}(R.) \cong \hom_D(M, \mathscr{F}). \tag{11}$$

More details can be found in [3, 20, 23, 30]. A remarkable consequence of (11) is that a linear system can be studied by means of M (that encodes the system equations) and by $\mathscr{F}$ (that is the functional space where the solutions are sought).[5] More importantly, built-in properties of the linear system $\ker_{\mathscr{F}}(R.)$ such as controllability, observability, and flatness can be characterized by module properties of M.

Next, we state basic module properties and definitions [25].

Definition 6 Let D be a domain and M a left D-module. The *dual module* $M^\star$ is the right D-module defined by $M^\star := \hom_D(M, D)$.

There is a canonical homomorphism between M and its bidual module $M^{\star\star}$:

$$\begin{aligned} \varepsilon : M &\longrightarrow M^{\star\star} = \hom_D(\hom_D(M, D), D) \\ m &\longmapsto \varepsilon(m), \;\; \text{with } \varepsilon(m)(f) := f(m), \;\; \forall f \in M^\star. \end{aligned} \tag{12}$$

[4] i.e., a bijective homomorphism of left D-modules

[5] The space $\mathscr{F}$ is also called the *signal space* in the behavioral approach [20].

Torsion modules play an important role in the theory of D-modules.

Definition 7 The *torsion submodule* $t(M)$ of a D-module M is defined by:

$$t(M) = \{m \in M \mid \exists\, d \in D \setminus \{0\} : \; d\, m = 0\}.$$

We recall that a left D-module M is *finitely generated* if $M = \sum_{i=1}^{r} D\, m_i$ with $m_i \in M$ and $r \in \mathbb{Z}_{\geq=0}$.

Definition 8 Let D be a domain and M a *finitely generated* left D-module.

1. M is *free* if there exists $r \in \mathbb{Z}_{\geq 0}$ such that $M \cong D^{1\times r}$.
2. M is *stably free* if there exist $r, \; s \in \mathbb{Z}_{\geq 0}$ such that $M \oplus D^{1\times s} \cong D^{1\times r}$.
3. M is *projective* if there exist $r \in \mathbb{Z}_{\geq 0}$ and a left D-module N such that:

$$M \oplus N \cong D^{1\times r}.$$

4. M is *reflexive* if the canonical homomorphism $\varepsilon : M \longrightarrow M^{\star\star}$ defined by (12) is an isomorphism.
5. M is *torsion-free* if $t(M) = 0$.
6. M is *torsion* if $t(M) = M$.

It is clear that a free module is stably free, and a stably free module is projective. If M is a finitely generated free module, then $M \cong M^{\star\star}$, so M is a reflexive module.

If M is a reflexive module, then M is torsion-free. To prove that, note that:

$$\ker(\varepsilon) = \left\{m \in M \mid \forall\, f \in M^{\star} : \; f(m) = 0\right\}.$$

If $m \in t(M)$, then there exists $d \in D \setminus \{0\}$ such that $d\, m = 0$, hence $d\, f(m) = f(d\, m) = f(0) = 0$ for all $f \in M^{\star}$, and so $f(m) = 0$, since D is a domain. That shows $t(M) \subseteq \ker(\varepsilon)$, and consequently, M is torsion-free whenever M is reflexive.

There is a strong relation between basic ideas in control theory, such as recognizing when a system is *parametrizable* or *flat* [3, 7, 23, 24], and the equivalence of the above Definition 8. Recall that if D is a *principal ideal domain*[6] (e.g., $D = \mathbb{K}[\partial; \mathrm{id}, 0]$ with $\mathbb{K}$ a field), then a torsion-free module is free (see for instance [25]).

Homological invariants concepts (defined in [18, 25]) of the ring of DTVD operators are characterized in the sequel. These invariants play an important role in the algebraic analysis approach [3, 23]. We start by recalling basic definitions of homological algebra [25].

Definition 9 Let $(M_i)_{i\in\mathbb{Z}}$ be a sequence of left (resp., right) D-modules. For $i \in \mathbb{Z}$, let $\delta_i \in \hom_D(M_i, M_{i-1})$.

[6] In a principal ideal domain D, every left (resp. right) ideal of D is generated by a single element of D, that is, of the form $D\, d$ (resp., $d\, D$) for a certain $d \in D$ [25].

1. The sequence $(M_i, \delta_i)_{i \in \mathbb{Z}}$ forms a *complex* if $\delta_i \circ \delta_{i+1} = 0$, that is, if we have $\text{im}\,(\delta_{i+1}) \subseteq \ker\,(\delta_i)$ for all $i \in \mathbb{Z}$. This complex is denoted by:

$$M_\bullet : \ \ldots \xrightarrow{\delta_{i+2}} M_{i+1} \xrightarrow{\delta_{i+1}} M_i \xrightarrow{\delta_i} M_{i-1} \xrightarrow{\delta_{i-1}} \ldots$$

2. The *defect of exactness* of the complex $M_\bullet$ at M_i is the left (resp., right) D-module defined by:
$$H_i(M_\bullet) = \ker\,(\delta_i) \,/\text{im}\,(\delta_{i+1})\,.$$
3. The complex $M_\bullet$ is said to be *exact at* M_i if we have $H_i(M_\bullet) = 0$, i.e., if we have $\ker\,(\delta_i) = \text{im}\,(\delta_{i+1})$, and is *exact* if:
$$H_i(M_\bullet) = 0, \ \forall\, i \in \mathbb{Z}.$$
We also say that this sequence is an *exact sequence*.
4. A *projective* (resp., *free*) *resolution* of a left D-module M is an exact sequence of the form
$$P_\bullet : \ \ldots \xrightarrow{\delta_3} P_2 \xrightarrow{\delta_2} P_1 \xrightarrow{\delta_1} P_0 \xrightarrow{\delta_1} M \longrightarrow 0,$$
where the P_i's are projective (resp., free) left D-modules. If $P_i = 0$ for $i \geq m+1$, then the *length* of the resolution $P_\bullet$ is said to be m. Similar definitions hold for right D-modules.
5. The *left projective dimension* of a left D-module M, denoted by $\text{lpd}_D(M)$, is the length of the shortest projective resolution of M (and similarly for the *right projective dimension* $\text{rpd}_D(M)$ of a right D-module M).
6. The *left* (resp., *right*) *global dimension* of D, denoted by $\text{lgd}\,(D)$ (resp., $\text{rgd}\,(D)$), is the supremum of $\text{lpd}_D(M)$ (resp., $\text{rpd}_D(M)$) over all the left (resp., right) D-modules M.

Theorem 3 (Auslander's theorem, Corollary 8.28 of [25]) *If A is a* Noetherian ring *(i.e., both left and right Noetherian ring), then:*
$$\text{lgd}\,(A) = \text{rgd}\,(A)\,.$$
We simply denote it by $\text{gd}\,(A)$.

Definition 10 A ring A is called a *left* (resp., *right*) *regular ring* if every finitely generated left (resp., right) A-module has finite projective dimension.

Theorem 4 *([18]) Let A be a ring with finite left (resp., right) global dimension* $\text{lgd}\,(A)$ *(resp.,* $\text{rgd}\,(A)$*) and $\sigma \in \text{End}(A)$. Then, we have:*

1. $\text{lgd}\,(A) \leq \text{lgd}\,(A[\partial; \sigma, \delta]) \leq \text{lgd}\,(A) + 1$
(resp., $\text{rgd}\,(A) \leq \text{rgd}\,(A[\partial; \sigma, \delta]) \leq \text{rgd}\,(A) + 1$*).*

2. *If* $\delta = 0$, *then* $\operatorname{lgd}(A[\partial;\sigma,\delta]) = \operatorname{lgd}(A) + 1$ *(resp.,* $\operatorname{rgd}(A[\partial;\sigma,\delta]) = \operatorname{rgd}(A) + 1$*).*
3. *If A is a semisimple Artinian ring (e.g., A is a field), then* $\operatorname{lgd}(A[\partial;\sigma,\delta]) = 1$ *(resp.,* $\operatorname{rgd}(A[\partial;\sigma,\delta]) = 1$*).*

4.2 Some Properties of the Ring of DTVD Operators

Using the results recalled in (4.1), we can prove the following results:

Proposition 2 *Let* $\mathscr{D} = A\left[\partial;\operatorname{id}_A,\frac{d}{dt}\right][S;\sigma,0]$ *be the ring of DTVD operators defined in Sect. 3. Assume that the function* $\ell : t \longmapsto t - h(t)$ *is bijective.*[7] *If A has finite left (resp., right) global dimension* $\operatorname{lgd}(A)$ *(resp.,* $\operatorname{rgd}(A)$*), then:*

$$\operatorname{lgd}(A) + 1 \leq \operatorname{lgd}(\mathscr{D}) \leq \operatorname{lgd}(A) + 2$$

(resp., $\operatorname{rgd}(A) + 1 \leq \operatorname{rgd}(\mathscr{D}) \leq \operatorname{rgd}(A) + 2$*). In particular, $\mathscr{D}$ is a left (resp., right) regular ring. If A is a semisimple Artinian ring,*[8] *then* $\operatorname{lgd}(\mathscr{D}) = 2$ *(resp.,* $\operatorname{rgd}(\mathscr{D}) = 2$*). Finally, if A is a Noetherian ring, then* $\operatorname{gd}(\mathscr{D}) = \operatorname{lgd}(\mathscr{D}) = \operatorname{rgd}(\mathscr{D})$.

This Proposition is a consequence of Theorem 4.

Example 7 Considering Example 6, if $\mathscr{D} = \mathscr{D}_1$ or $\mathscr{D} = \mathscr{D}_2$, then we have $\operatorname{gd}(\mathscr{D}) = 2$.

Definition 11 A ring A is said to be *projective stably free* if every finitely generated projective left/right A-module is stably free.

Theorem 5 ([27]) *If A is a left regular Noetherian ring and projective stably free (e.g., A is a field), then so is the ring* $\mathscr{D} = A\left[\partial;\operatorname{id}_A,\frac{d}{dt}\right][S;\sigma,0]$.

Proof It is a consequence of Corollary 12.3.3 of [18]. ■

Corollary 3 *Let* $\mathscr{D} = A\left[\partial;\operatorname{id}_A,\frac{d}{dt}\right][S;\sigma,0]$ *be the ring of DTVD operators defined in Sect. 3. Assume that the function* $\ell : t \longmapsto t - h(t)$ *is bijective. Let* $R \in \mathscr{D}^{q\times p}$ *and* $M = \mathscr{D}^{1\times p}/(\mathscr{D}^{1\times q}\,R)$ *be the left $\mathscr{D}$-module finitely presented by R. Then, M admits a free resolution of length less than or equal to* $\operatorname{lgd}(\mathscr{D}) + 1$.

Proof This can be proved as in Proposition 8 of [3]. ■

The reader may refer to [3, 23] and to the OREMODULES package [4] for an explicit way to compute free resolutions of finitely generated left D-modules based on Gröbner basis techniques.

We denote by $\operatorname{lKdim}(A)$ the *left Krull dimension* of the ring A. For more details, see [18] and the references therein.

---

[7] e.g., $h(t) = q\,t + h$ with $h \in \mathbb{R}_{\geq 0}$ and $0 < q < 1$.

[8] e.g., A is a field.

Theorem 6 (Proposition 6.5.4 of [18]) *If A is a left Noetherian ring, σ an automorphism of A and δ a σ-derivation, then:*

1. $\mathrm{lKdim}\,(A) \leq \mathrm{lKdim}\,(A[\partial;\, \sigma, \delta]) \leq \mathrm{lKdim}\,(A) + 1.$
2. $\mathrm{lKdim}\,(A[S;\, \sigma, 0]) = \mathrm{lKdim}\,(A).$
3. *If A is a left Artinian ring, then* $\mathrm{lKdim}\,(A[S;\, \sigma, 0]) = 1.$

Corollary 4 *Let $\mathscr{D} = A\left[\partial;\, \mathrm{id}_A, \frac{d}{dt}\right][S;\, \sigma, 0]$ be the ring of DTVD operators defined in Sect. 3, where the function $\ell : t \longmapsto t - h(t)$ is bijective, then:*

$$\mathrm{lKdim}\,(A) \leq \mathrm{lKdim}\,(\mathscr{D}) \leq \mathrm{lKdim}\,(A) + 1.$$

If D is a left Noetherian domain, then we can define its *total quotient field/field of fractions* $K := \{d^{-1}\, n \mid 0 \neq d,\, n \in D\}$ and the *rank* of a left D-module M is defined by $\mathrm{rank}_D(M) := \dim_K(K \otimes_D M)$, where $K \otimes_D M$ is the left K-vector space obtained by extending the scalars of M from D to K [25].

Theorem 7 (Theorems 11.1.14 and 11.1.17 of [18]) *If D is a left Noetherian domain, then any stably free left D-module M with* $\mathrm{rank}_D(M) \geq \mathrm{lKdim}(D) + 1$ *is free.*

5 Algebraic Analysis Approach

Let D be a Noetherian domain. In this Section, we briefly introduce the concept of *extension modules* [25]. Let $M = D^{1\times p}/(D^{1\times q}\, R)$ be the left D-module *finitely presented* by the system matrix $R \in D^{q\times p}$, which is associated with the linear system:

$$\ker_{\mathscr{F}}(R.) = \{\eta \in \mathscr{F}^p \mid R\,\eta = 0\}.$$

Then, we have the following exact sequence

$$0 \longrightarrow \ker_D(.R) \xrightarrow{\ i\ } D^{1\times q} \xrightarrow{\ .R\ } D^{1\times p} \xrightarrow{\ \pi\ } M \longrightarrow 0\,,$$

where $.R \in \hom_D(D^{1\times q}, D^{1\times p})$ is defined by $(.R)(\mu) = \mu\, R$ for all $\mu \in D^{1\times q}$, $\ker_D(.R) = \{\mu \in D^{1\times q} \mid \mu\, R = 0\}$ its kernel, i is the canonical injection and $\pi \in \hom_D(D^{1\times p}, M)$ is the canonical projection which sends $\lambda \in D^{1\times p}$ onto its *residue class* $\pi(\lambda) \in M$ (note that $\pi(\lambda) = \pi(\lambda')$ iff $\lambda - \lambda' \in D^{1\times q}\, R$). Since D is a left Noetherian ring, then the finitely generated left D-module $D^{1\times q}$ is Noetherian, and thus, $\ker_D(.R)$ is a finitely generated left D-module. Therefore, there exists a finite set of generators $\{\mu_j\}_{j=1,\ldots,r}$ of $\ker_D(.R)$ and we have $\ker_D(.R) = \mathrm{im}_D(.R_2) = D^{1\times r}\, R_2$ for $R_2 = (\mu_1^T\ \ldots\ \mu_r^T)^T \in D^{r\times q}$. Thus, we obtain the following long exact sequence of left D-modules:

$$0 \longrightarrow \ker_D(.R_2) \xrightarrow{\ i_2\ } D^{1\times r} \xrightarrow{\ .R_2\ } D^{1\times q} \xrightarrow{\ .R\ } D^{1\times p} \xrightarrow{\ \pi\ } M \longrightarrow 0.$$

Repeating the same reasoning as above for $\ker_D(.R_2)$ and so on, we finally obtain a *free resolution* of M

$$\dots \xrightarrow{.R_3} D^{1\times p_2} \xrightarrow{.R_2} D^{1\times p_1} \xrightarrow{.R_1} D^{1\times p_0} \xrightarrow{\pi} M \longrightarrow 0,$$

where $p_0 = p$, $p_1 = q$, $p_2 = r$ and $R_1 = R$. Applying the *contravariant left exact functor* $\hom_D(\cdot, \mathscr{F})$ [25] to the *truncated free resolution* of M defined by

$$\dots \xrightarrow{.R_3} D^{1\times p_2} \xrightarrow{.R_2} D^{1\times p_1} \xrightarrow{.R_1} D^{1\times p_0} \longrightarrow 0, \tag{13}$$

i.e., dualizing (13), we get the following complex of Abelian groups

$$\mathscr{F}^\bullet : \dots \xleftarrow{R_3.} \mathscr{F}^{p_2} \xleftarrow{R_2.} \mathscr{F}^{p_1} \xleftarrow{R_1.} \mathscr{F}^{p_0} \longleftarrow 0,$$

where $(R_i.)(\eta) = R_i\,\eta$ for all $\eta \in \mathscr{F}^{p_{i-1}}$ for $i \geq 1$. The defects of exactness of $\mathscr{F}^\bullet$ are characterizes by the *cohomology* groups of $\mathscr{F}^\bullet$ defined by:

$$\begin{cases} H^0(\mathscr{F}^\bullet) = \ker_{\mathscr{F}}(R_1.), \\ H^i(\mathscr{F}^\bullet) = \ker_{\mathscr{F}}(R_{i+1}.)/\mathrm{im}_{\mathscr{F}}(R_i.), \quad i \geq 1. \end{cases}$$

Using (11), we obtain $H^0(\mathscr{F}^\bullet) \cong \hom_D(M, \mathscr{F})$. More generally, an important result of homological algebra proves that the cohomologies $H^i(\mathscr{F}^\bullet)$ do not depend on the choice of the free resolution of M, i.e., up to isomorphism, they depend only on M and $\mathscr{F}$ [25]. They are then denoted by

$$\begin{cases} \mathrm{ext}^i_D(M, \mathscr{F}) = \hom_D(M, \mathscr{F}), \\ \mathrm{ext}^i_D(M, \mathscr{F}) = H^i(\mathscr{F}^\bullet), \quad i \geq 1, \end{cases}$$

and are called *extension Abelian groups*. In the case where $\mathscr{F} = D$, we can prove that the $\mathrm{ext}^i_D(M, D)$ inherit a right D-module structure. Similarly, if M is a right D-module, then the $\mathrm{ext}^i_D(M, D)$ inherit left D-module structures. For an implementation of the computation of the $\mathrm{ext}^i_D(M, D)$'s for certain Ore algebras, see the OREMODULES package [4].

Due to the homological nature of the main results of [3] (see also [23]), they can directly be applied to the ring of DTVD operators.

Theorem 8 *Let A be a regular Noetherian ring and $\mathscr{D} = A\left[\partial; \mathrm{id}_A, \frac{d}{dt}\right][S; \sigma, 0]$ the ring of DTVD operators with coefficients in A defined in Sect. 3. Assume that the function $\ell : t \longmapsto t - h(t)$ is bijective. Moreover, let $R \in \mathscr{D}^{q\times p}$, $M = \mathscr{D}^{1\times p}/(\mathscr{D}^{1\times q}\, R)$ and $N = \mathscr{D}^{q\times 1}/(R\, \mathscr{D}^{p\times 1})$ the so-called* Auslander transpose *of M. Then, we have:*

1. $t(M) \cong \mathrm{ext}^1_{\mathscr{D}}(N, \mathscr{D})$.
2. *M is torsion-free if and only if $\mathrm{ext}^1_{\mathscr{D}}(N, \mathscr{D}) = 0$.*

3. *M is reflexive if and only if* $\mathrm{ext}^1_{\mathcal{D}}(N, \mathcal{D}) = \mathrm{ext}^2_{\mathcal{D}}(N, \mathcal{D}) = 0$.
4. *M is projective if and only if* $\mathrm{ext}^i_{\mathcal{D}}(N, \mathcal{D}) = 0$ *for* $i = 1, \ldots, \mathrm{gd}(\mathcal{D})$.
5. *If A is a projective free ring, then M is projective left $\mathcal{D}$-module if and only if M is stably free left $\mathcal{D}$-module. Moreover, M is free when:*

$$\mathrm{rank}_{\mathcal{D}}(M) \geq \mathrm{lKdim}(\mathcal{D}) + 1.$$

Proof Assertions 1 and 2 are direct consequences of Theorem 5 of [3]. The assertion 3 (resp., 4) is a consequence of Theorem 6 (resp., Theorem 7) of [3]. Finally, let us prove 5. A stably free left $\mathcal{D}$-module is well-known to be projective. The converse is proved in Corollary 12.3.3 of [18]. ■

Within the algebraic analysis framework, the concept of an *injective cogenerator signal space* $\mathcal{F}$ plays a similar role as an algebraic *closed field* in algebraic geometry (think about the solutions of $x^2 + 1 = 0$ in $\mathbb{R}$). A non-trivial module $M = D^{1\times p}/(D^{1\times q}\, R)$ then defines a non-zero linear system/behavior $\ker_{\mathcal{F}}(R.)$. Moreover, a complete duality exists between linear systems/behaviors and finitely presented left modules [3, 20, 23, 30].

Definition 12 ([25])

1. A left D-module $\mathcal{F}$ is *cogenerator* if for every left D-module M and $m \in M \setminus \{0\}$, there exists $f \in \hom_D(M, \mathcal{F})$ such that $f(m) \neq 0$.
2. A left D-module $\mathcal{F}$ is *injective* if for all left D-modules M, $\mathrm{ext}^i_D(M, \mathcal{F}) = 0$ for all $i \geq 1$.

For a given ring D, it can be shown that an injective cogenerator left D-module $\mathcal{F}$ always exists (see, e.g., [25]).

According to [3, 7, 20, 22, 23, 30], we can state the following general definitions.

Definition 13 Let D be a Noetherian domain, $R \in D^{q\times p}$, $\mathcal{F}$ an injective cogenerator left D-module and $\ker_{\mathcal{F}}(R.)$ the linear system defined by R.

1. An *observable* $\psi(\eta)$ of $\ker_{\mathcal{F}}(R.)$ is a left D-linear combination of the system variables, i.e., $\psi(\eta) = \sum_{i=1}^{p} d_i\, \eta_i$, where $d_i \in D$ and $\eta = (\eta_1 \ \ldots \ \eta_p)^T \in \ker_{\mathcal{F}}(R.)$.
2. An observable $\psi(\eta)$ is called *autonomous* if it satisfies a left D-linear relation by itself, i.e., $d\, \psi(\eta) = 0$ for some $d \in D \setminus \{0\}$. It is called *free* it is not autonomous.
3. The linear system is said to be *controllable* if every observable is free.
4. The linear system is said to be *parametrizable* if there exists $Q \in D^{p\times m}$ such that

$$\ker_{\mathcal{F}}(R.) = \mathrm{im}_{\mathcal{F}}(Q.) = Q\, \mathcal{F}^m,$$

   i.e., if for every $\eta \in \ker_{\mathcal{F}}(R.)$, there exists $\xi \in \mathcal{F}^m$ such that $\eta = Q\, \xi$. Then, Q is called a *parametrization* and ξ a *potential*.
5. The linear system is said to be *flat* if there exists a parametrization $Q \in D^{p\times m}$ which admits a left inverse $T \in D^{m\times p}$, i.e. $T\, Q = I_p$. In other words, a flat system is a parametrizable system such that every component ξ_i of a potential ξ is an observable of the system. The potential ξ is then called a *flat output*.

The next Theorem explicitly characterizes the above definitions in terms of properties of modules.

Theorem 9 ([3, 7]) *Let $M = D^{1\times p}/(D^{1\times q}\, R)$ be the finitely presented left D-module associated with the linear system $\ker_{\mathscr{F}}(R.) = \{\eta \in \mathscr{F}^p \mid R\,\eta = 0\}$. With the hypotheses of Definition 13, we have:*

1. *The observables of the linear system are in one-to-one correspondence with the elements of M.*
2. *The autonomous elements of the linear system are in one-to-one correspondence with the torsion elements of M. The linear system is controllable if and only if M is torsion-free.*
3. *The linear system is parametrizable if and only if there exists a matrix $Q \in D^{p\times m}$ such that $M \cong D^{1\times p}\, Q$, i.e. M is a torsion-free left D-module. Then, the matrix Q is a parametrization, i.e. $\ker_{\mathscr{F}}(R.) = Q\,\mathscr{F}^m$. A parametrization Q can be computed by checking that $\mathrm{ext}^1_D(N, D) = 0$, where $N = D^{q\times 1}/(R\, D^{p\times 1})$.*
4. *The linear system is flat if and only if M is a free left D-module. Then, the bases of M are in one-to-one correspondence with the flat outputs of the linear system.*

Combining Theorems 8 and 9, the system properties listed in Definition 13 can be explicitly characterized in terms of module properties and in terms of the vanishing of the extension modules $\mathrm{ext}^i_D(N, D)$. Hence, the future implementation of Gröbner basis techniques for the ring D (see Theorem 2) in the OREMODULES package [4] will give us an effective way to check the system properties given in Definition 13.

Example 8 Let us consider the system $\dot{x}(t) = u(t - h(t))$ with $\ell : t \longmapsto t - h(t)$ bijective. Let $\mathscr{D}$ be a ring of DTVD operators, $R = (\partial \quad -S) \in \mathscr{D}^{1\times 2}$ the system matrix, $M = \mathscr{D}^{1\times 2}/(\mathscr{D}\, R)$ the left $\mathscr{D}$-module finitely presented by R and $N = \mathscr{D}/(R\,\mathscr{D}^{2\times 1})$ the Auslander transpose of M. Using the identity

$$\partial\, S = (1 - \dot{h})\, S\, \partial = S\, \left(\sigma^{-1}\left(1 - \dot{h}\right)\, \partial\right),$$

if we note

$$Q = \left(S \quad \sigma^{-1}\left(1 - \dot{h}\right)\, \partial\right)^T \in \mathscr{D}^{2\times 1},$$

then we have $\mathrm{im}_{\mathscr{D}}(Q.) = \ker_{\mathscr{D}}(R.)$. Thus, we obtain the free resolution:

$$0 \longleftarrow N \xleftarrow{\kappa} \mathscr{D} \xleftarrow{R.} \mathscr{D}^{2\times 1} \xleftarrow{Q.} \mathscr{D} \longleftarrow 0.$$

Dualizing this free resolution, we get the following complex

$$0 \longrightarrow \mathscr{D} \xrightarrow{.R} \mathscr{D}^{1\times 2} \xrightarrow{.Q} \mathscr{D} \longrightarrow 0,$$

which yields:

$$\begin{cases} \operatorname{ext}^1_{\mathscr{D}}(N,\mathscr{D}) = \ker_{\mathscr{D}}(.Q)/\operatorname{im}_{\mathscr{D}}(.R) \cong t(M), \\ \operatorname{ext}^2_{\mathscr{D}}(N,\mathscr{D}) = \mathscr{D}/(\mathscr{D}^{1\times 2}\, Q) = \mathscr{D}/\,(S,\,\partial) \neq 0. \end{cases}$$

We obtain that M is torsion-free but it is not a projective left $\mathscr{D}$-module since $\ker_{\mathscr{D}}(.Q) = \operatorname{im}_{\mathscr{D}}(.R)$ and $\operatorname{gd}(\mathscr{D}) = 2$.

If $\mathscr{F}$ is an injective left $\mathscr{D}$-module, then applying the contravariant exact functor $\hom_{\mathscr{D}}(\,\cdot\,,\mathscr{F})$ to the above complex results in the exact sequence of Abelian groups:

$$0 \longleftarrow \mathscr{F} \xleftarrow{R.} \mathscr{F}^2 \xleftarrow{Q.} \mathscr{F} \longleftarrow \ker_{\mathscr{F}}(Q.) \longleftarrow 0.$$

We get $\ker_{\mathscr{F}}(R.) = \operatorname{im}_{\mathscr{F}}(Q.)$, showing that Q is a parametrization of $\ker_{\mathscr{F}}(R.)$, so:

$$\begin{cases} x(t) = S\,\xi(t) = \xi(t - h(t)), \\ u(t) = \sigma^{-1}\left(1-\dot{h}\right)\,\partial\,\xi(t) = \left(1 - \dot{h}(\ell^{-1}(t))\right)\,\dot{\xi}(t). \end{cases}$$

Since σ is an automorphism of $A\left[\partial;\,\operatorname{id}_A,\,\frac{d}{dt}\right]$, then we can define *the skew Laurent polynomial ring* $E = A\left[\partial;\,\operatorname{id}_A,\,\frac{d}{dt}\right][S;\sigma,0][S^{-1};\sigma^{-1},0]$ [18]. Hence, $E \otimes_{\mathscr{D}} M$ is a free left E-module of rank 1 and x is a basis of $E \otimes_{\mathscr{D}} M$, i.e., the DTVD system is a S-*flat* system [7].

6 Conclusion

In this work, an algebraic analysis framework for the ring of differential time-varying delay operators was proposed via its realization as an Ore extension $\mathscr{D}$. An explicit construction of this ring was given, providing a novel algebraic approach for the study of linear differential varying time-delay systems. Homological algebraic properties of $\mathscr{D}$ were studied, and its global and Krull dimensions were analyzed.

Directions for future research include the possibility of $\mathscr{D}$ being an Auslander regular ring or a Cohen Macaulay ring [18]. Another important point would be to study the existence of an *involution* of $\mathscr{D}$ so that Gröbner bases can be calculated for right $\mathscr{D}$-modules (see [3] for more details). As a significant issue for the constructive aspects, Gröbner basis techniques for $\mathscr{D}$ will then be implemented.

References

1. A. Borel, *Algebraic D-Modules*, Perspectives in Mathematics (Academic Press, 1987)
2. C. Briat, O. Sename, J.f. Lafay, Design of LPV observers for LPV time-delay systems: an algebraic approach. Int. J. Control **84**(9), 1533–1542 (2011)
3. F. Chyzak, A. Quadrat, D. Robertz, Effective algorithms for parametrizing linear control systems over ore algebras. Appl. Algebr. Engrg. Comm. **16**, 319–376 (2005)

4. F. Chyzak, A. Quadrat, D. Robertz, OREMODULES*: A Symbolic Package for the Study of Multidimensional Linear Systems*. Lecture Notes in Control and Information Sciences, vol. 352, (Springer, 2007), pp. 233–264
5. F. Chyzak, B. Salvy, Non-commutative elimination in ore algebras proves multivariate identities. J. Symb. Comput. **26**, 187–227 (1998)
6. T. Cluzeau, A. Quadrat, Factoring and decomposing a class of linear functional systems. Linear Algebr. Appl. **428**, 324–381 (2008)
7. M. Fliess, H. Mounier, Controllability and observability of linear delay systems: an algebraic approach. ESAIM COCV **3**, 301–314 (1998)
8. E. Fridman, *Introduction to Time-Delay Systems: Analysis and Control*, Systems & Control: Foundations & Applications (Springer International Publishing, 2014)
9. C. Gallego, O. Lezama, Gröbner bases for ideals of σ-PBW extensions. Commun. Algebr. **39**, 50–75 (2010)
10. R.A. Johnson, Functional equations, approximations, and dynamic response of systems with variable time delay. IEEE Trans. Automat. Control **17**, 398–401 (1972)
11. M. Kashiwara, Algebraic study of systems of partial differential equations. Mémoires de la S. M. F., 1995, vol. 63, master's thesis (1970) (English translation)
12. M. Kashiwara, T. Kawai, T. Kimura, *Foundations of Algebraic Analysis*, vol. 37. (Princeton University Press, 1986)
13. V. Léchappé, Predictive control and estimation of uncertain systems with an input delay, Ph.D. Thesis, Ecole Centrale de Nantes (2015)
14. O. Lezama, C. Gallego, *Projective Modules and Gröbner Bases for Skew PBW Extensions*, Dissertationes Mathematicæ. Institute of Mathematics (Polish Academy of Sciences, 2017)
15. H. Li, *Noncommutative Gröbner Bases and Filtered-Graded Transfer*. 1795 (Springer, Lecture Notes in Mathematics, 2002)
16. J. Louisell, Delay differential systems with time-varying delay: new directions for stability theory. Kybernetika **37**, 239–251 (1972)
17. S.G. Margolis, J.J. O'Donnell, Rigorous treatments of variable time delays. IEEE Trans. Electron. Comput. **12**, 307–309 (1963)
18. J.C. McConnell, J.C. Robson, *Noncommutative Noetherian Rings* (American Mathematical Society, 2000)
19. S.-I. Niculescu, C.E. de Souza, L. Dugard, J.-M. Dion, Robust exponential stability of uncertain systems with time-varying delays. IEEE Trans. Automat. Control **43**, 743–748 (1998)
20. U. Oberst, Multidimensional constant linear systems. Acta Appl. Math **20**, 1–175 (1990)
21. J.W. Polderman, J.C. Willems, *Introduction to Mathematical Systems Theory: A Behavioral Approach*, TAM 26 (Springer, 1998)
22. J.-F. Pommaret, A. Quadrat, Algebraic analysis of linear multidimensional control systems. IMA J. Appl. Math. **16**, 275–297 (1999)
23. A. Quadrat, An introduction to constructive algebraic analysis and its applications. Les cours du CIRM, Journées Nationales de Calcul Formel **1**(2), 2010 (2010)
24. A. Quadrat, D. Robertz, Computation of bases of free modules over the Weyl algebras. J. Symb. Comput. **42**, 1113–1141 (2007)
25. J.J. Rotman, *An Introduction to Homological Algebra* (Springer, 2009)
26. J.P. Seddon, R.A. Johnson, The simulation of variable delay. IEEE Trans. Comput. **17**, 89–94 (1968)
27. J.-P. Serre, Sur les modules projectifs, Séminaire Dubreil. Algèbre et théorie des nombres, **14**(1) (1960–1961), talk:2. http://www.numdam.org/item/SD_1960-1961__14_1_A2_0/
28. E.I. Verriest, Stability of time varying systems with unknown bounded delays, in *Proceedings of the 33rd Conference on Decision and Control* (Florida, 1994)
29. L.A. Zadeh, Operational analysis of variable-delay systems, in *Proceedings of the I.* vol. 40: R.E, pp. 564–568 (1952)
30. E. Zerz, An algebraic analysis approach to linear time-varying systems. IMA J. Appl. Math. **23**, 113–126 (2006)

Practical Stability Preservation Under Sampling, Actuation Disturbance and Measurement Noise, for Globally Lipschitz Time-Delay Systems

Mario Di Ferdinando, Pierdomenico Pepe, and Emilia Fridman

Abstract In this chapter we deal with the problem of global exponential practical stability preservation under sampling, for globally Lipschitz time-delay systems with state feedback controllers affected by measurement noises and actuation disturbances. It is shown that, if the continuous-time closed-loop system at hand is globally exponentially stable and the maps describing the dynamics and the continuous-time state feedback are globally Lipschitz, then, under suitably fast sampling, the global exponential practical stability of the sampled-data closed-loop system is preserved even in the case of bounded measurement noises and bounded actuation disturbances.

1 Introduction

Sampled-data stabilization of linear, bilinear and nonlinear systems, even infinite dimensional ones, has been studied in the literature by many approaches such as: (i) the time-varying delay approach (see [11–13]); (ii) the approximate system discretization approach (see [14, 30, 35, 37, 38, 47–49]); (iii) the hybrid system approach (see [3, 4, 7, 19, 34, 36]); (iv) the stabilization in the sample-and-hold sense approach (see [5, 6, 9, 40–43]). The reader can refer to [20] for an interesting survey on the topic. The global (asymptotic, exponential) stability preservation

This work is supported in part by the Atheneum Project RIA 2018, and by the Italian Ministry of Education, University and Research MIUR, project FFABR 2017.

M. Di Ferdinando (✉) · P. Pepe
Department of Information Engineering, Computer Science, and Mathematics, University of L'Aquila, Via Vetoio (Coppito 1), 67100 L'Aquila, Italy
e-mail: mario.diferdinando@graduate.univaq.it

P. Pepe
e-mail: pierdomenico.pepe@univaq.it

E. Fridman
School of Electrical Engineering, Tel Aviv University, 69978 Tel Aviv, Israel
e-mail: emilia@eng.tau.ac.il

G. Valmorbida et al. (eds.), *Accounting for Constraints in Delay Systems*, Advances in Delays and Dynamics 12,
https://doi.org/10.1007/978-3-030-89014-8_6

under sampling, for delay-free systems has been extensively studied in [18, 22, 23, 29, 33]. In [4], the hybrid system approach is used in order to provide sufficient conditions for the computation of the maximum allowable sampling period by which global (asymptotic, exponential) stability of nonlinear delay-free systems is preserved under sampling.

In [1], for linear systems with globally Lipschitz perturbations as well as for systems described by semi-linear parabolic partial differential equations, sufficient conditions, ensuring the exponential stability preservation under sampling, are provided. Sufficient Lyapunov-like conditions for the global asymptotic stability preservation under sampling of control-affine time-varying systems with small delays in the input channel are given in [33]. Sampled-data stabilization of finite-dimensional nonlinear systems, with large delays in the input/output channels is extensively studied in [17, 24]. As far as nonlinear systems with state delays are concerned, in [40, 42], sufficient conditions, in terms of Lyapunov-Krasovskii functionals, are provided for the semi-global practical stability preservation (with arbitrarily small final target ball of the origin) by emulation of continuous-time state feedback controllers, for locally Lipschitz time-delay systems. In [8], using the same approach of [40, 42, 43], sufficient Lyapunov-like conditions are provided for the semi-global practical stability preservation by emulation of continuous-time dynamic output feedback controllers, for locally Lipschitz time-delay systems. Global exponential stability preservation results are provided in [45] for the class of nonlinear systems with state-delays, described by globally Lipschitz maps and admitting globally Lipschitz exponential stabilizers.

It is well known that actuation disturbances and observation errors can deteriorate the performances of controllers (see, for instance, [32, 52]), and the same, or even worse, kind of problems arises when the control law is applied by sampling and holding (see [31, 50, 51]). In [23] sufficient conditions ensuring input-to-state stability preservation, with respect to external disturbances, under suitably fast sampling, for nonlinear delay-free systems, are provided. In [41], concerning the stabilization in the sample-and-hold sense of nonlinear control-affine delay-free systems, an input-to-state stability redesign method is exploited, in order to attenuate the effects of bounded actuation disturbances and of suitably bounded observation errors. The same approach has been used for nonlinear control-affine time-delay systems in [9]. To our best knowledge, the problem related to the global exponential practical stability preservation under sampling for globally Lipschitz state feedback time-delay systems affected by bounded measurement noises and bounded actuation disturbances has never been addressed in the literature.

In this chapter, using the same reasonings in [45], under the assumption that the continuous-time closed-loop system at hand is globally exponentially stable and the maps describing the dynamics and the continuous-time state feedback are globally Lipschitz, the following result is proved: there exists a suitably fast sampling period such that, if the sampled-data controller is affected by bounded measurement noises and bounded actuation disturbances, the global exponential practical stability of the sampled-data closed-loop system is still ensured. Halanay's inequality (see [2, 15, 21]) and converse Lyapunov theorems (see [25, 28, 46]) are the main tools used to prove the results. As long as the continuous-time closed-loop system at hand is globally exponentially stable, since the existence of a Lyapunov-Krasovskii functional is ensured by the converse theorems, the proof of the existence of a suitably small sampling period does not require its explicit knowledge. If a Lyapunov-Krasovskii functional is explicitly known then a precise characterization of the sampling period can be always computed. Such characterization could well provide a conservative sampling period. On the other hand the results provided here are of the existence type and the provision of a non conservative sampling frequency is beyond the aims of this work. A numerical example is studied which validates the theoretical results.

Notation $\mathbb{N}$ denotes the set of nonnegative integer numbers, $\mathbb{R}$ denotes the set of real numbers, $\mathbb{R}^\star$ denotes the extended real line $[-\infty, +\infty]$, $\mathbb{R}^+$ denotes the set of nonnegative reals $[0, +\infty)$. The symbol $|\cdot|$ stands for the Euclidean norm of a real vector, or the induced Euclidean norm of a matrix. The essential supremum norm of an essentially bounded function is indicated with the symbol $\|\cdot\|_\infty$. For a positive integer n, for a positive real Δ (maximum involved time-delay): $\mathscr{C}^n$ denotes the space of the continuous functions mapping $[-\Delta, 0]$ into $\mathbb{R}^n$. For a continuous function $x : [-\Delta, c) \to \mathbb{R}^n$, with $0 < c \leq +\infty$, for any real $t \in [0, c)$, x_t is the function in $\mathscr{C}^n$ defined as $x_t(\tau) = x(t + \tau)$, $\tau \in [-\Delta, 0]$. For positive integers n, m, for a continuous map $f : \mathscr{C}^n \times \mathbb{R}^m \to \mathbb{R}^n$, and for a continuous functional $V : \mathscr{C}^n \to \mathbb{R}^+$, the derivative in Driver's form (see [39] and the references therein) $D^+V : \mathscr{C}^n \times \mathbb{R}^m \to \mathbb{R}^\star$, of the functional V, is defined, for $\phi \in \mathscr{C}^n$, $u \in \mathbb{R}^m$, as:

$$D^+V(\phi, u) = \limsup_{h \to 0^+} \frac{V(\phi_{h,u}) - V(\phi)}{h}, \tag{1}$$

where, for $0 \leq h < \Delta$, $\phi_{h,u} \in \mathscr{C}^n$ is defined, for $s \in [-\Delta, 0]$, as

$$\phi_{h,u}(s) = \begin{cases} \phi(s + h), & s \in [-\Delta, -h), \\ \phi(0) + (s + h) f(\phi, u), & s \in [-h, 0]. \end{cases}$$

Throughout the chapter, GES stands for globally exponentially stable or global exponential stability, RFDE stands for retarded functional differential equation.

2 Preliminaries

Let us consider a nonlinear time-delay system, described by the following RFDE (see [16, 27])

$$\begin{aligned} \dot{x}(t) &= f(x_t, u(t)), \quad t \geq 0 \ a.e., \\ x(\tau) &= x_0(\tau), \quad \tau \in [-\Delta, 0], \end{aligned} \tag{2}$$

where: $x(t) \in \mathbb{R}^n$, n is a positive integer; x_0, $x_t \in \mathscr{C}^n$; $\Delta > 0$ is the maximum involved time delay, which is assumed to be known; $u(t) \in \mathbb{R}^m$ is the input (the input signal is Lebesgue measurable and locally essentially bounded); m is a positive integer; f is a map from $\mathscr{C}^n \times \mathbb{R}^m$ to $\mathbb{R}^n$. It is assumed that $f(0, 0) = 0$.

For the reader's convenience, firstly, we recall important results which will be used as tools in the proofs of the main results. In particular, the following lemma provides necessary and sufficient conditions for the global exponential stability of nonlinear time-delay systems described by (2) in closed-loop with globally Lipschitz state feedbacks (see [25, 28, 46]). The following well-known definition concerns the GES of nonlinear time-delay systems described by (2) (see Definition 4.5, p. 150 of [26], as far as systems described by ODEs are concerned, see [46], as far as systems described by RFDEs are concerned).

Definition 1 (*see* [46]) The system described by (2) with $u(t) \equiv 0$ is said to be 0-GES if there exist positive reals M, λ such that, for any initial condition $x_0 \in \mathscr{C}^n$, the corresponding solution exists for all $t \geq 0$ and, furthermore, satisfies the inequality

$$|x(t)| \leq M e^{-\lambda t} \|x_0\|_\infty, \quad t \geq 0. \tag{3}$$

Notice that, in Definition 1, $M \geq 1$ is mandatory.

Lemma 1 *Let $k : \mathscr{C}^n \to \mathbb{R}^m$ be a globally Lipschitz map, satisfying $k(0) = 0$. Then, the continuous-time closed-loop system described by (2), with $u(t) = k(x_t)$, $t \geq 0$, is 0-GES if and only if there exist a globally Lipschitz functional $V : \mathscr{C}^n \to \mathbb{R}^+$, with L_V as Lipschitz constant, and positive reals α_i, $i = 1, 2, 3$, such that the following inequalities hold for any $\phi \in \mathscr{C}^n$*

$$\begin{aligned} \alpha_1 \|\phi\|_\infty \leq V(\phi) \leq \alpha_2 \|\phi\|_\infty, \\ D^+ V(\phi, k(\phi)) \leq -\alpha_3 \|\phi\|_\infty. \end{aligned} \tag{4}$$

The following lemma is a particular case of Theorem 3.2 in [21], as here needed for the forthcoming practical stability analysis. Theorem 3.2 in [21] extends the Halanay's inequality (see [2, 15]) to a more general case.

Lemma 2 (see Theorem 3.2 in [21]) *Let a, b, γ, r be positive reals, $a > b$. Let $z : [-r, +\infty) \to \mathbb{R}^+$ be a continuous function satisfying the inequality*

$$D^+ z(t) \leq -a z(t) + b \sup_{\theta \in [-r, 0]} z(t + \theta) + \gamma, \tag{5}$$

where $D^+ z(t) = \limsup_{h \to 0^+} \frac{z(t+h) - z(t)}{h}$ is the upper right-hand Dini derivative of the function z. Let λ be the positive real solution of the scalar equation

$$H(\bar{\lambda}) = \bar{\lambda} - 1 + \frac{b}{a} e^{\bar{\lambda} a r} = 0. \tag{6}$$

Let $\sigma = a - b$ and $N = e^{\lambda a r}$. Then the inequality

$$z(t) \leq \sup_{\theta\in[-r,0]} z(\theta)Ne^{-\lambda at} + \frac{\gamma}{\sigma} \tag{7}$$

holds for any $t \geq 0$.

We recall here the notion of partition of $[0, +\infty)$ (see [6, 43]) which is very helpful in sampled-data stabilization.

Definition 2 A partition $\pi = \{t_i,\ i = 0, 1, ...\}$ of $[0, +\infty)$ is a countable, strictly increasing sequence t_i, with $t_0 = 0$, such that $t_i \to +\infty$ as $i \to +\infty$. The diameter of π, denoted $diam\ (\pi)$, is defined as $\sup_{i\geq 0} t_{i+1} - t_i$. The dwell time of π, denoted $dwell\ (\pi)$, is defined as $\inf_{i\geq 0} t_{i+1} - t_i$. For any positive real $a \in (0, 1]$, $b > 0$, $\pi_{a,b}$ is any partition π with $ab \leq dwell\ (\pi) \leq diam\ (\pi) \leq b$.

The real $a \in (0, 1]$, in Definition 2, is introduced in order to allow non-uniform sampling in sampled-data stabilization.

We introduce the following assumption for the system described by (2).

Assumption 1 The map $f : \mathscr{C}^n \times \mathbb{R}^m \to \mathbb{R}^n$ is globally Lipschitz with Lipschitz constant L_f.There exists a globally Lipschitz feedback $k : \mathscr{C}^n \to \mathbb{R}^m$, with Lipschitz constant L_k, satisfying $k(0) = 0$, such that the continuous-time closed-loop system described by the RFDE

$$\dot{x}\ (t) = f\ (x_t, k(x_t)) \tag{8}$$

is 0-GES.

Notice that, taking into account Lemma 1, Assumption 1 ensures the existence of a globally Lipschitz functional $V : \mathscr{C}^n \to \mathbb{R}^+$, with L_V as Lipschitz constant, and positive reals α_i, $i = 1, 2, 3$, satisfying (4).

3 Main Results

The main result of this chapter is given by next Theorem 1. It is based on the results in [44] (see Theorem 8), [45] (see Theorem 5).

Theorem 1 *Let Assumption 1 hold. Let* α_1, α_2, α_3, L_V, L_k, L_f *be the positive constants provided in Lemma 1 and Assumption 1. Let* $a \in (0, 1]$. *Let* δ *be a positive real satisfying*

$$\delta < \min\left\{\Delta, \frac{\alpha_1\alpha_3}{\alpha_2 L_V L_f^2 L_k\,(1 + L_k)}\right\}. \tag{9}$$

Let λ *be the positive real solution of the scalar equation (with unknown variable* $\bar{\lambda}$*)*

$$\bar{\lambda} - 1 + \frac{\alpha_2 L_V L_f^2 L_k\,(1 + L_k)}{\alpha_1\alpha_3}\delta e^{\frac{\alpha_3\bar{\lambda}(\Delta+2\delta)}{\alpha_2}} = 0. \tag{10}$$

Let l be the positive integer such that

$$l a\delta \leq \Delta < (l+1)a\delta. \tag{11}$$

Let

$$\lambda^* = \frac{\alpha_3}{\alpha_2}\lambda. \tag{12}$$

Let r_1, r_2, r_3, be the following positive reals

$$\begin{aligned} r_1 &= \frac{\alpha_2}{\alpha_1}(2 + L_k L_f \delta)^{l+1} e^{(L_f + 2\lambda^*)(\Delta + 2\delta)}, \\ r_2 &= r_1 L_f \delta + \frac{\alpha_2 L_V L_f (1 + L_f L_k \delta)}{\alpha_1 \alpha_3 - \alpha_2 L_V L_f^2 L_k (1 + L_k)\delta}, \\ r_3 &= r_2 L_k. \end{aligned} \tag{13}$$

Then, for any partition $\pi_{a,\delta}$ (see Definition 1), for any initial state $x_0 \in \mathscr{C}^n$, for any sequence $e : \mathbb{N} \to \mathscr{C}^n$ (measurement noise) such that, for some positive real $\bar{e}$, $\|e_j\|_\infty \leq \bar{e}$, $j = 0, 1, 2, \ldots$, for any sequence $d : \mathbb{N} \to \mathbb{R}^m$ (actuation disturbance) such that, for some positive real $\bar{d}$, $|d_j| \leq \bar{d}$, $j = 0, 1, 2, \ldots$, the solution of the sampled-data closed-loop system described by

$$\dot{x}(t) = f\left(x_t, k(x_{t_j} + e_j) + d_j\right), \quad t_j \leq t < t_{j+1}, \quad j = 0, 1, 2, \ldots, \tag{14}$$

exists for all $t \geq 0$ and, furthermore, satisfies

$$|x(t)| \leq r_1 \|x_0\|_\infty e^{-\lambda^* t} + r_2 \bar{d} + r_3 \bar{e}. \tag{15}$$

Proof In the following, the structure of the proofs used in [44] (see Theorem 8), [45] (see Theorem 5) is suitably adapted in order to cope with actuation disturbances and measurement noises affecting the sampled-data controller. From the globally Lipschitz property of map f and the existence of a dwell-time for the partition $\pi_{a,\delta}$, it follows that the system described by (14) admits a unique locally absolutely continuous solution in $\mathbb{R}^+$. Let $x_t \in \mathscr{C}^n$ be the solution of (14). Let p be the positive integer such that, for the partition at hand, $t_p \leq \Delta < t_{p+1}$. Notice that, by (11), $p \leq l$. Firstly, let us consider the intervals $[t_j, t_{j+1}]$, $j = 0, 1, \ldots, p$. For $t \in [0, t_1]$, we have

$$\begin{aligned} &\|x_t\|_\infty \leq \\ &\|x_0\|_\infty + \sup_{\theta \in [-\Delta, 0],\, t+\theta \geq 0} \left(\|x_0\|_\infty + \int_0^{t+\theta} |f(x_\tau, k(x_0 + e_0) + d_0)| d\tau \right) \leq \\ &2\|x_0\|_\infty + \sup_{\theta \in [-\Delta, 0],\, t+\theta \geq 0} \int_0^{t+\theta} L_f(\|x_\tau\|_\infty + |k(x_0 + e_0) + d_0|) d\tau \leq \\ &2\|x_0\|_\infty + \sup_{\theta \in [-\Delta, 0],\, t+\theta \geq 0} \int_0^{t+\theta} L_f(\|x_\tau\|_\infty + L_k \|x_0 + e_0\|_\infty + |d_0|) d\tau \leq \\ &(2 + L_f L_k \delta)\|x_0\|_\infty + L_f L_k \delta \|e_0\|_\infty + L_f \delta |d_0| + \int_0^t L_f \|x_\tau\|_\infty d\tau. \end{aligned} \tag{16}$$

Let

$$c_1 = 2 + L_f L_k \delta, \qquad c_2 = L_f L_k \delta \bar{e} + L_f \delta \bar{d}. \tag{17}$$

By the Gronwall-Bellman Lemma (see Lemma A.1, pp. 651–652, in [26]) and taking into account (17), it follows from (16) that, for $t \in [0, t_1]$, the following inequalities/equality hold

$$\begin{aligned}&\|x_t\|_\infty \le ((2 + L_f L_k \delta)\|x_0\|_\infty + L_f L_k \delta \|e_0\|_\infty + L_f \delta |d_0|) e^{L_f t} \le \\ &(2 + L_f L_k \delta) e^{L_f t} \|x_0\|_\infty + (L_f L_k \delta \bar{e} + L_f \delta \bar{d}) e^{L_f t} = c_1 e^{L_f t} \|x_0\|_\infty + c_2 e^{L_f t}.\end{aligned} \tag{18}$$

By an induction reasoning, taking into account that, for any nonnegative integer y, the following inequalities hold

$$\begin{aligned}&\textstyle\sum_{i=0}^{y} c_1^i \le c_1^{y+1}, \\ &\|x_t\|_\infty \le c_1 e^{L_f (t - t_y)} \|x_{t_y}\|_\infty + c_2 e^{L_f (t - t_y)}, \ \ t \in [t_y, t_{y+1}],\end{aligned} \tag{19}$$

it follows that, for $t \in [t_j, t_{j+1}]$, $j = 0, 1, 2, ..., p$, the inequality holds

$$\|x_t\|_\infty \le c_1^{j+1} e^{L_f t} \|x_0\|_\infty + c_1^{j+1} c_2 e^{L_f t}. \tag{20}$$

From (20), recalling that $p \le l$ and taking into account (17), for any $t \in [0, t_{p+1}]$ the following inequality/equality holds

$$\begin{aligned}&\|x_t\|_\infty \le c_1^{l+1} e^{L_f (\Delta + 2\delta)} \|x_0\|_\infty + c_1^{l+1} c_2 e^{L_f (\Delta + 2\delta)} = \\ &(2 + L_f L_k \delta)^{l+1} e^{L_f (\Delta + 2\delta)} \|x_0\|_\infty + (2 + L_f L_k \delta)^{l+1} (L_f L_k \delta \bar{e} + L_f \delta \bar{d}) e^{L_f (\Delta + 2\delta)}.\end{aligned} \tag{21}$$

Now, let $w : \mathbb{R}^+ \to \mathbb{R}^+$ be the continuous function defined, for $t \in \mathbb{R}^+$, as $w(t) = V(x_t)$, with V the functional provided in Lemma 1 (for the system described by (8)). The following equalities hold for $t_j \le t < t_{j+1}$, $j = p + 1, p + 2, ...,$

$$\begin{aligned}&D^+ w(t) = \limsup_{h \to 0^+} \frac{1}{h} (V(x_{t+h}) - V(x_t)) = \\ &\limsup_{h \to 0^+} \frac{1}{h} (V(x_{t+h}) - V((x_t)_{h, k(x_{t_j} + e_j) + d_j}) + V((x_t)_{h, k(x_{t_j} + e_j) + d_j}) \\ &- V(x_t)),\end{aligned} \tag{22}$$

where, for $t \ge 0$, and any $v \in \mathbb{R}^m$, $(x_t)_{h,v} \in \mathscr{C}^n$ is defined (see the Notation section), for $h \in [0, \Delta)$, as

$$(x_t)_{h,v}(\theta) = \begin{cases} x_t(\theta + h), & \theta \in [-\Delta, -h), \\ x_t(0) + (\theta + h) f(x_t, v), & \theta \in [-h, 0]. \end{cases}$$

Now, for any positive real $h < \min\{t_{j+1} - t_j, \Delta\}$, the following equalities/inequalities hold (see [10, 39, 53])

$$\frac{1}{h}|V(x_{t+h}) - V((x_t)_{h,k(x_{t_j}+e_j)+d_j})| \leq \frac{L_V}{h}\|x_{t+h} - (x_t)_{h,k(x_{t_j}+e_j)+d_j}\|_\infty =$$

$$\frac{L_V}{h}\sup_{\theta\in[-\Delta,0]}|x_{t+h}(\theta) - (x_t)_{h,k(x_{t_j}+e_j)+d_j}(\theta)| =$$

$$\frac{L_V}{h}\sup_{\theta\in[-h,0]}|x(t+h+\theta) - x(t) - (\theta+h)f(x_t, k(x_{t_j}+e_j)+d_j)| \leq$$

$$\begin{aligned}&\frac{L_V}{h}\sup_{\theta\in[-h,0]}|x(t) + \textstyle\int_t^{t+h+\theta} f(x_\tau, k(x_{t_j}+e_j)+d_j)d\tau - x(t)\\ &-(\theta+h)f(x_t, k(x_{t_j}+e_j)+d_j)| =\\ &\frac{L_V}{h}\sup_{\theta\in(-h,0]}|(\theta+h)\left(\frac{1}{(\theta+h)}\textstyle\int_t^{t+h+\theta} f(x_\tau, k(x_{t_j}+e_j)+d_j)d\tau\right.\\ &\left.-f(x_t, k(x_{t_j}+e_j)+d_j)\right)| \leq\\ &L_V\sup_{\theta\in(-h,0]}|\left(\frac{1}{(\theta+h)}\textstyle\int_t^{t+h+\theta} f(x_\tau, k(x_{t_j}+e_j)+d_j)d\tau\right.\\ &\left.-f(x_t, k(x_{t_j}+e_j)+d_j)\right)|.\end{aligned} \tag{23}$$

From (23), taking into account of the continuity of the map f and of the solution $x_\tau \in \mathscr{C}^n$, $\tau \in \mathbb{R}^+$ (see Lemma 2.1, p. 40, in [16]) the limit follows

$$\limsup_{h\to 0^+}\frac{1}{h}(V(x_{t+h}) - V((x_t)_{h,k(x_{t_j}+e_j)+d_j})) = 0. \tag{24}$$

From (22) and (24) it follows that

$$\begin{aligned}&D^+w(t) = \limsup_{h\to 0^+}\frac{1}{h}(V((x_t)_{h,k(x_{t_j}+e_j)+d_j}) - V(x_t)) =\\ &\limsup_{h\to 0^+}\frac{1}{h}(V((x_t)_{h,k(x_{t_j}+e_j)+d_j}) - V((x_t)_{h,k(x_t)}) + V((x_t)_{h,k(x_t)})\\ &-V(x_t)).\end{aligned} \tag{25}$$

From (25) and by the use of Lemma 1 we obtain

$$\begin{aligned}&D^+w(t) \leq \limsup_{h\to 0^+}\frac{1}{h}(V((x_t)_{h,k(x_{t_j}+e_j)+d_j}) - V((x_t)_{h,k(x_t)}))+\\ &\limsup_{h\to 0^+}\frac{1}{h}(V((x_t)_{h,k(x_t)}) - V(x_t)) =\\ &D^+V(x_t, k(x_t)) + \limsup_{h\to 0^+}\frac{1}{h}(V((x_t)_{h,k(x_{t_j}+e_j)+d_j}) - V((x_t)_{h,k(x_t)})) \leq\\ &-\alpha_3\|x_t\|_\infty + \limsup_{h\to 0^+}\frac{1}{h}(V((x_t)_{h,k(x_{t_j}+e_j)+d_j}) - V((x_t)_{h,k(x_t)})).\end{aligned} \tag{26}$$

Moreover, we have

$$
\begin{aligned}
&\limsup_{h\to 0^+} \frac{1}{h}|V((x_t)_{h,k(x_{t_j}+e_j)+d_j}) - V((x_t)_{h,k(x_t)})| \leq \\
&\limsup_{h\to 0^+} \frac{L_V}{h}\left(\sup_{\theta\in[-\Delta,0]} |(x_t)_{h,k(x_{t_j}+e_j)+d_j}(\theta) - (x_t)_{h,k(x_t)}(\theta)|\right) = \\
&\limsup_{h\to 0^+} \frac{L_V}{h}(\sup_{\theta\in[-h,0]} |x(t) + (\theta+h)f(x_t, k(x_{t_j}+e_j)+d_j) - x(t) \\
&-(\theta+h)f(x_t, k(x_t))|) = \\
&\limsup_{h\to 0^+} \frac{L_V}{h}\left(\sup_{\theta\in[-h,0]}(\theta+h)|f(x_t, k(x_{t_j}+e_j)+d_j) - f(x_t, k(x_t))|\right) \leq \\
&L_V L_f(L_k\|x_{t_j}+e_j-x_t\|_\infty + |d_j|) \leq L_V L_f(L_k\|x_{t_j}-x_t\|_\infty + L_k\bar{e} + \bar{d}).
\end{aligned} \tag{27}
$$

From (26), taking into account (27), the following inequality holds

$$
D^+w(t) \leq -\alpha_3\|x_t\|_\infty + L_V L_f(L_k\|x_{t_j}-x_t\|_\infty + L_k\bar{e} + \bar{d}). \tag{28}
$$

Now, let us consider any $j \geq p+1$. Let $\hat{t}_1 = \max_{i\in\mathbb{N}}\{t_i \in \pi_{a,\delta} : t_i \leq t_j - \Delta\}$ and let $\hat{t}_2, \hat{t}_3, \ldots, \hat{t}_s$, s positive integer, be the set of all sampling times, in increasing order, in the interval $(t_j-\Delta, t_j]$. Taking into account (11) and since $\hat{t}_s = t_j$ it must be $2 \leq s \leq l+2$. Then, for $t_j \leq t < t_{j+1}$, we have

$$
\begin{aligned}
&\|x_{t_j} - x_t\|_\infty = \sup_{\theta\in[-\Delta,0]} |x(t_j+\theta) - x(t+\theta)| \leq \\
&\sup_{q=1,2,\ldots,s-1} \sup_{\theta\in[\hat{t}_q-t_j, \hat{t}_{q+1}-t_j]} |x(t_j+\theta) - x(t+\theta)| \leq \\
&\sup_{q=1,2,\ldots,s-1} \sup_{\theta\in[\hat{t}_q-t_j, \hat{t}_{q+1}-t_j]} \left| \int_{t_j+\theta}^{t+\theta} (L_f(1+L_k)\sup_{\beta\in[t_j+\theta,t+\theta]} \|x_\beta\|_\infty \right. \\
&\left. + L_f L_k\bar{e} + L_f\bar{d})d\tau \right| \leq \\
&\sup_{q=1,2,\ldots,s-1} \sup_{\theta\in[\hat{t}_q-t_j, \hat{t}_{q+1}-t_j]} L_f(1+L_k)\delta \sup_{\beta\in[t_j+\theta,t+\theta]} \|x_\beta\|_\infty \\
&+ L_f L_k\delta\bar{e} + L_f\delta\bar{d} \leq \\
&L_f(1+L_k)\delta \sup_{\beta\in[\hat{t}_1,t]} \|x_\beta\|_\infty + L_f L_k\delta\bar{e} + L_f\delta\bar{d} \leq \\
&\qquad \frac{L_f(1+L_k)\delta}{\alpha_1} \sup_{\beta\in[\hat{t}_1,t]} w(\beta) + L_f L_k\delta\bar{e} + L_f\delta\bar{d}.
\end{aligned} \tag{29}
$$

From (28), taking into account (29) and setting $w(\theta) = w(0)$ for $\theta \in [-2\delta, 0]$, by the use of Lemma 1, the following inequality holds for all $t \geq t_{p+1}$

$$D^+w(t) \leq -\frac{\alpha_3}{\alpha_2}w(t) + \frac{L_V L_f^2 L_k(1+L_k)\delta}{\alpha_1} \sup_{\theta\in[-\Delta-2\delta,0]} w(t+\theta) \\ +(\delta L_V L_f^2 L_k^2 + L_V L_f L_k)\bar{e} + (\delta L_V L_f^2 L_k + L_V L_f)\bar{d}. \tag{30}$$

From (30), by the use of Lemma 2 in the interval $[t_{p+1}, +\infty)$ and taking into account (12), we obtain

$$w(t) \leq \sup_{\theta\in[-\Delta-2\delta,0]} w(t_{p+1}+\theta)e^{\lambda^*(\Delta+2\delta)}e^{-\lambda^*(t-t_{p+1})} \\ +\frac{\alpha_1\alpha_2(L_V L_f L_k + L_V L_f^2 L_k^2\delta)}{\alpha_1\alpha_3 - \alpha_2 L_V L_f^2 L_k\,(1+L_k)\,\delta}\bar{e} + \frac{\alpha_1\alpha_2(L_V L_f + L_V L_f^2 L_k\delta)}{\alpha_1\alpha_3 - \alpha_2 L_V L_f^2 L_k\,(1+L_k)\,\delta}\bar{d}. \tag{31}$$

From (31), by Lemma 1 and taking into account (13), (21), for $t \in [t_{p+1}, +\infty)$, the following inequalities hold

$$\|x_t\|_\infty \leq \frac{\alpha_2}{\alpha_1}(2+L_f L_k\delta)^{l+1}e^{(L_f+\lambda^*)(\Delta+2\delta)}\|x_0\|_\infty e^{-\lambda^*(t-t_{p+1})} \\ +\frac{\alpha_2}{\alpha_1}(2+L_f L_k\delta)^{l+1}e^{(L_f+\lambda^*)(\Delta+2\delta)}(L_f L_k\delta\bar{e} + L_f\delta\bar{d})e^{-\lambda^*(t-t_{p+1})} \\ +\frac{\alpha_2(L_V L_f L_k + L_V L_f^2 L_k^2\delta)}{\alpha_1\alpha_3 - \alpha_2 L_V L_f^2 L_k\,(1+L_k)\,\delta}\bar{e} + \frac{\alpha_2(L_V L_f + L_V L_f^2 L_k\delta)}{\alpha_1\alpha_3 - \alpha_2 L_V L_f^2 L_k\,(1+L_k)\,\delta}\bar{d}. \tag{32}$$

From (32), taking into account that $t_{p+1} \leq \Delta + \delta$ and that, for $t \in [0, t_{p+1}]$, the right-hand side of the inequality in (32) is greater than the right-hand side of the equality in (21), we can conclude that, for $t \in [0, +\infty)$, the following inequality holds

$$\|x_t\|_\infty \leq \frac{\alpha_2}{\alpha_1}(2+L_f L_k\delta)^{l+1}e^{(L_f+\lambda^*)(\Delta+2\delta)}\|x_0\|_\infty e^{-\lambda^*(t-\delta-\Delta)} \\ +\frac{\alpha_2}{\alpha_1}(2+L_f L_k\delta)^{l+1}e^{(L_f+\lambda^*)(\Delta+2\delta)}(L_f L_k\delta\bar{e} + L_f\delta\bar{d})e^{-\lambda^*(t-\delta-\Delta)} \\ +\frac{\alpha_2(L_V L_f L_k + L_V L_f^2 L_k^2\delta)}{\alpha_1\alpha_3 - \alpha_2 L_V L_f^2 L_k\,(1+L_k)\,\delta}\bar{e} + \frac{\alpha_2(L_V L_f + L_V L_f^2 L_k\delta)}{\alpha_1\alpha_3 - \alpha_2 L_V L_f^2 L_k\,(1+L_k)\,\delta}\bar{d}. \tag{33}$$

From (33), we obtain

$$\|x_t\|_\infty \leq \frac{\alpha_2}{\alpha_1}(2 + L_f L_k \delta)^{l+1} e^{(L_f+2\lambda^*)(\Delta+2\delta)} \|x_0\|_\infty e^{-\lambda^* t}$$

$$+\frac{\alpha_2}{\alpha_1}(2 + L_f L_k \delta)^{l+1} e^{(L_f+2\lambda^*)(\Delta+2\delta)} (L_f L_k \delta \bar{e} + L_f \delta \bar{d})$$

$$+\frac{\alpha_2(L_V L_f L_k + L_V L_f^2 L_k^2 \delta)}{\alpha_1\alpha_3 - \alpha_2 L_V L_f^2 L_k (1 + L_k)\delta}\bar{e} + \frac{\alpha_2(L_V L_f + L_V L_f^2 L_k \delta)}{\alpha_1\alpha_3 - \alpha_2 L_V L_f^2 L_k (1 + L_k)\delta}\bar{d} =$$

$$r_1 \|x_0\|_\infty e^{-\lambda^* t} + r_1(L_f L_k \delta \bar{e} + L_f \delta \bar{d}) + \frac{\alpha_2(L_V L_f L_k + L_V L_f^2 L_k^2 \delta)}{\alpha_1\alpha_3 - \alpha_2 L_V L_f^2 L_k (1 + L_k)\delta}\bar{e} \tag{34}$$

$$+\frac{\alpha_2(L_V L_f + L_V L_f^2 L_k \delta)}{\alpha_1\alpha_3 - \alpha_2 L_V L_f^2 L_k (1 + L_k)\delta}\bar{d} = r_1 \|x_0\|_\infty e^{-\lambda^* t} + r_2 \bar{d} + r_3 \bar{e}.$$

The proof of the theorem is complete. ∎

Remark 1 Notice that the sampling period δ computed in (9), as well as the positive reals λ^*, r_1, r_2, r_3 computed in (12), (13), depend neither on the upper bound for the actuation disturbance $\bar{d}$, nor on the upper bound for the measurement noise $\bar{e}$.

Next corollary readily follows from Theorem 1, and therefore the proof is omitted.

Corollary 1 *Let Assumption 1 hold. Let $a \in (0, 1]$. Then, there exist positive reals $\delta, \lambda^*, r_1, r_2, r_3$ such that, for any partition $\pi_{a,\delta}$ (see Definition 2), for any initial state $x_0 \in \mathscr{C}^n$, for any sequence $e : \mathbb{N} \to \mathscr{C}^n$ (measurement noise) such that, for some positive real $\bar{e}$, $\|e_j\|_\infty \leq \bar{e}$, $j = 0, 1, 2, \ldots$, for any sequence $d : \mathbb{N} \to \mathbb{R}^m$ (actuation disturbance) such that, for some positive real $\bar{d}$, $|d_j| \leq \bar{d}$, $j = 0, 1, 2, \ldots$, the solution of the sampled-data closed-loop system described by*

$$\dot{x}(t) = f\left(x_t, k(x_{t_j} + e_j) + d_j\right), \quad t_j \leq t < t_{j+1}, \quad j = 0, 1, 2, \ldots, \tag{35}$$

exists for all $t \geq 0$ and, furthermore, satisfies

$$|x(t)| \leq r_1 \|x_0\|_\infty e^{-\lambda^* t} + r_2 \bar{d} + r_3 \bar{e}. \tag{36}$$

Remark 2 Notice that the existence of a Lyapunov-Krasovskii functional for the continuous-time closed-loop system described by (8) is guaranteed by Lemma 1, as long as this system is globally exponentially stable. Then, the preservation of the global exponential practical stability for the sampled-data closed-loop system described by (14) is ensured, as long as Assumption 1 holds. Furthermore, if a globally Lipschitz Lyapunov-Krasovskii functional is explicitly known, then a precise characterization of the sufficiently small sampling period is provided by using the involved Lipschitz constants and the lower and upper bounds related to the Lyapunov-Krasovskii functional (see (9) in Theorem 1).

Remark 3 According to our experience, the upper bound of the sampling period provided in (9) could be conservative. The source of such conservatism may be the use of Lipschitz constants of many involved functions as well as lower and upper bounds of Lyapunov-Krasovskii functionals and derivatives (see (9)). The reduction of the conservativeness of the provided sampling period may require different proof techniques which, anyway, require further and not easy developments (that are beyond the aims of this chapter) to cope with the case of nonlinear systems with state delays. The study of the reduction of the conservativeness of the provided sampling period is left for future investigations. The hybrid system approach seems to be a very promising tool in such direction (see [4], in the case of systems described by ordinary differential equations). On the other hand, the results provided in this Chapter are of the existence type, and the study of the conservativeness of the provided sampling frequency is beyond the aims of this work.

Remark 4 We highlight that, when $|d_j| = |e_j| = 0$, for any $j = 0, 1, ...,$ the condition (15) implies the global exponential stability property of the sampled-data closed-loop system described by (14) (see Theorem 1 and Corollary 1).

4 Example

Let us consider the system described by the following scalar RFDE (see [45])

$$\begin{aligned} \dot{x}(t) &= -x(t) + \tanh(x(t) + x(t-\Delta) + u(t)), \\ x(\tau) &= x_0(\tau), \quad \tau \in [-\Delta, 0], \end{aligned} \tag{37}$$

where $x(t) \in \mathbb{R}$, $\Delta = 0.1$, $u(t) \in \mathbb{R}$. As in [45], let us consider $k : \mathscr{C}^1 \to \mathbb{R}$ defined for $x_t \in \mathscr{C}^1$ as

$$k(x_t) = -x(t) - x(t-\Delta). \tag{38}$$

Moreover, we have that $L_f = 3$ and $L_k = 2$. The continuous-time closed-loop system described by (37), (38) is globally exponentially stable. Then, Theorem 1 can be applied for the system at hand. As in [45], according to Theorem 1 let us choose $V : \mathscr{C}^1 \to \mathbb{R}^+$ defined, for $\phi \in \mathscr{C}^1$, as $V(\phi) = |\phi(0)|$. Then we have $\alpha_1 = \alpha_2 = \alpha_3 = L_V = 1$. From (9), we obtain $\delta < 18.5$ms. In performed simulations, uniform sampling is used, and the sampling period has been chosen equal to 18ms. The actuator disturbance has been chosen as $d(j) = 0.003 sin\left(t_j\right)$, $j = 0, 1, ...$ As far as the observation error is concerned, we considered $e_j \in \mathscr{C}^1$, with $e_j(0) = e_j(-\Delta) = \tilde{e}$, and $\tilde{e}$ taken from the interval $[-0.01, 0.01]$, by emulation of the uniform probability density function. The initial state is chosen constant in $[-\Delta, 0]$ and equal to 0.1. The practical stability property of the sampled-data closed-loop system is validated by the plots in Fig. 1.

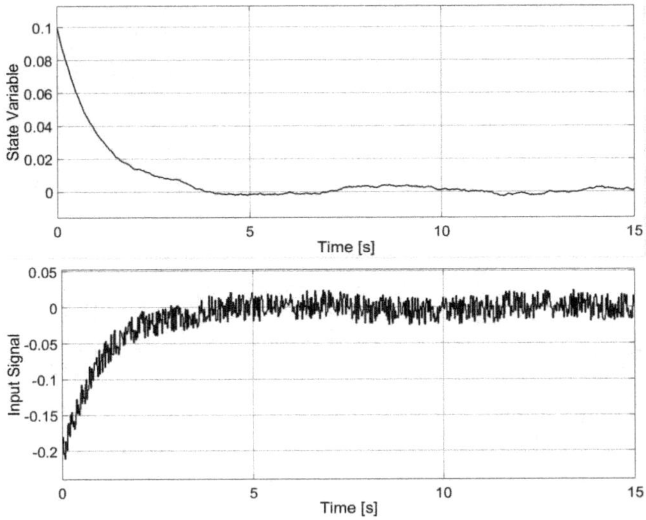

Fig. 1 State variable and input signal with uniform sampling period $\delta = 18$[ms]

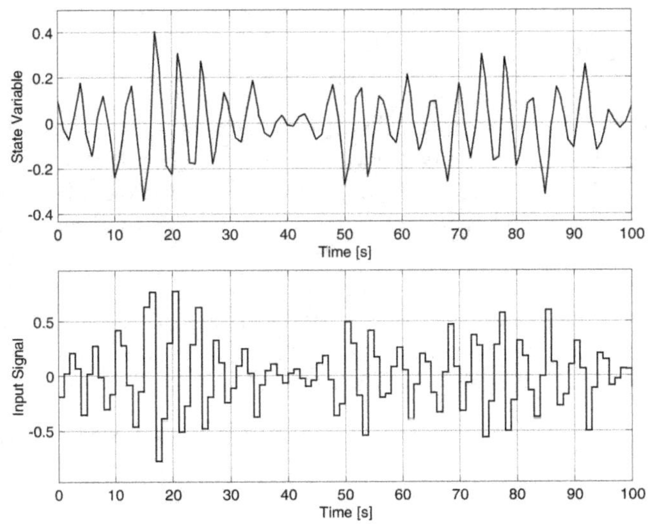

Fig. 2 State variable and input signal with uniform sampling period $\delta = 1$[s]

Further simulations have been performed using larger sampling periods. The simulations show that the performances of the sampled-data controller deteriorates, leading to large oscillations of the system state variable (see Fig. 2), with sampling periods equal to 1[s]. Stability is lost with sampling period equal to 2.5[s] leads to instability (see Fig. 3). Simulations fully validate the theoretical results.

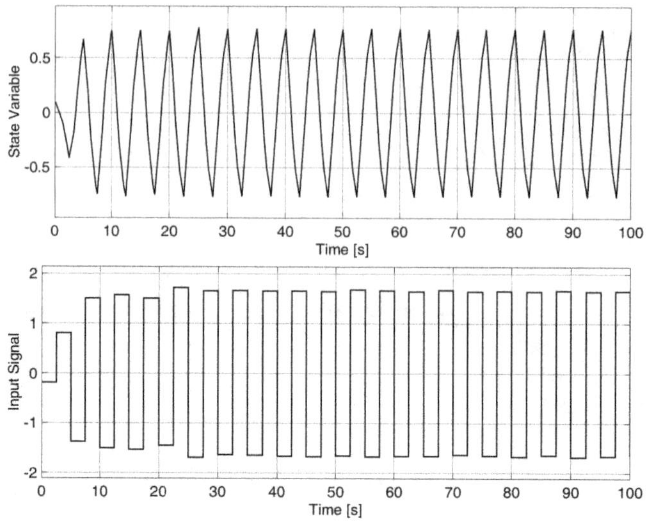

Fig. 3 State variable and input signal with uniform sampling period $\delta = 2.5[s]$

5 Conclusion

In this chapter, it has been shown that if the continuous-time closed-loop system at hand is globally exponentially stable and the maps describing the dynamics and the continuous-time state feedback are globally Lipschitz, then there exists a suitably fast sampling period such that, if the sampled-data controller is affected by bounded measurement noises and bounded actuation disturbances, the global exponential practical stability of the sampled-data closed-loop system is preserved. The results provided here are of the existence type and the study of the conservativeness of the provided sampling frequency is beyond the aims of this work. Future developments will concern methodologies which may allow for less conservative sampling frequencies.

References

1. T. Ahmed-Ali, E. Fridman, F. Giri, L. Burlion, F. Lamnabhi-Lagarrigue, Using exponential time-varying gains for sampled-data stabilization and estimation. Automatica **67**, 244–251 (2016)
2. C.T.H. Baker, E. Buckwar, Exponential stability in pth mean of solutions, and of convergent Euler-type solutions, of stochastic delay differential equations. J. Comput. Appl. Math. **184**, 404–427 (2005)
3. C. Briat, Convex conditions for robust stability analysis and stabilization of linear aperiodic impulsive and sampled-data systems under dwell-time constraints. Automatica **49**, 3449–3457 (2013)

4. D. Carnevale, A.R. Teel, D. Nesic, A Lyapunov proof of an improved maximum allowable transfer interval for networked control systems. IEEE Trans. Automat. Control **52**, 892–897 (2007)
5. F.H. Clarke, Discontinuous feedback and nonlinear systems, in *Plenary Lecture at IFAC Conference on Nonlinear Control Systems* (2010)
6. F.H. Clarke, Y.S. Ledyaev, E.D. Sontag, A.I. Subbotin, Asymptotic controllability implies feedback stabilization. IEEE Trans. Automat. Control **42**, 1394–1407 (1997)
7. J.M. Gomes da Silva, I. Queinnec, A. Seuret, S. Tarbouriech, Regional stability analysis of discrete-time dynamic output feedback under aperiodic sampling and input saturation. IEEE Trans. Automat. Control **61**, 4176–4182 (2016)
8. M. Di Ferdinando, P. Pepe, Sampled-Data emulation of dynamic output feedback controllers for nonlinear time-delay systems. Automatica **99**(1), 120–131 (2019)
9. M. Di Ferdinando, P. Pepe, Robustification of sample-and-hold stabilizers for control-affine time-delay systems. Automatica **83**, 141–154 (2017)
10. R.D. Driver, Existence and stability of solutions of a delay-differential system. Arch. Ration. Mech. Anal. **10**, 401–426 (1962)
11. E. Fridman, Introduction to time-delay systems: analysis and control. *Birkhauser* (2014)
12. E. Fridman, A refined input delay approach to sampled-data control. Automatica **46**, 421–427 (2010)
13. E. Fridman, A. Seuret, J.P. Richard, Robust sampled-data stabilization of linear systems: an input delay approach. Automatica **40**, 1441–1446 (2004)
14. L. Grune, D. Nesic, Optimization based stabilization of sampled-data nonlinear systems via their approximate discrete-time models. SIAM J. Control Optim. **42**, 98–122 (2003)
15. A. Halanay, *Differential Equations, Stability, Oscillations, Time Lags* (Academic Press, New York, 1966)
16. J.K. Hale, S.M. Verduyn Lunel, *Introduction to Functional Differential Equations* (Springer Verlag, New York, 1993)
17. W.P.M.H. Heemels, A.R. Teel, N. Van De Wouw, D. Nesic, Networked control systems with communication constraints: tradeoff between transmission intervals, delays and performance. IEEE Trans. Autom. Control **55**, 1781–1796 (2010)
18. G. Herrmann, S.K. Spurgeon, C. Edwards, Discretization of sliding mode based control schemes, in *38th IEEE Conference on Decision and Control*, vol. 5, pp. 4257–4262 (1999)
19. J. Hespanha, A model for stochastic hybrid systems with application to communication networks. Nonlinear Anal. Spec. Issue Hybrid Syst. **62**, 1353–1383 (2005)
20. L. Hetel, C. Fiter, H. Omran, A. Seuret, E. Fridman, J.P. Richard, S.I. Niculescu, Recent developments on the stability of systems with aperiodic sampling: an overview. Automatica **76**, 309–335 (2017)
21. L.V. Hien, V.N. Phat, H. Trinh, New generalized Halanay inequalities with applications to stability of nonlinear non-autonomous time delay systems. Nonlinear Dyn. **82**, 563–575 (2015)
22. P. Hsu, S. Sastry, The effect of discretized feedback in a closed loop system, in *26th IEEE Conference on Decision and Control*, vol. 26, pp. 1518–1523 (1987)
23. I. Karafyllis, K. Kravaris, Necessary and sufficient Lyapunov-like conditions for robust nonlinear stabilization. Int. J. Robust Nonlinear Control **19**, 1105–1128 (2009)
24. I. Karafyllis, M. Krstic, Nonlinear stabilization under sampled and delayed measurements, and with inputs subject to delay and zero-order-hold. IEEE Trans. Automat. Control **57**, 1141–1154 (2012)
25. I. Karafyllis, P. Pepe, Z.P. Jiang, Global output stability for systems described by retarded functional differential equations: Lyapunov characterizations. Eur. J. Control **6**, 516–536 (2008)
26. H.K. Khalil, *Nonlinear Systems*, 3rd edn. (Prentice Hall, Upper Saddle River, NJ, 2000)
27. V. Kolmanovskii, A. Myshkis, *Introduction to the Theory and Applications of Functional Differential Equations* (Kluwer Academic, Dordrecht, 1999)
28. N.N. Krasovskii, *Stability of Motion* (Stanford University Press, 1963)
29. D. S. Laila, D. Nesic, A. Astolfi, Sampled-data control of nonlinear systems, in *Lecture Notes in Control and Information Sciences*, ed. by A. Loria, F. Lamnabhi-Lagarrigue, E. Panteley (Eds.), vol. 328, pp. 91–137 (2006)

30. D.S. Laila, D. Nesic, A.R. Teel, Open and closed loop dissipation inequalities under sampling and controller emulation. Eur. J. Control **18**, 109–125 (2002)
31. Y.S. Ledyaev, E.D. Sontag, A Lyapunov characterization of robust stabilization. Nonlinear Anal. Ser. A: Theory Methods **37**, 813–840 (1999)
32. M. Malisoff, E.D. Sontag, Asymptotic controllability and input-to-state stabilization: the effect of actuator errors, in *Optimal Control, Stabilization and Nonsmooth Analysis*, vol. 301. Lecture Notes in Control and Information Sciences, pp. 855–171 (2004)
33. F. Mazenc, M. Malisoff, T.H. Dinh, Robustness of nonlinear systems with respect to delay and sampling of the controls. Automatica **49**, 1925–1931 (2013)
34. P. Naghshtabrizi, J. Hespanha, A.R. Teel, On the robust stability and stabilization of sampled-data systems: a hybrid system approach, in *Proceedings of the 45th Conference on Decision and Control* (2006)
35. D. Nesic, A.R. Teel, A framework for stabilization of nonlinear sampled-data systems based on their approximate discrete-time models. IEEE Trans. Automat. Control **49**, 1103–1122 (2004)
36. D. Nesic, A.R. Teel, Input-to-state stability of networked control systems. Automatica **40**, 2121–2128 (2004)
37. D. Nesic, A.R. Teel, P.V. Kokotovic, Sufficient conditions for stabilization of sampled-data nonlinear systems via discrete-time approximations. Syst. Control Lett. **45**, 259–270 (1999)
38. H. Omran, L. Hetel, J.P. Richard, F. Lamnabhi-Lagarrigue, Stability analysis of bilinear systems under a periodic sampled-data control. Automatica **50**, 1288–1295 (2014)
39. P. Pepe, On Lyapunov-Krasovskii functionals under Caratheodory conditions. Automatica J. IFAC **43**, 701–706 (2007)
40. P. Pepe, Stabilization in the sample-and-hold sense of nonlinear retarded systems. SIAM J. Control Optim. **52**, 3053–3077 (2014)
41. P. Pepe, Robustification of nonlinear stabilizers in the sample-and-hold sense. J. Franklin Inst. **42**, 4107–4128 (2015)
42. P. Pepe, On stability preservation under sampling and approximation of feedbacks for retarded systems. SIAM J. Control Optim. **54**, 1895–1918 (2016)
43. P. Pepe, On control Lyapunov-Razumikhin functions, Nonconstant delays, Nonsmooth feedbacks, and nonlinear sampled-data stabilization. IEEE Trans. Autom. Control **62**, 5604–5619 (2017)
44. P. Pepe, E. Fridman, On global exponential stability preservation under sampling for globally lipschitz delay-free and retarded systems, in 13th *IFAC Workshop on Time-Delay Systems Istanbul, IFAC-PapersOnLine*, vol. 49, pp. 41–46 (2016)
45. P. Pepe, E. Fridman, On global exponential stability preservation under sampling for globally Lipschitz time-delay systems. Automatica **82**, 295–300 (2017)
46. P. Pepe, I. Karafyllis, Converse Lyapunov-Krasovskii theorems for systems described by neutral functional differential equations in Hale's form. Internat. J. Control **86**, 232–243 (2013)
47. R. Postoyan, T. Ahmed-Ali, F. Lamnabhi-Lagarrigue, Robust backstepping for the Euler approximate model of sampled-data strict-feedback systems. Automatica **45**, 2164–2168 (2009)
48. A. Seuret, A novel stability analysis of linear systems under asynchronous samplings. Automatica **48**, 177–182 (2012)
49. A. Seuret, C. Briat, Stability analysis of uncertain sampled-data systems with incremental delay using looped-functionals. Automatica **55**, 274–278 (2015)
50. E.D. Sontag, Clocks and insensitivity to small measurement errors. ESAIM Control Optim. Calc. Var. **4**, 537–557 (1999)
51. E.D. Sontag, Stability and stabilization: discontinuities and the effect of disturbances. Math. Phys. Sci. **528**, 551–598 (1999)
52. E.D. Sontag, Smooth stabilization implies coprime factorization. IEEE Trans. Autom. Control **34**, 435–443 (1989)
53. T. Yoshizawa, Stability theory by Liapunov's second method, *Publications of the Mathematical Society of Japan* (1966)

Numerical Methods for Analysis and Design

15 Years or So of Pseudospectral Collocation Methods for Stability and Bifurcation of Delay Equations

Alessia Andò, Dimitri Breda, Davide Liessi, Stefano Maset, Francesca Scarabel, and Rossana Vermiglio

Abstract Stability and bifurcation analyses of delay equations represent fundamental challenges in many applications, emerging in important fields like, e.g., control theory and population dynamics. Exact approaches are usually unattainable due to the infinite dimension of the associated state space, so that resorting to numerical methods is unavoidable. This work is a survey on the use of collocation techniques to address the problems above from the numerical standpoint. As such it summarizes the main contributions in this context of the research of the authors in the last 15 years or so. Methods for the linearized stability analysis of equilibria and periodic

CDLab—Computational Dynamics Laboratory (http://www.cdlab.uniud.it) and of INdAM Research group GNCS. Dimitri Breda and Rossana Vermiglio are supported by the project PSD_2015_2017_DIMA_PRID_2017_ZANOLIN "SIDIA—SIstemi DInamici e Applicazioni" (UNIUD).

A. Andò · D. Breda (✉) · D. Liessi · R. Vermiglio
Department of Mathematics, Computer Science and Physics, University of Udine,
via delle scienze 206, 33100 Udine, Italy
e-mail: dimitri.breda@uniud.it

A. Andò
e-mail: alessia.ando@gssi.it

D. Liessi
e-mail: davide.liessi@uniud.it

R. Vermiglio
e-mail: rossana.vermiglio@uniud.it

S. Maset
Department of Mathematics and Geosciences, University of Trieste,
via Valerio 12/1, 34127 Trieste, Italy
e-mail: maset@units.it

F. Scarabel
Department of Mathematics and Statistics, University of Helsinki,
P.O. Box 68, 00014 Helsinki, Finland
e-mail: francesca.scarabel@manchester.ac.uk

G. Valmorbida et al. (eds.), *Accounting for Constraints in Delay Systems*,
Advances in Delays and Dynamics 12,
https://doi.org/10.1007/978-3-030-89014-8_7

solutions, as well as for the bifurcation analysis of general nonlinear systems are illustrated, together with an essential overview of relevant problems and potential developments.

1 Introduction

Let d_X, d_Y be positive integers, $\tau > 0$ and consider the Banach spaces of functions

$$X := L^1([-\tau, 0]; \mathbb{R}^{d_X}), \qquad \|\varphi\|_X := \int_{-\tau}^{0} |\varphi(\theta)|d\theta$$

and

$$Y := C([-\tau, 0]; \mathbb{R}^{d_Y}), \qquad \|\psi\|_Y := \max_{\theta\in[-\tau,0]} |\psi(\theta)|$$

for $|\cdot|$ a norm in either $\mathbb{R}^{d_X}$ or $\mathbb{R}^{d_Y}$. This work concerns systems of the form

$$\begin{cases} x(t) = F(x_t, y_t) \\ y'(t) = G(x_t, y_t) \end{cases} \tag{1}$$

for $F : X \times Y \to \mathbb{R}^{d_X}$ and $G : X \times Y \to \mathbb{R}^{d_Y}$ smooth, autonomous, in general nonlinear functionals. Above, $x_t \in X$ denotes the X-component of the state of the dynamical system on the state space $X \times Y$ associated to (1), defined through the translation

$$x_t(\theta) := x(t + \theta), \quad \theta \in [-\tau, 0]. \tag{2}$$

The same notation is adopted for the Y-component. We refer to (1) as a *coupled* or *delay equation*. Let us remark that Lipschitz continuity of F and G ensures well-posedness (see [30, 33]).

Our interest is in questions about stability and bifurcations of mainly equilibria and periodic orbits of (1). These are undoubtedly important problems in view of applications, as often witnessed in major fields like control theory and population dynamics (see [39, 46] to name just a couple of starting references).

The solution to the above problems is in general unattainable analytically since the state space has infinite dimension. Resorting to numerical methods is thus unavoidable. This survey aims at illustrating the diverse uses of collocation techniques that have been devised by the authors since [9] to address the above issues. In particular, we refer to a framework known as *pseudospectral collocation* approach, that we describe in its generality in Sect. 1 in order to highlight its main features related to the context of interest, and to introduce the necessary notations and ingredients.

The rest of the work is organized as follows. Section 2 deals with the problem of stability of equilibria, addressed through the numerical approximation of the characteristic roots. We call the *infinitesimal generator approach* the relevant collocation method, introduced in Sect. 2 and, for ease of exposition, detailed only for the differential part of (1). Section 3 deals with the problem of stability of periodic solutions, addressed through the numerical approximation of the characteristic multipliers. We call the *solution operator approach* the relevant collocation method, introduced in Sect. 3 and, for ease of exposition, detailed only for the non-differential part of (1). In view of continuation and bifurcation, Sect. 4 deals with the approximation of (1) with *ordinary differential equations* (ODEs). We call the *nonlinear approach* the relevant collocation method, introduced in Sect. 4. In Sect. 5 we briefly discuss some directions of extension and generalization. Finally, we give some comments on future developments in Sect. 6, together with a summary overview.

1.1 Pseudospectral Collocation

Many problems in mathematics involve functions, but most such problems are not solvable exactly. Notorious instances are differential equations and definite integrals. One way to find a reasonable approximation is that of substituting the function with its interpolating polynomial at a given set of nodes discretizing the domain of interest. Then the pseudospectral collocation approach can be summarized as *do to the interpolating polynomial as you would do to the given function.* For an explanation of the term *pseudospectral* see [25, Sect. 1.7].

To recall the basics and fix a notation, let us restrict to the scalar case of a space of functions $Z([a, b]; \mathbb{R})$. This choice does not preclude generality as extension to the vector-valued case is straightforward. The space Z is assumed to be regular enough for point-wise evaluation to make sense.

Let n be a positive integer and consider the discretization of $[a, b]$ through $n + 1$ distinct points, say

$$a \leq t_{n,0} < t_{n,1} < \cdots < t_{n,n} \leq b. \tag{3}$$

Correspondingly, introduce $Z_n := \mathbb{R}^{n+1}$ as the finite-dimensional counterpart of Z. Define then the *restriction* operator $R_n : Z \to Z_n$ as

$$R_n f := (f(t_{n,0}), f(t_{n,1}), \ldots, f(t_{n,n}))^T$$

and the *prolongation* operator $P_n : Z_n \to Z$ as

$$P_n F := p_n \tag{4}$$

for p_n the polynomial of degree at most n interpolating the components of $F = (F_0, F_1, \ldots, F_n)^T$ at the points (3), i.e., satisfying

$$p_n(t_{n,i}) = F_i, \quad i = 0, 1, \ldots, n. \tag{5}$$

The polynomial p_n uniquely exists. Another matter is the way to represent it, which is not unique. A convenient one, both theoretically and numerically [7], uses the basis of Lagrange polynomials $\{\ell_0, \ell_1, \ldots, \ell_n\}$ associated to the points (3), i.e.,

$$\ell_{n,j}(t) = \prod_{\substack{k=0 \\ k \neq j}}^{n} \frac{t - t_{n,k}}{t_{n,j} - t_{n,k}}, \quad j = 0, 1, \ldots, n, \ t \in [a, b]. \tag{6}$$

Then

$$P_N F = \sum_{j=0}^{n} \ell_{n,j} F_j.$$

Indeed, to verify (5) it is enough to observe that the Lagrange polynomials satisfy the cardinal property $\ell_{n,j}(t_{n,i}) = \delta_{i,j}$ for $\delta_{i,j}$ the Kronecker's delta.

Remark 1 In what follows we will use lower case letters for functions and upper case letters for their finite-dimensional counterparts, i.e., the vectors of their values at the given points. Moreover, we allow the subscript indicating the discretization index n to be flexible, in the sense that it may refer to an indexing starting either from 0 or 1, leading to a finite dimension $n + 1$ or n, respectively.

The idea behind the pseudospectral collocation approach is to obtain convergence of p_n to f as $n \to \infty$. In this sense polynomial interpolation is an excellent technique when the interpolated function is smooth and the interpolation nodes are suitably chosen. Examples of this good behavior are illustrated in Fig. 1, where pseudospectral collocation is compared to classical methods of finite order. To the left, the derivative of an analytic function is approximated with finite differences (first order) and with pseudospectral collocation as

$$f'(t) \approx (P_n F)'(t) \tag{7}$$

for a given $t \in (a, b)$ and $F := (f(t_{n,0}), f(t_{n,1}), \ldots, f(t_{n,n}))^T$. To the right, the definite integral is approximated with the trapezoid rule (second order) and with pseudospectral collocation as

$$\int_a^b f(t)dt \approx \int_a^b (P_n F)(t)dt.$$

In practice, ensured the right choice of collocation nodes, the pseudospectral collocation approach can exploit the smoothness of the function, giving rise to errors decaying in general as $O(n^{-k})$ for f of class C^k, or as $O(n^{-k})$ for every $k \in \mathbb{N}$ for f of class C^∞, or even as $O(n^{-n})$ for f analytic. The latter two are known as *convergence of infinite order* or *spectral accuracy*, [48]. Opposite, a method of finite order, say m, applied to a function f of class C^k with $k > m$ will perform with an error decaying as $O(n^{-m})$ independently of k.

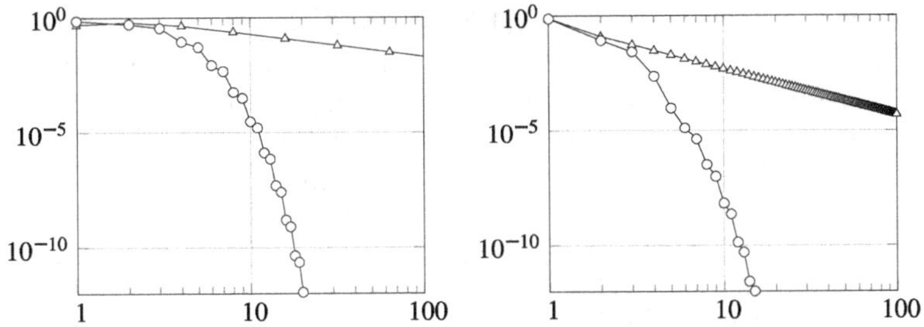

Fig. 1 Errors of pseudospectral collocation (∘) and finite-order methods (△) for approximating $f'(0)$ (left) and $\int_{-1}^{1} f(t)dt$ (right) for $f(t) = e^{-t^2}$; the collocation points are Chebyshev extrema, see Sect. 2.1

In the problems of interest for this work we have the freedom to select the appropriate nodes. Thus the attention is focused on the degree of smoothness of the concerned functions. As we will see, we approximate mainly spectral values of linear operators, so that at the core we deal with eigenfunctions, which are in general smooth. This motivates the use of pseudospectral collocation techniques. In fact, most applications require to tackle the stability problem under parameters' variation or uncertainty (e.g., bifurcation analysis, stability charts), so that computations of spectra must be repeated many times. Therefore low-dimensional approximations are very welcome. As a general drawback, error analyses and proofs of convergence can be rather cumbersome compared to those for methods of finite order.

With respect to the above, let us anticipate that by no means the proposed methods are intended to be applied for the sake of time-integration or, in other words, simulation of solutions of related initial value problems. Indeed, there is no guarantee that such solutions are smooth, as due in general to the presence of *breaking points* when delay is involved, [6].

2 Equilibria and Characteristic Roots

For ease of exposition, let us consider problems of the form (1) restricted to the differential part alone, i.e.,

$$y'(t) = G(y_t) \tag{8}$$

for $G : Y \to \mathbb{R}^{d_Y}$ a smooth, autonomous, in general nonlinear functional. Extension to coupled equations will be commented on at the end of the section. Equation (8) is called a *retarded functional differential equation*: the right-hand side G depends on the history y_t and it prescribes the (right-hand) derivative of the solution at the current time. In the applications of interest, (8) is usually a *delay differential* or *integro-differential equation*, meaning that the dependence on the past is described in terms of a finite number of discrete delays, i.e., terms like $y(t - \tau_k)$ for some $\tau_k \in (0, \tau]$, $k = 1, \ldots, p$, or, respectively, distributed delays, i.e., integrals of terms involving $y(t+\theta)$ for θ in some interval in $[-\tau, 0]$. Let us assume a constant function $\bar{y} \in Y$ to be an equilibrium of (8), i.e., satisfying $G(\bar{y}) = 0$. The study of the local stability properties of $\bar{y}$ can be addressed by resorting to the principle of linearized stability (see, e.g., [33, Sect. VII.6, Corollary 5.12]). Thus the problem is translated into studying the stability of the null solution of the linearized equation

$$y'(t) = L_G y_t, \tag{9}$$

where the linear functional $L_G : Y \to \mathbb{R}^{d_Y}$ is the Fréchet derivative of G at $\bar{y}$. For the class of problems we have in mind this functional has the form

$$L_G \psi = A_Y \psi(0) + B_Y \psi(-\tau) + \int_{-\tau}^{0} C_Y(\theta)\psi(\theta)d\theta$$

for $A_Y, B_Y \in \mathbb{R}^{d_Y \times d_Y}$ and $C_Y : [-\tau, 0] \to \mathbb{R}^{d_Y \times d_Y}$ smooth, or more general versions with any finite number of discrete and distributed delays.

The stability problem for (9) is usually formulated in terms of *characteristic roots*, i.e., solutions $\lambda \in \mathbb{C}$ of the characteristic equation

$$\det(\lambda I_{d_Y} - L_G e^{\lambda \cdot}) = 0. \tag{10}$$

In particular, $\bar{y}$ is locally asymptotically stable if and only if all the characteristic roots have negative real part, unstable as soon as one root falls to the right of the imaginary axis. Equation (10) is difficult to obtain in general and, moreover, its numerical solution may give rise to ill-conditioning, [9]. For these reasons in [19] it is proposed to exploit the following alternative characterization.

Remark 2 In (10) $L_G e^{\lambda \cdot}$ is a shorthand for the matrix $(L_G e^{\lambda \cdot} e_1 | \cdots | L_G e^{\lambda \cdot} e_{d_Y})$, $e_1, \ldots, e_{d_Y}$ being the canonical vectors in $\mathbb{R}^{d_Y}$.

Let $T_G(t) : Y \to Y$ be the *solution operator* associated to (9), i.e.,

$$T_G(t)\psi = y_t, \quad t \geq 0,$$

for y the solution of the initial value problem for (9) with initial condition $y_0 = \psi \in Y$. The family $\{T_G(t)\}_{t \geq 0}$ is a strongly continuous semigroup, [34]. Its *infinitesimal generator* is the linear unbounded operator $\mathscr{A}_G : \mathscr{D}(\mathscr{A}_G) \subseteq Y \to Y$ given by

$$\mathscr{A}_G \psi = \psi', \quad \psi \in \mathscr{D}(\mathscr{A}_G) = \left\{\psi \in Y \ : \ \psi' \in Y \text{ and } \psi'(0) = L_G \psi\right\}. \tag{11}$$

It can be proved that $\mathscr{A}_G$ has only point spectrum $\sigma(\mathscr{A}_G)$, whose elements are all eigenvalues of finite type and

$$\lambda \in \sigma(\mathscr{A}_G) \Leftrightarrow \det(\lambda I_{d_Y} - L_G e^{\lambda \cdot}) = 0,$$

see, e.g., [33, Chap. IV, Corollary 3.3] or [38, Chap. 7, Lemma 2.1].

In the following section we show how to use the pseudospectral collocation approach to discretize $\mathscr{A}_G$ into a finite-dimensional linear operator. The idea is to use the eigenvalues of the latter to approximate the characteristic roots, in a sense that will be clarified later on.

2.1 Infinitesimal Generator Approach

Let M be a positive integer and let

$$\theta_{M,i} = \frac{\tau}{2}\left(\cos\left(\frac{i\pi}{M}\right) - 1\right), \quad i = 0, 1, \ldots, M, \tag{12}$$

be the Chebyshev extremal nodes in $[-\tau, 0]$. Denote the corresponding restriction and prolongation operators as introduced in Sect. 1 respectively by $R_M : Y \to Y_M$ and $P_M : Y_M \to Y$ for $Y_M = \mathbb{R}^{d_Y(M+1)}$ the finite-dimensional counterpart of Y.

Define the linear finite-dimensional operator $\mathscr{A}_{G,M} : Y_M \to Y_M$ as

$$\begin{cases} [\mathscr{A}_{G,M}\Psi]_0 = L_G(P_M\Psi) \\ [\mathscr{A}_{G,M}\Psi]_i = [P_M\Psi]'(\theta_i), \quad i = 1, \ldots, M. \end{cases}$$

The first component above (index 0) mimics the derivative action of the generator at $\theta_{M,0} = 0$. As such it is given through the boundary condition characterizing the domain in (11) applied to the collocation polynomial $P_M\Psi$. The remaining components (index $i = 1, \ldots, M$) are given by the exact derivative of the collocation polynomial at the relevant nodes, according to the general derivative action in (11).

To get a matrix representation of $\mathscr{A}_{G,M}$ it is enough to choose a basis to represent the collocation polynomial. By using the Lagrange basis (see (6)), the resulting matrix reads

$$\mathscr{A}_{G,M} = \begin{pmatrix} L_G\ell_{M,0} & L_G\ell_{M,1} & \cdots & L_G\ell_{M,M} \\ d_{M,1,0} & d_{M,1,1} & \cdots & d_{M,1,M} \\ \vdots & \vdots & \ddots & \vdots \\ d_{M,M,0} & d_{M,M,1} & \cdots & d_{M,M,M} \end{pmatrix} \in \mathbb{R}^{d_Y(M+1)\times d_Y(M+1)} \tag{13}$$

for

$$d_{M,i,j} := \ell'_{M,j}(\theta_{M,i})I_{d_Y}, \quad i = 1, \ldots, M, \ j = 0, 1, \ldots, M, \tag{14}$$

and matrices $L_G\ell_{M,j}$, $j = 0, 1, \ldots, M$, according to Remark 2. Notice that the block-structure of $\mathscr{A}_{G,M}$ represents a perturbation of the Chebyshev differentiation matrix along the first row, which is motivated exclusively by the boundary condition in $\mathscr{D}(\mathscr{A}_G)$: this is the only part truly affected by the specific equation (9). Indeed, one can always rescale time to set $\tau = 1$, and hence fix nodes and (14) consequently. In this sense most entries of $\mathscr{A}_{G,M}$ can be computed independently of (9) and, moreover, explicitly (see, e.g., [48]).

A complete error analysis has been developed in [19], based on comparing (10) with the characteristic equation relevant to $\mathscr{A}_{G,M}$ via Rouché's Theorem. A softened version of the final result follows.

Theorem 1 *Let $\lambda \in \sigma(\mathscr{A}_G)$ have algebraic multiplicity ν. Then there exists $M \in \mathbb{N}$ sufficiently large such that $\mathscr{A}_{G,M}$ has ν eigenvalues $\lambda_{M,1}, \ldots, \lambda_{M,\nu}$ counted with multiplicities satisfying*

$$\max_{i=1,\ldots,\nu} |\lambda_{M,i} - \lambda| \leq C_2 \left(\frac{C_1}{M!}\right)^{M/\nu} \tag{15}$$

for C_1 and C_2 constants depending on λ but independent of M.

The constant C_1 is proportional to $|\lambda|$, hence approximations to eigenvalues of smaller magnitude converge faster. The issue is delicate since for stability one is rather interested in rightmost roots (see [52] for some work in this direction). Another due comment regards the automatic choice of M, in view of keeping errors below a desired tolerance. Although some indications are given in [19], on the one hand a precise answer is not available yet, on the other hand it is commonly verified that M of the order of few tens is enough to get machine accuracy for simple roots when the relevant eigenfunctions are not highly oscillatory (the role of multiplicities will be briefly accounted for in Sect. 3).

A recent full exposition of the infinitesimal generator approach can be found in [23]. There, possible variations of the above basic formulation include the choice of different sets of collocation nodes and the treatment of distributed delays. Concerning the latter, a suitable quadrature formula has to be adopted to compute the terms $L_G \ell_{M,j}$ in (13), which amounts to an additional error term inside the parentheses in (15), [23, Sect. 5.3.3]. A piecewise collocation is also advised according to the possible presence of multiple delays, leading to a reduced computational effort and an increased efficiency, [23, Sects. 5.2 and 5.4]. The associated MATLAB code `eigAM` is freely available at http://www.cdlab.uniud.it/software.

As an example of application of the infinitesimal generator approach for (8), let us consider the delay differential equation

$$\begin{cases} v'(t) = v(t) - \frac{v(t)^3}{3} - w(t) + \mu[v(t-\tau) - v_0] \\ w'(t) = \rho[v(t) + a - bw(t)], \end{cases} \tag{16}$$

where v_0 is a real root of $v - v^3/3 - (v + a)/b$, $a, b, \mu, \rho \in \mathbb{R}$ and $\tau > 0$. The system models recurrent neural feedback and has been studied in [26]. A nontrivial equilibrium is given by

$$(\bar{v}, \bar{w}) = \left(v_0, \frac{v_0 + a}{b}\right). \tag{17}$$

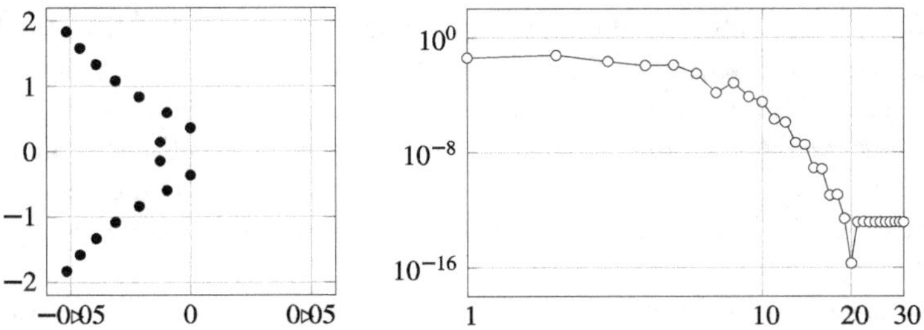

Fig. 2 Characteristic roots for (16) with $a = 0.7$, $b = 0.8$, $\rho = 0.08$ and $\tau = 25$ linearized around (17), computed with `eigAM` using $M = 30$ (left) and absolute error on the rightmost imaginary couple for increasing M (right)

The relevant linearized system (9) reads

$$\begin{cases} v'(t) = (1 - \bar{v}^2)v(t) - w(t) + \mu v(t - \tau) \\ w'(t) = \rho v(t) - b\rho w(t). \end{cases}$$

By using the infinitesimal generator approach we are able to find a Hopf bifurcation for $a = 0.7$, $b = 0.8$, $\rho = 0.08$, $\tau = 25$ and $\mu \approx -0.4996$. Figure 2 shows the relevant characteristic roots computed with `eigAM` (left) and the absolute error on the rightmost imaginary couple for increasing M (right). Observe that the error decays with the expected spectral accuracy.

Finally, the infinitesimal generator approach has been extended also to coupled equations (1), with two different but computationally equivalent formulations [15, 16]. Beyond the necessary changes to take into account for different state spaces, which non-trivially affect the proof of convergence, the delicate point is a diverse treatment of the boundary condition in the domain of the generator. Indeed, the condition regarding the X-component prescribes the value of the function at 0 rather than that of its derivative. Consequently, a direct collocation cannot be employed as it is in the first row of (13), but rather the collocated condition is solved externally to remove one degree of freedom, reflecting into a perturbation of the whole matrix. More on this will be detailed in Sect. 4.

3 Periodic Solutions and Floquet Multipliers

For ease of exposition, let us consider problems of the form (1) restricted to the non-differential part alone, i.e.,

$$x(t) = F(x_t) \tag{18}$$

for $F : X \to \mathbb{R}^{d_X}$ a smooth, autonomous, in general nonlinear functional. Extension to coupled equations will be commented on at the end of the section. Equation (18) is called a *retarded functional equation*: the right-hand side F depends on the history x_t and it prescribes the solution at the current time. In the applications of interest, (18) is usually a *renewal* or *Volterra integral equation*, meaning that the dependence on the past is described by integrals of terms involving $x(t+\theta)$ for θ in some interval in $[-\tau, 0]$.

Let us assume a function $t \mapsto \bar{x}(t) \in \mathbb{R}^{d_X}$ to be a periodic solution of (18) of (minimal) period Ω, i.e., $\bar{x}(t+\Omega) = \bar{x}(t)$ for all t. The study of the local stability properties of $\bar{x}$ can be addressed by resorting to the principle of linearized stability and to the Floquet theory as extended to (18). We assume the latter to hold although it is not yet fully developed (see [18, 41]). Thus the problem is translated into studying the stability of the null solution of the linearized problem

$$x(t) = L_F(t)x_t, \tag{19}$$

where the linear functional $L_F(\cdot) : X \to \mathbb{R}^{d_X}$ is the Fréchet derivative of F at $\bar{x}$, periodic in t of period Ω. For the class of problems we have in mind this functional has the form

$$L_F(t)\varphi = \int_{-\tau}^{0} C_X(t,\theta)\varphi(\theta)d\theta$$

for $C_X(t,\cdot) : [-\tau, 0] \to \mathbb{R}^{d_X \times d_X}$ smooth and periodic in t of period Ω.

Contrary to the *autonomous* case treated in Sect. 2, there is no direct formulation of the stability problem for the null solution of (19) in terms of characteristic equations. Rather one can introduce the two-parameters solution operator $T_F(t,s) : X \to X$ associated to (19), i.e.,

$$T_F(t,s)\varphi = x_t, \quad t \geq s,$$

for x the solution of the initial value problem for (19) with initial condition $x_s = \varphi \in X$, $s \in \mathbb{R}$. The family $\{T_F(t,s)\}_{t \geq s}$ is a strongly continuous *evolution family*, [29, 33]. Correspondingly, one can define *monodromy operators*

$$\mathcal{M}_F(s) := T_F(s+\Omega, s), \tag{20}$$

which advance the evolution of one period. It can be proved that these are compact operators. The Floquet theory states that the spectrum $\sigma(\mathcal{M}_F)$ of $\mathcal{M}_F(s)$ is independent of s and its elements $\mu \in \mathbb{C}$ are eigenvalues of finite type, known as *characteristic* or *Floquet multipliers*. Moreover, taking into account that a *trivial* multiplier $\mu = 1$ is always present due to linearization without affecting stability, the null solution of (19) is asymptotically stable if and only if all the non-trivial multipliers are inside the unit circle (and the trivial one is simple), unstable as soon as one multiplier falls outside. Eventually, in the absence of non-trivial multipliers on the unit circle, the periodic solution $\bar{x}$ of (18) inherits the same stability properties.

In the following section we show how to use the pseudospectral collocation approach to discretize $\mathcal{M}_F := \mathcal{M}_F(s)$ with a finite-dimensional linear operator. The idea is again to use the eigenvalues of the latter to approximate the characteristic multipliers. But first it is convenient to reformulate $\mathcal{M}_F$ as follows.

Let us introduce the spaces $X^+ := L^1([0, \Omega]; \mathbb{R}^{d_X})$ and $X^\pm := L^1([-\tau, \Omega]; \mathbb{R}^{d_X})$ besides X. Define the operator $V : X \times X^+ \to X^\pm$ as

$$[V(\varphi, w)](t) := \begin{cases} w(t), & t \in (0, \Omega], \\ \varphi(t), & t \in [-\tau, 0], \end{cases}$$

and the operator $\mathcal{F}_s : X^\pm \to X^+$ as

$$[\mathcal{F}_s v](t) := L_F(s+t)v_t, \quad t \in [0, \Omega].$$

The latter corresponds to the right-hand side of (19) with time shifted by s to set 0 as initial time. Then $\mathcal{M}_F$ can be recast as

$$\begin{cases} \mathcal{M}_F\varphi = V(\varphi, w^*)_\Omega \\ w^* = \mathcal{F}_s V(\varphi, w^*), \end{cases} \tag{21}$$

where the subscript Ω in the first equation refers to the translation (2) and w^* denotes the fixed point solution of the second equation. Indeed, it is not difficult to see that w^* is the solution in $[0, \Omega]$ of the initial value problem for (19) with initial condition $x_0 = \varphi \in X$. Consequently, $V(\varphi, w^*)$ is the solution in $[-\tau, \Omega]$ and $V(\varphi, w^*)_\Omega$ is the state after one period, which is in fact $\mathcal{M}_F\varphi$.

3.1 Solution Operator Approach

The approach presented in this section assumes $\Omega \geq \tau$ for simplicity of presentation. In the case $\Omega < \tau$ it is necessary to adopt a piecewise strategy, whose details are discussed in [18].

Let M be a positive integer and let $\theta_{M,i}$, $i = 0, 1, \ldots, M$, be the Chebyshev extremal nodes in $[-\tau, 0]$ given in (12). Denote the corresponding restriction and prolongation operators as introduced in Sect. 1 respectively by $R_M : \tilde{X} \to X_M$ and $P_M : X_M \to X$ for $X_M = \mathbb{R}^{d_X(M+1)}$ the finite-dimensional counterpart of X. Above R_M is defined on a suitable subspace $\tilde{X} \subset X$ where point-wise evaluation makes sense (see [18, Sect. 3.2] for relevant comments). Similarly, let N be a positive integer and let

$$t_{N,i} = \frac{\Omega}{2}\left(1 - \cos\left(\frac{(2i-1)\pi}{2N}\right)\right), \quad i = 1, \ldots, N,$$

be the Chebyshev zeros in $(0, \Omega)$. Denote the corresponding restriction and prolongation operators as introduced in Sect. 1 respectively by $R_N^+ : \tilde{X}^+ \to X_N^+$ and $P_N^+ : X_N^+ \to X^+$ for $X_N^+ = \mathbb{R}^{d_X N}$ the finite-dimensional counterpart of X^+. The subspace $\tilde{X}^+ \subset X^+$ has the same role as $\tilde{X}$ above.

Define the linear finite-dimensional operator $\mathscr{M}_{F,M,N} : X_M \to X_M$ as

$$\begin{cases} \mathscr{M}_{F,M,N}\Phi = R_M V(P_M\Phi, P_N^+ W^*)_\Omega \\ W^* = R_N^+ \mathscr{F}_s V(P_M\Phi, P_N^+ W^*). \end{cases} \tag{22}$$

Notice that (21) is discretized by applying the operators V and $\mathscr{F}_s$ to the collocation polynomials $P_M\Phi$ in $[-\tau, 0]$ and $P_N^+ W^*$ in $(0, \Omega]$ and by restricting the images through R_M and R_N^+. Both images are continuous, so the restrictions are well defined.

To get a matrix representation of $\mathscr{M}_{F,M,N}$ is not as immediate as for $\mathscr{A}_{G,M}$ in Sect. 2. Details are elaborated in [18], where the construction of the matrix is explicitly described for the sake of implementation. Also a complete error analysis has been developed in [18], based on comparing $\mathscr{M}_F$ with a finite-rank infinite-dimensional version of $\mathscr{M}_{F,M,N}$. Convergence in norm of the latter to $\mathscr{M}_F$ guarantees the convergence of the relevant multipliers and (generalized) eigenspaces thanks to the general results in [27]. Notice that the structure of the convergence analysis is completely different from (and notably more difficult than) that of the infinitesimal generator approach, due to the absence of a characteristic equation for $\mathscr{M}_F$. A softened version of the final result follows.

Theorem 2 *Let $\mu \in \sigma(\mathscr{M}_F)$ have algebraic multiplicity ν and ascent l, with (generalized) eigenfunctions of class C^p. Then there exists $N \in \mathbb{N}$ sufficiently large such that, for $M \geq N$, $\mathscr{M}_{F,M,N}$ has ν eigenvalues $\mu_{M,N,1}, \ldots, \mu_{M,N,\nu}$ counted with multiplicities satisfying*

$$\max_{i=1,\ldots,\nu} |\mu_{M,N,i} - \mu| = o\left(N^{\frac{1-p}{l}}\right).$$

The degree of smoothness p of the (generalized) eigenfunctions is not yet known in case of problems of the form (18), for relevant comments see [41, Chap. 9]. Nevertheless they are expected to be infinitely smooth, provided the right-hand side is smooth. Indeed, this is confirmed experimentally by the infinite order of convergence typically obtained, see the example below.

The eigenfunctions are instead known to be analytic functions for problems of the form (8), for which the solution operator approach was originally developed, [22]. The main differences, beyond the state space, reside in the definition of the operator V, which reconstructs the solution in $[0, \Omega]$ by integrating the derivative, and in the fact that the right-hand side $\mathscr{F}_s$ furnishes indeed the derivative, being the equation differential. A recent full exposition for (8) can be found in [23]. Similarly to the infinitesimal generator approach, it is explained there how to treat the possible presence of distributed delays, amounting again to an additional error term. The associated MATLAB code `eigTMN` is freely available at http://www.cdlab.uniud.it/software.

The original solution operator approach for (8) has been revisited in [41], where it has been extended also to coupled problems of the form (1), requiring to deal with the delicate interplay between the different regularization properties of the X- and Y-components, fundamental in the convergence analysis, [41, Chap. 7]. The associated MATLAB code `eigTMNc` is freely available at http://www.cdlab.uniud.it/software.

In view of this recent extension, as an example of application of the solution operator approach for (1), let us consider the coupled equation

$$\begin{cases} x(t) = -\frac{1}{2}\Big[\int_0^{\frac{7}{2}\pi} x(t-\sigma)d\sigma - \int_0^{\frac{\pi}{2}} \ln(y(t-\sigma))d\sigma\Big] \\ y'(t) = -\ln\Big(y\Big(t-\frac{\pi}{2}\Big)\Big)y(t), \end{cases} \tag{23}$$

which has been constructed in [41] so that it has the periodic solution

$$(\bar{x}(t), \bar{y}(t)) = \left(\sin t, \mathrm{e}^{\sin t}\right). \tag{24}$$

The linearization of (23) around this solution reads

$$\begin{cases} x(t) = -\frac{1}{2}\Big[\int_0^{\frac{7}{2}\pi} x(t-\sigma)d\sigma - \int_0^{\frac{\pi}{2}} \frac{y(t-\sigma)}{\bar{y}(t-\sigma)}d\sigma\Big] \\ y'(t) = -\ln\Big(\bar{y}\Big(t-\frac{\pi}{2}\Big)\Big)y(t) - \frac{y(t-\frac{\pi}{2})}{\bar{y}(t-\frac{\pi}{2})}\bar{y}(t). \end{cases}$$

Figure 3 shows the relevant Floquet multipliers computed with `eigTMNc` for $M = N = 55$ (left) and the errors on the two dominant multipliers, which are both 1, for increasing $M = N$ (right). Observe that the error plot exhibits an infinite order of convergence.

It is worth noticing that the solution operator approach can be used to compute also the characteristic roots for equilibria. In this case the same collocation framework used to discretize the monodromy operators (20) can be equally adopted for problems

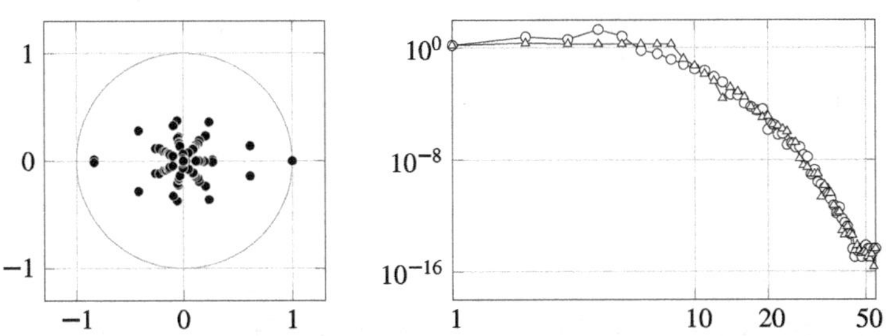

Fig. 3 Floquet multipliers (with respect to the unit circle) for (23) linearized around (24), computed with `eigTMNc` using $M = N = 55$ (left) and absolute errors on the two dominant multipliers for increasing $M = N$

(19) in the autonomous case $L_F(t) \equiv L_F$ to discretize the solution operator $T_F(h) := T_F(h, 0)$ advancing the solution of any given and fixed step $h > 0$. The same clearly holds for T_G in case of (9), or for coupled problems. Indeed, the spectral mapping theorem ensures the exponential relation $\mu = e^{\lambda h}$ between $\mu \in \sigma(T_F(h))$ and the characteristic roots λ. Notice, however, that the convergence results in Theorems 1 and 2 highlight the different roles of algebraic multiplicities and ascents. This issue is currently under further investigation by some of the authors, given also the recent emerging interest in the effects of multiplicities in control of time-delay systems (see, e.g., [8, 45]).

4 Continuation and Bifurcation Analysis

In applications the study of the dynamics of a given model is seldom restricted to only equilibria or periodic solutions, but it rather aspires to explore the general behavior under parameters' variation or uncertainty, in the framework of continuation and bifurcation analyses.

DDE-BIFTOOL [2] and Knut [3] are the most evolved tools available for problems with delays, but still they cannot handle the generality of (1). Thus the method in [13] is proposed in the direction of filling this gap. The idea is to apply the pseudospectral collocation approach directly to the original nonlinear problem, without resorting to linearization. In this sense the method can be seen as an extension to the nonlinear case of the infinitesimal generator approach of Sect. 2.1. The result is a nonlinear ODE, which can be recovered quite easily and can be treated with one of the many consolidated continuation packages for bifurcation analysis of ODEs (e.g., AUTO [1], MatCont [4], XPPAUT [5]). We summarize this method in Sect. 4.1. First we need to recall some theoretical background. The initial value problem for (1) is well-posed for $t \geq 0$ as soon as an initial condition $(x_0, y_0) = (\varphi, \psi) \in X \times Y$ is imposed. This allows to define a dynamical system on the state space $X \times Y$, with state (x_t, y_t) defined through the solution $(x(t), y(t))$ according to (2). The relevant nonlinear evolution family $\{T_{F,G}(t)\}_{t \geq 0}$ for $T_{F,G}(t) : X \times Y :\rightarrow X \times Y$ given by

$$T_{F,G}(t)(\varphi, \psi) = (x_t, y_t), \quad t \geq 0,$$

is generated by the nonlinear operator $\mathscr{A}_{F,G} : \mathscr{D}(\mathscr{A}_{F,G}) \subseteq X \times Y :\rightarrow X \times Y$ given by

$$\mathscr{A}_{F,G}(\varphi, \psi) = (\varphi', \psi')$$

with domain

$$\mathscr{D}(\mathscr{A}_{F,G}) = \big\{(\varphi, \psi) \in X \times Y : (\varphi', \psi') \in X \times Y \\ \text{and } (\varphi(0), \psi'(0)) = (F(\varphi, \psi), G(\varphi, \psi))\big\}. \tag{25}$$

It is important to remark that the *nonlinear* domain condition involves the value at 0 for the X-component and the value of the derivative at 0 for the Y-component, due respectively to the non-differential and differential nature of the relevant equations. This difference plays a key role in the discretization developed below. Eventually, it is possible to describe the evolution of the state by the abstract Cauchy problem

$$\begin{cases} \frac{d}{dt}(u(t), v(t)) = \mathscr{A}_{F,G}(u(t), v(t)), & t \geq 0, \\ (u(0), v(0)) = (\varphi, \psi). \end{cases} \tag{26}$$

Indeed, the solution is $(u(t), v(t)) = (x_t, y_t)$ as soon as $(\varphi, \psi) \in \mathscr{D}(\mathscr{A}_{F,G})$ (otherwise the equivalence passes through the definition of a *mild* solution, [34]).

In the following section we show how to use the pseudospectral collocation approach to discretize the abstract differential equation above with an ODE.

4.1 Nonlinear Approach

Let M be a positive integer and let $\theta_{M,i}$, $i = 0, 1, \ldots, M$, be the Chebyshev extremal nodes in $[-\tau, 0]$ given by (12). Correspondingly, define also the following submatrices of the relevant differentiation matrix:

$$d_M := \begin{pmatrix} d_{M,1,0} \\ \vdots \\ d_{M,M,0} \end{pmatrix} \in \mathbb{R}^M, \qquad D_M := \begin{pmatrix} d_{M,1,1} & \cdots & d_{M,1,M} \\ \vdots & \ddots & \vdots \\ d_{M,M,1} & \cdots & d_{M,M,M} \end{pmatrix} \in \mathbb{R}^{M \times M}$$

for

$$d_{M,i,j} := \ell'_{M,j}(\theta_{M,i}), \quad i = 1, \ldots, M, \quad j = 0, 1, \ldots, M.$$

Since F and G in (1) have finite-dimensional ranges, we need to introduce only a prolongation operator P_M. This is done according to (4), allowing P_M to act on vectors of blocks of size either d_X or d_Y.

For all $t \geq 0$ and $i = 0, 1, \ldots, M$, define the following approximations to the states at the nodes:

$$U_i(t) \approx u(t)(\theta_i) = x_t(\theta_i) \in \mathbb{R}^{d_X}, \qquad V_i(t) \approx v(t)(\theta_i) = y_t(\theta_i) \in \mathbb{R}^{d_Y}.$$

Set also

$$U(t) := (U_1(t), \ldots, U_M(t))^T \in \mathbb{R}^{d_X M}, \qquad V(t) := (V_1(t), \ldots, V_M(t))^T \in \mathbb{R}^{d_Y M}.$$

Recall now that the boundary condition for the X-component in (25) concerns the value of the state at 0. Thus its discretization

$$U_0(t) = F(P_M(U_0(t), U(t)), P_M(V_0(t), V(t))) \tag{27}$$

is solved to recover the unknown $U_0(t) \in \mathbb{R}^{d_X}$. If it cannot be solved explicitly, some nonlinear root finder has to be employed. Having done this, (1) or, equivalently, the abstract differential equation in (26), is reduced to the ODE in $\mathbb{R}^{d_X M + d_Y(1+M)}$

$$\begin{cases} U'(t) = (D_M \otimes I_{d_X})U(t) + (d_M \otimes I_{d_X})U_0(t) \\ V_0'(t) = G(P_M(U_0(t), U(t)), P_M(V_0(t), V(t))) \\ V'(t) = (D_M \otimes I_{d_Y})V(t) + (d_M \otimes I_{d_Y})V_0(t). \end{cases} \tag{28}$$

Notice that only (27) and the second in (28) are affected by the right-hand sides of (1), inheriting their nonlinearities. The rest of (28), which amounts to the most, is linear and can be defined automatically based just on M, d_X and d_Y (recall that τ can always be rescaled to 1). Of course, if needed, the discretization can be further specialized by choosing a different index M for the X- and Y-components of the state, this generalization being straightforward. Finally, as mentioned in Sect. 2.1, similar comments apply regarding the presence in the right-hand sides of integrals to be approximated and the possibility of implementing a piece-wise strategy.

The analysis in [13] shows that there is a one-to-one correspondence between the equilibria of (1) and those of (28). In particular, let $(\bar{x}, \bar{y}) \in X \times Y$ satisfy

$$\begin{cases} F(\bar{x}, \bar{y}) = \bar{x} \\ G(\bar{x}, \bar{y}) = 0 \end{cases}$$

and, without confusion, let $(\bar{x}, \bar{y}) \in \mathbb{R}^{d_X} \times \mathbb{R}^{d_Y}$ denotes also the value of the function $(\bar{x}, \bar{y}) \in X \times Y$. Then $(\bar{x}, \ldots, \bar{x}, \bar{y}, \bar{y}, \ldots, \bar{y}) \in \mathbb{R}^{d_X M + d_Y(1+M)}$ is an equilibrium for (28). The opposite direction holds as well. Moreover, the matrix at the right-hand side of the linearization of (28) (and of (27)) at this equilibrium is the same that is obtained by applying the infinitesimal generator approach of Sect. 2.1 (extended according to [16]) to the linearization of (1) at the same equilibrium, as to say that linearization and discretization commute. Consequently, a bifurcation analysis of the equilibria of (28) performed, e.g., by MatCont, returns spectrally accurate results based on Theorem 1. The convergence of the similar approach to the bifurcation of limit cycles has not been rigorously justified at the moment, but it is largely supported by numerous experiments, an example of which is illustrated below. See also [14] for the case of renewal equations (18).

Let us consider the delay differential equation

$$y'(t) = -\sin y(t - \tau). \tag{29}$$

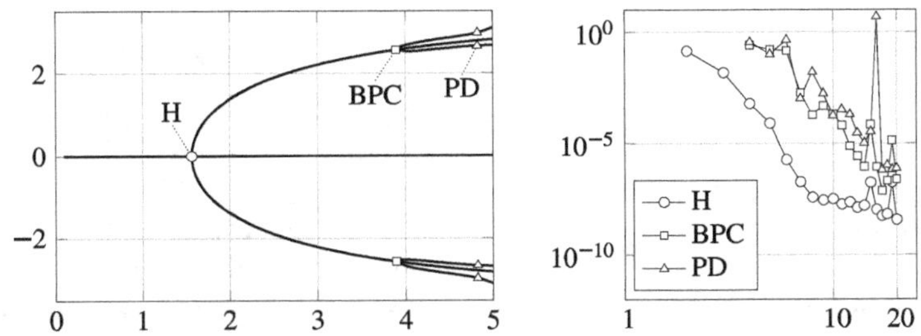

Fig. 4 Bifurcation diagram of (29) computed with MatCont and the nonlinear approach using $M = 10$ (left) and absolute errors on the bifurcation points for increasing M (right)

The relevant bifurcation properties are described in detail in [47], where (29) is presented as one of the simplest delay problems with chaotic behavior. The trivial equilibrium $\bar{y} = 0$, which exists for all $\tau > 0$, undergoes a Hopf bifurcation (H) at $\tau = \pi/2$, after which a stable limit cycle appears. The latter destabilizes through a pitchfork bifurcation (BPC) at $\tau \approx 3.894$, beyond which there are two coexisting stable limit cycles. On this branch of limit cycles, a period doubling (PD) occurs at $\tau \approx 4.828$. Figure 4 shows the bifurcation diagram obtained by applying MatCont to the ODE approximating (29) via the nonlinear approach using $M = 10$ (left) and the absolute error in the computed bifurcation points H, BPC and PD for increasing M. Observe the typical trend of spectral convergence until the tolerance 10^{-6} (set as default in MatCont) is reached.

5 Extensions

The collocation approaches presented in the previous sections have been or can be suitably extended in diverse directions. A brief account follows.

5.1 *Infinite Delay*

In some models there is no bound to the delay. This happens for instance when the model involves a probability density function (e.g., exponential or Gamma distributions). In this case, the state (x_t, y_t) is defined on the unbounded domain $(-\infty, 0]$ and we talk about *infinite delay*. Fixing a suitable $\rho > 0$, the initial value problem for (1) with initial condition $(x_0, y_0) = (\varphi, \psi)$ is well-posed on $X_\rho \times Y_\rho$ with

$$X_\rho := \left\{ \varphi : (-\infty, 0] \to \mathbb{R}^{d_X} \; : \; \int_{-\infty}^{0} e^{\rho\theta} |\varphi(\theta)| d\theta < +\infty \right\}$$

and

$$Y_\rho := \left\{ \psi : (-\infty, 0] \to \mathbb{R}^{d_Y} \ : \lim_{\theta \to -\infty} \psi(\theta) = 0, \ \sup_{\theta \le 0} e^{\rho\theta} |\psi(\theta)| < +\infty \right\}$$

provided with the corresponding weighted norms [31]. Note that, with this definition, constant functions belong to the state space, and the principle of linearized stability holds for equilibria. In [37] the nonlinear approach presented in Sect. 4.1 has been adapted to the case of infinite delay. Given a discretization index M, the collocation nodes in $(-\infty, 0]$ are chosen as the points corresponding to the extrema of the Laguerre polynomial of degree $M + 1$, plus the node $\theta = 0$. Recall that Laguerre polynomials are orthogonal with respect to the exponential weight $e^{\rho\theta}$, and therefore they appear particularly suitable to this context. The structure of the resulting ODE is very similar to (28), where most of the system is defined via the differentiation matrix relevant to the nodes, and the nonlinearities enter only through F and G. One-to-one correspondence of equilibria holds, and numerical tests suggest spectral convergence to the eigenvalues λ with $\Re\lambda > -\rho$ of the infinitesimal generator of the linearized problem.

5.2 *Lyapunov Exponents*

In the case of generic solutions of (8), i.e., when linearization leads to nonautonomous problems, chaos can be detected via Lyapunov exponents. These can be defined through the evolution family as the logarithm of the eigenvalues of the operator $\Lambda : Y \to Y$ given by

$$\Lambda := \lim_{k\to\infty} [T_G(s + k\tau, s)^H T_G(s + k\tau, s)]^{1/2k},$$

[10]. The above, indeed, reminds of the long-time average growth of the norm of the evolution family along a uniform sequence of time instants $\{t_k\}_{k\geq 0}$ with step τ, viz. $t_k = s + k\tau$. The mentioned norm is relevant to an inner product, hence the state space of (8) has to be changed to the Hilbert space $Y = L^2([-\tau, 0]; \mathbb{R}^{d_Y})$. Beyond giving a proper sense to that notion of orthogonality that is fundamental in the computation of the exponents, this change allows to raise to infinite dimension the framework of QR techniques commonly employed in the case of ODEs. This corresponds to the method developed in [24], which employs generalized Fourier projections (see also [28]). The above need for changing state space vanishes if one resorts instead to the nonlinear approach of Sect. 4.1, thus applying standard schemes for computing the exponents to the resulting ODE, as described in [12].

5.3 Uncertain Parameters

In practical applications a crucial question is the analysis of the effect on stability of uncertainty in the specification of model constants and parameters. The uncertain data are often modeled as random quantities in a suitable probabilistic framework and, in the last few years, the Polynomial Chaos expansions (see, e.g., [51]) have been successfully applied to problems of uncertainty propagation and quantification in the random model solutions, [40]. Recently, within this context, retarded functional differential equations with uncertain parameters have been introduced and analyzed. Moreover the uncertainty quantification and the sensitivity analysis of the random numerical spectral abscissa have been performed by combining the infinitesimal generator approach with suitable stochastic numerical methods, [49, 50].

5.4 Other Classes of Equations

The solution operator approach of Sect. 3.1 is in principle more tunable than the infinitesimal generator approach of Sect. 2.1. In fact, while the existence of a generator is not always guaranteed, that of a solution operator is always available once the original problem is well posed.

In this sense, the reformulation (21), as abstracted to a general linear evolution operator $T_{F,G}(s+h,s)$ for $h>0$ and $s\in\mathbb{R}$, is prone to be adapted to many classes of evolution problems. It is sufficient to define properly the operator V that constructs the solution in finite time once the required ingredients are provided. These are usually represented by an initial or boundary function (the first argument of V) together with a rule to advance the solution in time (through the second argument of V). In the case of problems concerning delay, the former is the history function on the delay interval $[-\tau,0]$. In the case of hyperbolic partial differential equations (PDEs), for instance, it is the function provided on a suitable boundary of the domain of the "space" variable. The second ingredient of V is instead provided by integrating the derivative in case of (9) or by the solution itself in case of (19). In either case the key elements are given by the right-hand side of the equation (the operator $\mathscr{F}_s$), giving rise to the fixed point problem at the second line of (21). Once this reformulation is available, the formal discretization follows straightforwardly as in (22) through restrictions and prolongations suitably defined on the concerned spaces. Of course this setting must be conformed to the specific class of equations through the definitions of the operators V and $\mathscr{F}_s$ as well as to the regularity of the spaces where these are defined. A similar general approach has been largely discussed in [42–44] for the numerical solution of boundary value problems, but it has never been addressed adequately for approximations of spectral values.

Alternatively, whenever available, the structure of a derivative operator with non-local boundary conditions characterizing its domain can be favorably exploited by following the principles outlined in Sect. 2 for the infinitesimal generator of problems with delay. On the one hand, the derivative action is always treated according to (7), giving rise to blocks depending on the underlying differentiation matrix. On the other hand, the boundary condition is discretized by applying it to the collocation polynomial, giving rise to a perturbation of the above differentiation parts. Variations of this approach have been discussed in [20], where similar schemes have been proposed to compute characteristic eigenvalues of mixed-type functional differential equations (i.e., problems containing both delays and advances), age-structured population models (hyperbolic PDEs) and also delay equations of neutral type (delays in the derivative of the solution). A similar methodology has been applied in [21] also to PDEs with delay, in particular reaction-diffusion problems with delay in the reaction part. Here, the pseudospectral collocation approach is used at the end to reduce to a matrix the eigenvalue problem obtained by previously applying a spectral method to the original formulation.

6 Overview and Future

Beyond what summarized in the previous section, there are other directions to be explored in the "to-do" list of the authors and collaborators, in view of extending existing collocation approaches, or developing new ones.

In particular, we mention the class of problems concerning *state-dependent* delays, for which a nonlinear approach with variable mesh seems a reasonable approach, worthy of being investigated [36]. Moreover, the class of neutral problems, in which delays affect also the derivative of the unknown function, has not been adequately considered so far, and the abstract approach developed in [44] for boundary value problems may serve as good starting point for generalizing the solution operator approach of Sect. 3.1. With regards to the computation of Lyapunov exponents, proper

Table 1 IG—infinitesimal generator approach, SO—solution operator approach, LE—Lyapunov exponents, NL—nonlinear approach; ✓—done, ✍—ongoing, ✗—not done

Class of problems	IG	SO	LE	NL
Retarded functional differential	✓ [19]	✓ [22]	✓ [12, 24]	✓ [13]
Retarded functional	✓ [15, 16]	✓ [18]	✍	✓ [13]
Coupled	✓ [15, 16]	✓ [41]	✍	✓ [13]
Neutral functional differential	✓ [20]	✗	✗	✗
Mixed functional differential	✓ [20]	✗	✗	✗
Partial retarded functional differential	✓ [21]	✗	✗	✗
Age-structured population	✓ [11, 17]	✗	✗	✗
Infinite delay	✓ [37]	✗	✗	✓ [37]
State-dependent	✗	✗	✗	✍

extensions of [24] should be devised for both renewal and coupled equations, for which analogous tools completely lack.

There are also other possibilities of exploiting collocation approaches, beyond those regarding the approximation of spectra of operators or the generic nonlinear approach of Sect. 4.1. We refer in particular to the computation of periodic solutions, for which the consolidated piece-wise and adaptive strategy behind AUTO and MatCont, adopted also in DDE-BIFTOOL [35], can be potentially extended to more general classes of problems with delay. This is rather important in view of analyzing the stability of such solutions, in the framework of continuation. Finally, as far as the latter is concerned, collocation can favorably contribute also to the treatment of complex models, as is the emerging case of structured populations where vital rates are implicitly defined through external conditions, often involving ODEs (see, e.g., [32]). The latter can be suitably collocated and thus included in the continuation structure in order to improve the overall performance and efficiency.

Eventually, in order to give a panoramic view of the "state-of-the-art" so far, we refer to Table 1 below, which may also serve as a list of relevant references.

References

1. AUTO. http://indy.cs.concordia.ca/auto/
2. DDE-BIFTOOL. http://ddebiftool.sourceforge.net/
3. Knut. http://rs1909.github.io/knut/
4. MatCont. https://sourceforge.net/projects/matcont/
5. XPPAUT. http://www.math.pitt.edu/~bard/xpp/xpp.html
6. A. Bellen, M. Zennaro, *Numerical Methods for Delay Differential Equations* Numerical mathemathics and scientifing computing series. (Oxford University Press, 2003)
7. J.-P. Berrut, L.N. Trefethen, Barycentric Lagrange interpolation. SIAM Rev. **46**, 501–517 (2004)
8. I. Boussada, S.-I. Niculescu, Tracking the algebraic multiplicity of crossing imaginary roots for generic quasipolynomials: a Vandermonde-based approach. IEEE Trans. Automat. Control **61**, 1601–1606 (2016)
9. D. Breda, *Numerical Computation of Characteristic Roots for Delay Differential Equations*, Ph.D. thesis, Ph.D. in Computational Mathematics, Università di Padova (2004)
10. D. Breda, Nonautonomous delay differenti.al equations in Hilbert spaces and Lyapunov exponents. Diff. Int. Equ. **23**, 935–956 (2010)
11. D. Breda, C. Cusulin, M. Iannelli, S. Maset, R. Vermiglio, Stability analysis of age-structured population equations by pseudospectral differencing methods. J. Math. Biol. **54**, 701–720 (2007)
12. D. Breda, S. Della Schiava, Pseudospectral reduction to compute Lyapunov exponents of delay differential equations. Disc. Contin. Dyn. S.-B **23**, 2727–2741 (2018)
13. D. Breda, O. Diekmann, M. Gyllenberg, F. Scarabel, R. Vermiglio, Pseudospectral discretization of nonlinear delay equations: new prospects for numerical bifurcation analysis. SIAM J. Appl. Dyn. Sys. **15**, 1–23 (2016)
14. D. Breda, O. Diekmann, D. Liessi, F. Scarabel, Numerical bifurcation analysis of a class of nonlinear renewal equations. Electron. J. Qual. Theory Differ. Equ. **65**, 1–24 (2016)
15. D. Breda, O. Diekmann, S. Maset, R. Vermiglio, A numerical approach for investigating the stability of equilibria for structured population models. J. Biol. Dyn. **7**, 4–20 (2013)

16. D. Breda, P. Getto, J. Sánchez Sanz, R. Vermiglio, Computing the eigenvalues of realistic Daphnia models by pseudospectral methods. SIAM J. Sci. Comput. **37**, 2607–2629 (2015)
17. D. Breda, M. Iannelli, S. Maset, R. Vermiglio, Stability analysis of the Gurtin-MacCamy model. SIAM J. Numer. Anal. **46**, 980–995 (2008)
18. D. Breda, D. Liessi, Approximation of eigenvalues of evolution operators for linear renewal equations. SIAM J. Numer. Anal. **56**, 1456–1481 (2018)
19. D. Breda, S. Maset, R. Vermiglio, Pseudospectral differencing methods for characteristic roots of delay differential equations. SIAM J. Sci. Comput. **27**, 482–495 (2005)
20. D. Breda, S. Maset, R. Vermiglio, Pseudospectral approximation of eigenvalues of derivative operators with non-local boundary conditions. Appl. Numer. Math. **56**, 318–331 (2006)
21. D. Breda, S. Maset, R. Vermiglio, Numerical approximation of characteristic values of partial retarded functional differential equations. Numer. Math. **113**, 181–242 (2009)
22. D. Breda, S. Maset, R. Vermiglio, Approximation of eigenvalues of evolution operators for linear retarded functional differential equations. SIAM J. Numer. Anal. **50**, 1456–1483 (2012)
23. D. Breda, S. Maset, R. Vermiglio, *Stability of Linear Delay Differential Equations—A Numerical Approach with MATLAB*. SpringerBriefs in Control, Automation and Robotics (Springer, New York, 2015)
24. D. Breda, E.S. Van Vleck, Approximating Lyapunov exponents and Sacker-Sell spectrum for retarded functional differential equations. Numer. Math. **126**, 225–257 (2014)
25. H. Brunner, Collocation methods for Volterra integral and related functional differential equations, no. 15, in *Cambridge monographs on applied and computational mathematics* (Cambridge University Press, Cambridge, 2004)
26. A.M. Castelfranco, H.W. Stech, Periodic solutions in a model of recurrent neural feedback. SIAM J. Appl. Math. **47**, 573–588 (1987)
27. F. Chatelin, *Spectral Approximation of Linear Operators* (Academic Press, New York, 1983)
28. M.D. Chekroun, M. Ghil, H. Liu, S. Wang, Low-dimensional Galerkin approximations of nonlinear delay differential euqations. Discrete Contin. Dyn. Syst. **36**, 4133–4177 (2016)
29. C. Chicone, Y. Latushkin, Evolution semigroups in dynamical systems and differential equations, no. 70, in *Mathematical Surveys and Monographs* (American Mathematical Society, Providence, 1999)
30. O. Diekmann, P. Getto, M. Gyllenberg, Stability and bifurcation analysis of Volterra functional equations in the light of suns and stars. SIAM J. Math. Anal. **39**, 1023–1069 (2008)
31. O. Diekmann, M. Gyllenberg, Equations with infinite delay: blending the abstract and the concrete. J. Differ. Equ. **252**, 819–851 (2012)
32. O. Diekmann, M. Gyllenberg, J.A.J. Metz, S. Nakaoka, A.M. de Roos, Daphnia revisited: local stability and bifurcation theory for physiologically structured population models explained by way of an example. J. Math. Biol. **61**, 277–318 (2010)
33. O. Diekmann, S.A. van Gils, S.M. Verduyn Lunel, H.-O. Walther, Delay equations—functional, complex and nonlinear analysis, no. 110 in *Applied Mathematical Sciences* (Springer, New York, 1995)
34. K. Engel, R. Nagel, One-Parameter semigroups for linear evolution equations, no. 194, in *Graduate Texts in Mathematics* (Springer, New York, 1999)
35. K. Engelborghs, T. Luzyanina, K.J. in 't Hout, D. Roose, Collocation methods for the computation of periodic solutions of delay differential equations. SIAM J. Sci. Comput. **22**, 1593–1609 (2001)
36. PH. Getto, M. Gyllenberg, Y. Nakata, F. Scarabel, Stability analysis of a state-dependent delay differential equation for cell maturation: analytical and numerical methods. J Math Biol. **79**(1), 281–328. (2019). https://doi.org/10.1007/s00285-019-01357-0
37. M. Gyllenberg, F. Scarabel, R. Vermiglio, Equations with infinite delay: numerical bifurcation analysis via pseudospectral discretization. Appl. Math. Comput. **333**, 490–505 (2018)
38. J.K. Hale, Theory of functional differential equations, no. 99, in *Applied Mathematical Sciences*, 1st edn. (Springer, New York, 1977)
39. H. Inaba, *Age-Structured Population Dynamics in Demography and Epidemiology* (Springer, New York, 2017)

40. O. Le Maître, O. Knio, *Spectral Methods for Uncertainty Quantification with Applications to Computational Fluid Dynamics* (Springer, New York, 2010)
41. D. Liessi, Pseudospectral methods for stability of periodic solutions of delay models, Ph.D. thesis, Ph.D. in Computer Science, Mathematics and Physics (Università di Udine, 2018)
42. S. Maset, The collocation method in the numerical solution of boundary value problems for neutral functional differential equations. Part I: Convergence results. SIAM J. Numer. Anal. **53**, 2771–2793 (2015). https://doi.org/10.1137/130935550
43. S. Maset, The collocation method in the numerical solution of boundary value problems for neutral functional differential equations. Part II: Differential equations with deviating arguments. SIAM J. Numer. Anal. **53**, 2794–2821 (2015). https://doi.org/10.1137/140979022
44. S. Maset, An abstract framework in the numerical solution of boundary value problems for neutral functional differential equations. Numer. Math. **133**, 525–555 (2016)
45. W. Michiels, I. Boussada, S.-I. Niculescu, An explicit formula for the splitting of multiple eigenvalues for nonlinear eigenvalue problems and connections with the linearization for the delay eigenvalue problem. SIAM J. Matrix Anal. Appl. **38**, 599–620 (2017)
46. W. Michiels, S.-I. Niculescu, Stability, control, and computation for time-delay systems. An eigenvalue based approach, no. 27, in *Advances in Design and Control*, 2nd edn. (SIAM, Philadelphia, 2014)
47. J.C. Sprott, A simple chaotic delay differential equation. Phys. Lett. A **366**, 397–402 (2007)
48. L.N. Trefethen, *Spectral Methods in MATLAB* Software-Environment-Tools series. (SIAM, Philadelphia, 2000)
49. R. Vermiglio, Polynomial chaos expansions for the stability analysis of uncertain delay differential equations. SIAM/ASA J. Uncertain. Quantif. **5**, 278–303 (2017)
50. R. Vermiglio, A. Zamolo, Sensitivity analysis for stability of uncertain delay differential equations using polynomial chaos expansions. 698–699
51. N. Wiener, The homogeneous chaos. Am. J. Math. **60**, 897–936 (1938)
52. Z. Wu, W. Michiels, Reliably computing all characteristic roots of delay differential equations in a given right half plane using a spectral method. J. Comput. Appl. Math. **236**, 2499–2514 (2012)

Sensitivity Analysis for Stability of Uncertain Delay Differential Equations Using Polynomial Chaos Expansions

Rossana Vermiglio and Andrea Zamolo

Abstract In applications, the analysis of the effect of uncertain parameters on the stability of linear(ized) delay differential equations is a crucial question. Such uncertainties are often modelled as random quantities in a suitable probabilistic framework and, as a result, the spectral abscissa, which determines the stability of the deterministic delay differential equations and depends on the parameters, is a random variable. Within this context, Polynomial Chaos Expansions reveal to be a powerful tool to represent random quantities and to perform uncertain quantification and sensitivity analysis. Here we present a numerical method, called the Stochastic Infinitesimal Generator approach, to efficiently compute the expansion coefficients of the numerical spectral abscissa, showing its application to variance-based global sensitivity analysis through the computation of the Sobol' indices. The technique implements the non intrusive spectral projection methods and makes use also of the so-called Padua points to reduce the computation cost. Some numerical examples are given.

The work of R.V. was partially supported by INdAM GNCS 2017 project "Analisi e sviluppo di metodologie numeriche per certi tipi non classici di sistemi dinamici" and is part of the project SiDiA, "Sistemi Dinamici e Applicazioni," PRID 2017. R.V. is member of the INdAM Research group GNCS.

R. Vermiglio (✉)
CDLab - Computational Dynamics Laboratory, Department of Mathematics, Computer Science and Physics, University of Udine, via delle Scienze 206, 33100 Udine, Italy

Member of the INdAM Research group GNCS, Rome, Italy
e-mail: rossana.vermiglio@uniud.it

A. Zamolo
Department of Mathematics, Computer Science and Physics, University of Udine, via delle Scienze 206, 33100 Udine, Italy

G. Valmorbida et al. (eds.), *Accounting for Constraints in Delay Systems*,
Advances in Delays and Dynamics 12,
https://doi.org/10.1007/978-3-030-89014-8_8

1 Introduction

In the past decades delay differential equations (DDEs) have been proposed to model systems with retarded arguments in several application fields, namely control theory and population dynamics. Nowadays there exists an extensive literature on both theoretical and numerical aspects (see for instance [1, 2, 23, 26, 34, 35, 40, 50] and the references therein). In applications a crucial question concerns the local stability of equilibria, which, according to the principle of linearized stability, can be inferred from the stability analysis of the zero solution of the linearized system. Since DDEs generate infinite-dimensional dynamical systems, efficient numerical methods are needed, and various approaches have been proposed [5, 7, 25, 33, 38]. In [6, 8] the pseudospectral collocation technique has been employed to discretize the infinitesimal generator (IG) of the semigroup of solution operators associated with linear DDEs, whose spectral abscissa, i.e. the real part of the rightmost eigenvalue, determines the (un)stability of the zero solution. Specifically, if the spectral abscissa is negative the zero-solution is asymptotically stable, whereas it is unstable if the spectral abscissa is positive. The so-called IG approach turns the original eigenvalues problem for the operator IG into the computation of the eigenvalues of a matrix. The outstanding feature of the IG approach is the spectral accuracy of the approximations of the eigenvalues. But in applications there is another important aspect to take into account: the model constants and parameters are often uncertain and, as a result, affect the stability properties since the spectral abscissa can change sign. Such uncertain data are often modelled as random quantities in a suitable probabilistic framework. Within this context, in the last few years the Polynomial Chaos Expansions (PCEs) have received a great attention, since they provide an efficient way to represent random variables and, complemented with suitable numerical methods, to analyse and quantify the effect of data uncertainty in the model response, revealing to be an efficient alternative to Monte-Carlo (MC) simulations. The implementation of MC methods is usually straight forward, but it provides random estimates with limited accuracy. The use of PCEs allows one to exploit the possible regularity of the model solution with respect to the random parameters to accelerate the estimation of expected values and statistical moments. The interested readers are referred to the books [36, 58] and the references therein.

In this context, linear DDEs with uncertain parameters (UDDEs) have been recently introduced and analysed in order to investigate the effect of the uncertain parameters on the stability of the zero solution. To this aim a numerical method, which combines the spectral IG-approach and the Stochastic Collocation (SC) method, has been derived to compute the deterministic coefficients of the PCE of the random numerical spectral abscissa [52]. Once the expansion coefficients are known, the uncertainty quantification (UQ) can be performed through the estimation of the statistical moments. The SC methods are non intrusive, since they require deterministic solvers only, but they suffer of the curse of dimensionality, i.e. the number of model resolutions increases exponentially with the number of independent random parameters. Therefore sparse grids, such as Smolyak grids, have been introduced to reduce the complexity [36, 43]. Such grids have been considered in [27] for the UQ of spectral abscissa of UDDEs. Beyond UQ another important issue is the sensitivity analysis (SA), i.e. the quantification of the effect of random data or of a group of them onto the variance of the model solution. In the literature on this topic the Sobol' indices have received much attention [44, 46], and PCE-based methods have been proposed to compute them in efficient way [20, 49].

This paper presents a brief and self-contained overview of the PCEs and UDDEs in order to show the potential of PCE-based methods not only for UQ but also for the SA of the random numerical abscissa associated with linear(ized) UDDEs in the whole domain of the uncertain parameters. We introduce a new variant of the methodology proposed in [52], which couples the IG approach to non intrusive spectral projection (NISP) methods. We call it Stochastic IG (SIG) approach. The implementation of NISP methods requires to use multivariate integration formulas. To limit the computational cost with respect to Gaussian tensored formulas, we propose interpolation formulas based on the so called Padua points, which can be applied when the number of uncertain parameters is small, namely two or three, and uniformly distributed. The Padua points of the square lie on a particular Lissajous curve and are optimal for total degree polynomial interpolation in two variables [10, 11, 15–17]. The Lissajous curve technique has been extend to dimension three in [12] and to higher dimensions in [13], whereas a new cubature formula is developed in [22]. So far the Padua points have not been used for the PCE-based methods for UQ and SA, and their use represents a novelty within this context.

The paper is organized as follows. In Sect. 2 we present the notations and we briefly discuss the background material on PCEs and their application to the computation of Sobol' sensitivity indices. In particular we present the NISP methods and we introduce the Padua points. In Sect. 3 we introduce the UDDEs, summarising the relevant results in view of the definition of the SIG approach in Sect. 4. The experimental analysis of the SIG approach for the SA is given in Sect. 5, where we also compare the numerical results obtained by using Gaussian tensored and Padua integration formulas on some test problems from [3, 18, 28, 41]. We use the Matlab codes developed in [59]. Finally in Sect. 6 we draw the conclusions and future developments.

2 Polynomial Chaos Expansions

In this section we set the probabilistic framework and we briefly introduce PCEs, along with their application to SA via Sobol' sensitivity indices. For easy notation we consider random variables only, but the same representations and approximations can be extended to complex-valued random variables, random vectors and stochastic processes.

Let $(\Omega, \Sigma, \mathscr{P})$ be a complete probability space, where Ω is the event space, Σ is the σ-algebra, and $\mathscr{P}$ is the probability measure. We denote by $L^2(\Omega, \Sigma, \mathscr{P})$ the Hilbert space of square-integrable random variables on it.

The Wiener's theory states that a random variable $Y \in L^2(\Omega, \Sigma, \mathscr{P})$ can be expressed as a convergent series of polynomials in a sequence of Gaussian random variables $\{\xi_i\}_{i=1}^{\infty}$ of the form

$$Y(\omega) = y_0 H_0 + \sum_{\alpha_1=1}^{\infty} y_{\alpha_1} H_1(\xi_{\alpha_1}(\omega)) + \sum_{\alpha_1=1}^{\infty} \sum_{\alpha_2=1}^{\alpha_2} y_{\alpha_1,\alpha_1} H_2(\xi_{\alpha_1}(\omega), \xi_{\alpha_2}(\omega)) + \dots, \tag{1}$$

where H_j's are the multivariate Hermite polynomials of degree j, which are mutually orthogonal with regards to the Gaussian measure associated with $\{\xi_i\}_{i=1}^{\infty}$ [14, 31, 57]. The series (1) is the *PCE* of Y and its deterministic coefficients, simply named *PC coefficients*, capture the probabilistic description of Y. In applications, the expansions are restricted to a finite number M of random variables, which leads to PCEs of finite dimension. The so-called stochastic dimension M is determined by the number of the sources of uncertainty in the model. Moreover it is also necessary to truncate the PCE to order P, so that it results also finite. Specifically all the infinite sums in (1) are replaced by sums over M dimensions with polynomials of degree less or equal to P. The resulting truncated expansion of order P and dimension M converges in the mean square sense as both M and P go to infinity [14]. As already pointed out above, Hermite polynomials were first considered. Nowadays PCEs have been generalized to different families of orthogonal polynomials, as well as to piecewise polynomial functions and wavelets, and to situations involving dependent random variables. For convergence results of generalized PCEs we refer to [24] and to the books [31, 36, 58] for further details and references.

Let $\xi = (\xi_1, \dots, \xi_M)$ be a vector of M independent standard random variables defined on $(\Omega, \Sigma, \mathscr{P})$ with joint probabilistic density function p_ξ, i.e.

$$p_\xi(\xi) = \prod_{i=1}^{M} p_i(\xi_i), \tag{2}$$

where $p_i(\xi_i)$ is the probability density function of ξ_i. We denote by Ξ the support of ξ and by $L_2 := L^2(\Xi, \mathscr{B}_\Xi, p_\xi)$ the Hilbert space of square-integrable random variables defined on the probability space $(\Xi, \mathscr{B}_\Xi, p_\xi)$, where $\mathscr{B}_\Xi$ is the Borel set of Ξ. The inner product on L_2 is given by

$$\langle Y, Z \rangle := \int_\Xi Y(\xi) Z(\xi) p_\xi(\xi) d\xi,$$

and $\|Y\| := \sqrt{\langle Y, Y \rangle}$ is the induced norm.

The random variable $Y \in L^2(\Omega, \Sigma, \mathscr{P})$ can be approximated by the truncated expansion Y_P given by

$$Y_P(\omega) = \sum_{\alpha \in \mathscr{A}_P} y_\alpha \Psi_\alpha(\xi(\omega)), \tag{3}$$

where $\{\Psi_\alpha(\xi)\}$ are the multivariate orthogonal polynomials induced by the density function p_ξ of the random vector ξ, $\alpha = (\alpha_1, ..., \alpha_M) \in \mathbb{N}_0^M$ is a M-dimensional multi-index, $|\cdot|$ is a suitable norm on $\mathbb{N}_0^M$, and $\mathscr{A}_P = \{\alpha \in \mathbb{N}_0^M : |\alpha| \leq P\}$. Let n_P be the cardinality of $\mathscr{A}_P$. By using the $|\cdot|_1$ we have that $n_P = \binom{M+P}{P}$, whereas $n_p = (P+1)^M$ for the $|\cdot|_\infty$. In both cases n_p rapidly increases with respect to the stochastic dimension M and the expansion degree P. Other truncation sets $\mathscr{A}_P$ have been proposed in literature [36, 37, 58].

The polynomial basis $\{\Psi_\alpha\}$ is built via the tensor product of one-dimensional polynomials, i.e.

$$\Psi_\alpha(\xi) = \Psi_{\alpha_1}(\xi_1) \cdots \Psi_{\alpha_M}(\xi_M), \; \alpha \in \mathbb{N}_0^M, \tag{4}$$

where Ψ_{α_i} is the orthogonal polynomial w.r.t. $p_i(\xi_i)$ of degree α_i. For example we choose Legendre polynomials for uniformly distributed random variables and Laguerre polynomials for gamma distribution. Note that the tensored construction of the PC basis (4) also allows to consider random variables ξ_i associated with different probability laws $p_i(\xi_i)$.

Now we can operate in the stochastic space L_2 and approximate, for instance, the expectation and the variance of Y as follows

$$\begin{aligned} E[Y] = \int_\Omega Y(\omega) d\mathscr{P}(\omega) \quad &\approx \sum_{\alpha \in \mathscr{A}_P} y_\alpha \int_\Xi \Psi_\alpha(\xi) p_\xi(\xi) d\xi = y_0, \\ Var[Y] = E[(Y - E[Y])^2] \approx &\sum_{\alpha \in \mathscr{A}_P} |y_\alpha|^2 \int_\Xi \Psi_\alpha^2(\xi) p_\xi(\xi) d\xi. \end{aligned} \tag{5}$$

PCEs coupled with suitable numerical methods to compute the chaos coefficients provide a convenient framework to quantify the propagation of data uncertainties in the model solution. Once the uncertain input data are considered as a random vector X with components in $L^2(\Omega, \Sigma, \mathscr{P})$ and represented by its PCE, one can tackle the propagation of uncertainty induced by X into the random solution $Y = g(\xi)$ through mathematical model $\mathscr{M}$, i.e. $\mathscr{M}(Y, X) = 0$, by determining the PC coefficients $\{y_\alpha\}$ of Y by suitable numerical methods.

2.1 Computation of the Chaos Coefficients

Several methods have been proposed to compute the chaos coefficients (see e.g. [36, 58]). Here we use the NISP methods, which, thank to the expression

$$y_\alpha = \frac{\langle Y, \Psi_\alpha \rangle}{\langle \Psi_\alpha, \Psi_\alpha \rangle}, \; \alpha \in \mathscr{A}_P, \tag{6}$$

reduce their computation to a numerical integration problem. Given a M-dimensional numerical integration formula with quadrature nodes ξ_β and weights w_β, $\beta \in \mathscr{Q}$, we get

$$y_\alpha \approx \frac{1}{\langle \Psi_\alpha, \Psi_\alpha \rangle} \sum_{\beta \in \mathscr{Q}} w_\beta Y(\xi_\beta) \Psi_\alpha(\xi_\beta), \; \alpha \in \mathscr{A}_P, \tag{7}$$

Note that $\langle \Psi_\alpha, \Psi_\alpha \rangle$ depend only on the basis and they are known analytically for classical orthogonal polynomials. Therefore we need only to evaluate the numerators in the left-hand side of (7), and the computational cost is given by the total number Q of model evaluations, where Q denotes the cardinality of the set of multi-indices $\mathscr{Q}$. M-dimensional numerical integration formulas of tensored type, simply called tensored formulas, can be constructed by using 1-dimensional integration formula with $q + 1$ nodes in each directions. For instance, if the random parameters $\xi_i, i = 1, \ldots M$, are identically distributed, we can consider along all the directions the same integration formula with nodes ξ_j and weights w_j, $j = 0, \ldots, q$, and define $\xi_\beta = (\xi_{\beta_1}, \cdots, \xi_{\beta_M})$, $w_\beta = w_{\beta_1} \times \cdots \times w_{\beta_M}$, for $\beta \in \mathscr{Q} = \{\beta \in \mathbb{N}_0^M : |\beta|_\infty \leq q\}$. Note that one can choose different numbers of nodes in each direction. Moreover the tensored construction can easily extend to the case of random parameters ξ_i with different distributions p_i, simply by using different integrations rules along the different integration directions. The 1-dimensional Gaussian integration formulas, which exactly integrate polynomials of degree $\leq 2q + 1$, are closely related to orthogonal polynomials, and so Gaussian tensored formulas for NISP methods is a natural choice. But they suffer of curse of dimensionality since $Q = (1 + q)^M$ and so they are used for moderate M only. To overcome this problem, sparse grid formulas, such as Smolyak's ones, have been derived [43]. Here we propose M-dimensional integration formulas of interpolation type based on Padua points, which are well suited for uniform distributions and $M = 2, 3$, even if the recent paper [13] suggests the possibility to consider higher dimensions. For $M = 2$ the Padua points are the side contacts and self-intersections of the Lissajous curve. For interpolation of degree m, the number of Padua points is $Q = \frac{(m+1)(m+2)}{2}$. Moreover there are two points on consecutive vertices and $2n - 1$ on the edges of the square, whereas the remaining ones are inside. In Fig. 1 we plot the Padua points for $m = 10$ (left) and $m = 11$ (right). The construction has been also extended to dimension $M = 3$, and in Fig. 2 we plot $Q = \frac{(m+4)(m+5)}{2}$ Padua points for $m = 4$ (left) and $m = 9$ (right). The interested reader is referred again to [10–12, 15–17] for more details and convergence results. The paper [13] gives the possibility to exploit the NISP approach with Padua points for higher dimensions M in the future.

2.2 Sobol Indices and PCEs

Recently the so called PCE-based methods have been also proposed for the global SA, which aims to completely understand the respective impacts of the different uncertainty sources in Y in the whole domain of the uncertain parameters. In this context Sobol' decompositions and Sobol' indices provide a way to characterise and measure the induced variabilities in the model response [44–46] and the orthogonality

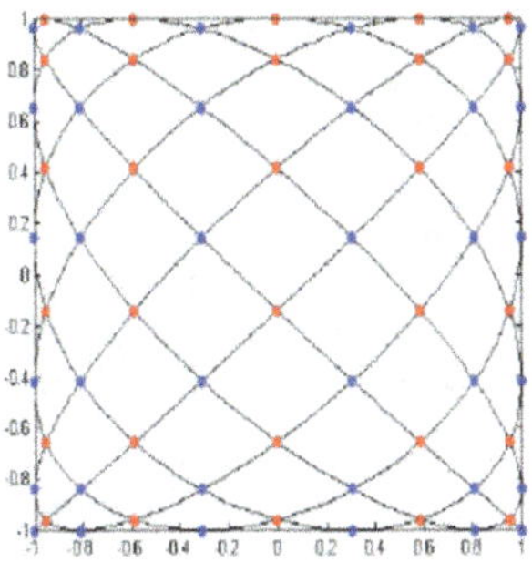

Fig. 1 Padua points: $Q = 66$ (left) and $Q = 78$ (right) in the square $[-1, 1]^2$

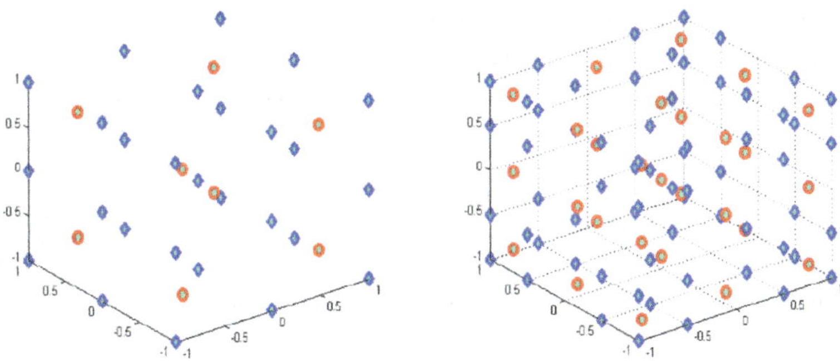

Fig. 2 Padua points: $Q = 36$ (left) and $Q = 91$ (right) in the cube $[-1, 1]^3$

and the tensored construction of the PC basis (4) allows one to compute the sensitivity indices in an efficient way [20, 36, 49]. In what follows, we recall the definition of Sobol' indices and we briefly describe the PCE-based method to compute them.

Any square integrable function $g(\xi)$, $\xi \in [0, 1]^M$, has a unique orthogonal decomposition, known as Sobol' decomposition, of the form

$$g(\xi) = \sum_{\sigma \subseteq \{1,\ldots,M\}} g_\sigma(\xi_\sigma), \tag{8}$$

where $\sigma = \{i_1, \ldots, i_m\}$ and $\xi_\sigma = \{\xi_{i_1}, \ldots, \xi_{i_m}\}$ and $g_\emptyset = E[Y]$ [44]. The integral of each g_σ over any of its independent variables is zero, i.e.:

$$\int_0^1 g_\sigma(\xi_\sigma) d\xi_j = 0, \;\; j \in \sigma.$$

From (4), we get that the decomposition of the chaos polynomial Ψ_α involves a single term, namely

$$\Psi_\alpha(\xi) = \Psi_{\sigma_\alpha}(\xi_{\sigma_\alpha}), \;\; \sigma_\alpha = \{i \;s.t.\; \alpha_i \neq 0\}.$$

If $\xi \sim \mathscr{U}[0,1]^M$ the decomposition of the random variable $Y = g(\xi)$ can be obtained from its Legendre PCE. For instance we have that

$$g_\sigma(\xi_\sigma) = \sum_{\alpha \in \mathscr{A}_\sigma} y_\alpha \Psi_\alpha(\xi)$$

where $\mathscr{A}_\sigma := \{\alpha \in \mathbb{N}_0^M : \alpha_i > 0, i \in \sigma, \alpha_i = 0, i \notin \sigma\}$. Due to orthogonality nature of (8), the total variance of Y can be decomposed as

$$Var[Y] = \sum_{\sigma \subseteq \{1,\dots,M\}, \sigma \neq \emptyset} Var_\sigma[Y]$$

where $Var_\sigma[Y] = \langle g_\sigma, g_\sigma \rangle$ is the contribution to $Var[Y]$ of the interaction of the set of random parameters $\{\xi_i \ s.t. \ i \in \sigma\}$. To compare the impact of the different sources of uncertainty we use the Sobol' sensitivity indices, which are given by

$$\mathscr{S}_\sigma = \frac{Var_\sigma[Y]}{Var[Y]}, \quad \sigma \subseteq \{1, \dots, M\}$$

Each $\mathscr{S}_\sigma$ measures which amount of the total variance $Var[Y]$ is due to the uncertainties in ξ_σ, and it can be approximated by using (3) and

$$g_{\sigma,P}(\xi_\sigma) = \sum_{\alpha \in \mathscr{A}_\sigma \cap \mathscr{A}_P} y_\alpha \Psi_\alpha(\xi) \tag{9}$$

[49]. Generalized Sobol' decompositions can be derived for more general independent variables [51] as well as for dependent ones [19].

3 Uncertain Delay Differential Equations

In what follows, motivated by the PC theory, we define DDEs with uncertain parameters and we summarise the fundamental results on stability. A detailed exposition of the subject can be found in [52].

We denote the independent variable *time* by $t \in \mathbb{R}$ and the *dimension* of the system by $d \geq 1$. The space $\mathbb{R}^d$ is equipped with the Euclidean norm $|y|$, $y = (y_1, \dots, y_d)^T \in \mathbb{R}^d$.

Bearing in mind the probabilistic setting of Sect. 2, we consider a vector $\xi = (\xi_1, \dots, \xi_M)$ of M independent random variables on Ω with joint probabilistic density function (2) with support Ξ, and we introduce the Hilbert space $\mathscr{Y}$ of $\mathbb{R}^d$-valued functions on Ξ with all components in L_2 equipped with the inner product

$$\langle Y, Z\rangle_{\mathcal{Y}} := \int_{\Xi} Y(\xi)^T Z(\xi) p_\xi(\xi) d\xi,$$

and the induced norm $\|Y\|_{\mathcal{Y}} := \sqrt{\langle Y, Y\rangle_{\mathcal{Y}}}$.

Definition 1 We define *uncertain DDE* (UDDE) of stochastic dimension M the following parameter dependent DDE

$$\frac{d}{dt} Y(t,\xi) = F(Y(t,\xi), Y_t(\cdot,\xi), \xi)\; t \geq 0,\; a.e.\, \xi \in \Xi, \tag{10}$$

where

- $\tau > 0$ is the maximum delay,
- $Y_t(\theta,\xi) := Y(t+\theta,\xi), \theta \in [-\tau, 0]$, is the *history segment* living on the *history space* $\mathcal{H} := L^2([-\tau, 0] \times \Xi; \mathbb{R}^d) \sim L^2([-\tau, 0]; \mathcal{Y})$ equipped with the inner product

$$\langle \Phi, \Psi\rangle_{\mathcal{H}} = \int_{-\tau}^{0} \langle \Phi(\theta,\cdot), \Psi(\theta,\cdot)\rangle_{\mathcal{Y}}\, d\theta,\; \Phi, \Psi \in \mathcal{H},$$

- $F : \mathcal{Y} \times \mathcal{H} \times \Xi \to \mathcal{Y}$ is sufficiently smooth function.

Notice that also the delays can be a random quantities, which are assumed to be essentially bounded by τ in Definition 1 (see [52] for more details and some numerical examples along this line).

The UDDE (10) can be reformulated as an abstract DDE in $\mathcal{Y}$. As a consequence, for a given the initial condition

$$Y_0(\cdot,\xi) = \Phi,\; \Phi \in \mathcal{Y}, \tag{11}$$

we can define classical and mild solutions, and prove the well-posedness of the initial problem (10)–(11) by means of classic results on the theory of DDEs with initial data in L^p-spaces [1, 21]. Moreover the local stability properties of a steady state solution $\overline{Y} \in \mathcal{Y}$ of (10) can be linked to the mean-square stability of the zero solution of the linearized system at $\overline{Y}$, i.e.

$$\frac{d}{dt} Y(t,\xi) = B(\xi) Y(t,\xi) + L(\xi) Y_t(\cdot,\xi), \quad t \geq 0, \tag{12}$$

where $B(\xi) := DF_1(\overline{Y}(\xi), \overline{Y}(\xi), \xi)$, $L(\xi) := DF_2(\overline{Y}(\xi), \overline{Y}(\xi), \xi)$, and $DF_1(\mathrm{y}, \varphi, \xi)$, $DF_2(\mathrm{y}, \varphi, \xi)$ are the derivative of F at $(\mathrm{y}, \varphi, \xi)$ with respect to y and φ respectively. Finally the norm stability of the zero-solution of linear UDDE (12) is described by the *stability indicator*

$$s := ess\text{-}sup_{\xi\in\Xi}\, Re(\sigma(A(\xi)) = ess\text{-}sup_{\xi\in\Xi} s(A(\xi)), \tag{13}$$

where $s(A(\xi))$ is the spectral abscissa of the operator $A(\xi) : D(A(\xi)) = \{(y, \varphi) \in \mathbb{R}^d \subset \mathscr{Y} \to \mathscr{Y}$ associated with a realization of (12), namely

$$A(\xi)(y, \varphi) = (B(\xi)y + L(\xi)\varphi, \tfrac{d}{d\theta}\varphi),$$
$$D(A(\xi)) = \{(y, \varphi) \in \mathbb{R}^d \times W^{1,2}([-\tau, 0]; \mathbb{R}^d) \ : \ \varphi(0) = y\} \tag{14}$$

where

$$W^{1,2}([-\tau, 0]; \mathbb{R}^d) := \{\varphi \in L^2([-\tau, 0]; \mathbb{R}^d) : \begin{array}{l} \textit{there exits } \psi \in L^2([-\tau, 0]; \mathbb{R}^d) \\ \textit{s.t. } \varphi(\theta) = \varphi(0) + \int_0^\theta \psi(s)ds \end{array} \}$$

is the (vector-valued) Sobolev space and $\frac{d}{d\theta}\varphi := \psi$ is the weak derivative [1, p. 236].

In particular we have that the zero-solution is mean-square asymptotically stable if $s < 0$. Note that $S(\xi) := s(A(\xi))$, for the random vector ξ, defines the *random spectral abscissa* and that the stability indicator (13) srepresents its L_∞ norm. Moreover $S \in L_2$ with $\|S\| \le |s| \left(\int_\Xi p_\xi(\xi)d\xi \right)^{1/2}$ and therefore it admits either the PCE (1) or the Sobol's decomposition (8).

4 The Stochastic IG Approach

In order to estimate the stability indicator (13) associated with the linear UDDE (12) and analyse the effect induced on it by the uncertain parameters, we apply the Stochastic IG (SIG) method, which combines the IG approach for deterministic linear DDEs with PCE-methods [52]. First, by applying the IG approach, we construct the random matrix $A_N(\xi)$ obtained by the spectral discretization of the infinite-dimensional operator (14), where N is the discretization index. Due to the spectral accuracy in the approximation of the eigenvalues, N is of the order of few tenths [6, 8, 56]. Second, since the random *numerical spectral abscissa* $S_N(\xi) := s(A_N(\xi)) \in L_2$, we compute the PC coefficients of the truncated PCE at order P of S_N, i.e.

$$S_{N,P} = \sum_{\alpha \in \mathscr{A}_P} s_\alpha \Psi_\alpha(\xi), \tag{15}$$

with $\mathscr{A}_P = \{\alpha \in \mathbb{N}_0^M : |\alpha|_1 \le P\}$.

Once the coefficients are known, we can determine the statistical moments for UQ and the Sobol' indices for SA, as shown in Sect. 2. In NISP methods the formulas (7) need the Q deterministic values $S_N(\xi_\beta) = s(A_N(\xi_\beta))$, $\beta \in \mathscr{Q}$, and each of them requires to solve a matrix eigenvalue problem of dimension $d(N + 1)$. Therefore the total number Q of the integration points estimates the main contribution of the computation cost of the SIG-approach. We remark that, with no additional evaluations of S_N, one can easily obtain also the approximation $\tilde{s}$ of the stability indicator (13) as follows

$$\tilde{s} = max_{\beta \in \mathscr{Q}} S_N(\xi_\beta). \tag{16}$$

Hereafter we refer to $\tilde{s}$ as the *numerical stability indicator.*

The PC coefficients can be also obtained through the resolution of a Least Squares (LS) minimization problem

$$argmin \frac{1}{R} \sum_{\rho=1}^{R} (\sum_{\alpha \in \mathscr{A}_P} s_\alpha \Psi_\alpha(\xi_\rho) - S_N(\xi_\rho))^2,$$

where $\{\xi_\rho\}_{\rho=1}^R$ are a selected set of *minimization* points. The interested reader can find in the paper [49] a detailed presentation of the approach and of construction of the R minimization points. In Sect. 5 we present the basic ideas and we experimentally compare LS method and NISP method using a test equation.

We remark that also the Stochastic Collocation (SC) method, in which the set of deterministic values is used together with interpolation schemes, are non intrusive approaches. They have been considered to approximate S_N by interpolating on grids of tensored type and sparse Smolyak's grids in [27, 52] respectively. It is worthwhile to recall that NISP method with Gaussian tensored formulas with $q = P$ for (15) with $\mathscr{A}_P = \{\alpha \in \mathbb{N}_0^M : |\alpha|_\infty \leq P\}$ reduces to the SC-method at the Gaussian nodes.

Finally we underline that the regularity of S_N affects the rate of convergence of both the truncation error and integration error, and as a consequence it plays an important role on the choice of the parameter P as well as the quadrature formulas. We refer to [9, 42, 53–55] sand the reference therein for theoretical insights. In the forthcoming section we experimentally perform this analysis on some numerical examples.

Remark 1 The SIG describes a methodology to investigate the effect of uncertain parameters on stability of linear UDDEs, based in the IG-approach for DDEs. Since also the spectral abscissa $S \in L_2$, in order to possibly reduce the total computation cost, we can also approximate the expansion coefficients by using different values of N to estimate $S(\xi_\beta)$ for $\xi_\beta, \beta \in \mathscr{Q}$. Indeed, due to the spectral accuracy, N of the order of tenths ensures an accuracy of machine precision. So here we choose to present the approach with N fixed.

5 Numerical Results

This section is specifically devoted to illustrate the SIG approach outlined in Sect. 4 through its application to some examples of UDDEs. The aim is not only to show the potential of PCE-based methods for UQ and SA of the random numerical spectral abscissa S_N, but also to experimentally compare NISP methods with Gaussian tensored formulas and Padua integration formulas. All the uncertainties considered are uniformly distributed, therefore we use Legendre polynomials. Moreover we always fix $N = 20$.

In the first two examples we consider elementary linear UDDEs, paying particular attention to how the regularity of S_N affects the choice of the parameter P and the accuracy of the integration formulas (7). Here we always choose $q = P$ for Gaussian tensored formulas. We refer to [29] for some insights on the role of the regularity in PCE-based method for random numerical abscissa. As first step, in order to determine a suitable P, we experimentally analyse the convergence rate by comparing the expectation and the variance w.r.t. some reference values. Note that for that four digits of accuracy it is sufficient to estimate the Sobol' indices. In the last example we consider a nonlinear DDE and we study the effect of uncertain parameters on the stability of an equilibrium by analysing the numerical spectral abscissa of the resulting linearized UDDE. In this case we also include the numerical results obtained by applying the LS-methods and MC-methods to underline that, contrary to MC-method, PCE-methods usually take advantage from the possible smoothness of the model output with respect to the random inputs. In particular for the choice of the minimization points in the LS-method, we follow the same strategy in [49]. Let C be the matrix with coefficients $C_{\rho,\alpha} = \Psi_\alpha(\xi_\rho)$, $\rho = 1, ..., R$, $\alpha \in \mathscr{A}_P$. The basic idea consists in selecting among the M-uplets of the full-tensored Gaussian integration points, the points which have greater weight. After ordering them according to their increasing norm, the R minimization points are obtained iteratively added until the *information matrix* $C^T M$ becomes invertible. All the codes are developed in Matlab [59].

5.1 Oscillator with Feedback Delay

In [28] the authors consider an oscillator with delay controlled by static feedback in a probabilistic framework, which can described by the following UDDE of stochastic dimension $M = 2$

$$\begin{cases} Y_1'(t,\xi) = Y_2(t,\xi) \\ Y_2'(t,\xi) = -\nu(\xi_1)^2 Y_1(t,\xi_1) - 2\nu\varepsilon(\xi_2)Y_2(t,\xi) + U(t-1,\xi) \\ U(t,\xi) = K_1 Y_1(t,\xi) + K_2 Y_2(t,\xi), \end{cases} \tag{17}$$

where the angular frequency ν and the damping ratio ε are uniform independent random variables given by $\nu(\xi_1) = 0.1\xi_1 + 1 \sim \mathscr{U}[0.9, 1.1]$, $\varepsilon(\xi_2) = 0.05\xi_2 + 0.15 \sim \mathscr{U}[0.1, 0.2]$, with $\xi = (\xi_1, \xi_2)$, $\xi_i \sim \mathscr{U}[-1, 1]$, $i = 1, 2$. For each realisations ξ, $Y_1(t,\xi)$ is the normalized position, $Y_2(t,\xi)$ is the velocity, $U(\cdot,\xi)$ is a "control" force that acts with a delay $\tau = 1$ and K_1, K_2 are control variables. From now on, we fix $K_1 = 0.58$, $K_2 = 0.02$. The Fig. 3 shows the spectral abscissa $s(A_{20})$ as a function of $\nu \in [0.9, 1.1]$ for each realizations $\varepsilon^i = 0.1 + 0.02 \cdot i$, $i = 0, \ldots, 5$. Since the spectral abscissa is negative, we can conclude that the zero-solution is asymp-

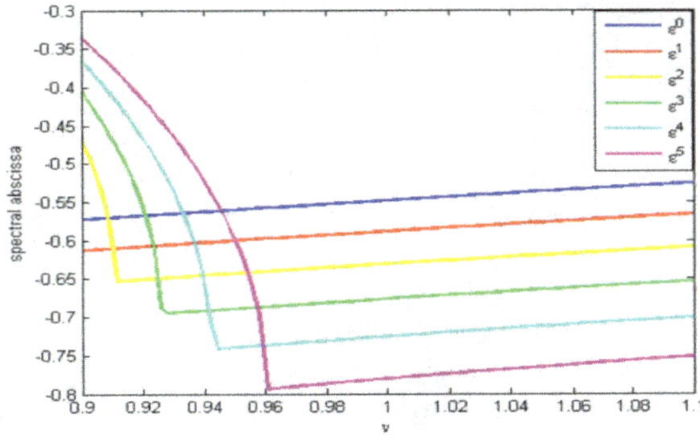

Fig. 3 Numerical spectral abscissa $s(A_{20})$ of (17) as a function of ν for $\varepsilon^i = 0.1 + 0.02 \cdot i$, $i = 0, \dots, 5$

totically stable for $\nu \in [0.9, 1.1]$ and ε^i, $i = 0, \dots, 5$. Note that these functions are continuous but non-differentiable or non-Lipschitz continuous. As a consequence we expect a slow convergence w.r.t. P and the numerical results highlight this fact by using either tensored Gaussian formulas or Padua integration formulas.

In Tables 1 and 2 we show the expectation $E_P := E[S_{20,P}]$, the variance $V_P := Var[S_{20,P}]$ as well as their relative errors rounded to 5 digits. The reference values obtained by tensored Gaussian formulas with $P = 40$, are $E_{40} = -6.3084800048655874e - 01$, $V_{40} = 6.677006595280072e - 03$ rounded to 16 digits. In Table 1 the results are obtained by applying the SIG approach with Gaussian tensored formulas with $Q = (P + 1)^2$ points, whereas in Table 2 the Padua integration formulas with $Q = (P + 4)(P + 5)/2$ points have been used. Note that by using the available $Q = 41^2$ evaluations of the numerical spectral abscissa at the tensored Gaussian points, we obtain that the numerical stability indicator (16) rounded to 5 digits is $\tilde{s} = -3.3611e - 01$, confirming the robustness of the system, under the effect of uncertainty on the parameters.

In Tables 3 and 4 we show the Sobol' indices related to $S_{20,P}$ respectively for Gaussian and Padua integration points. The results experimentally confirm a slow convergence in both cases, but we reach the accuracy needed to compare the respective impact of the different sources of uncertainty also for small P, namely $P = 10$. Note that the set $\{\nu, \varepsilon\}$ mainly influences the stability of the system (17). In fact $\mathscr{S}_{\nu,\varepsilon}$ represents about 45% of the total variability, whereas $\mathscr{S}_\nu$ and $\mathscr{S}_\varepsilon$ contribute with about 30% and 24% respectively. It is worthwhile to recall that Padua integration

Table 1 PCE-based expectation and variance of $S_{20,P}$ for the UDDE (17) varying P computed by using $Q = (P + 1)^2$ Gaussian integration points

P	E_P	$\lvert E_P - E_{40}\rvert/\lvert E_{40}\rvert$	V_P	$\lvert V_P - V_{20}\rvert/\lvert V_{40}\rvert$
10	$-6.3097e - 01$	$2.01e - 04$	$6.7029e - 03$	$1.19e - 02$
20	$-6.3087e - 01$	$3.39e - 05$	$6.6519e - 03$	$3.76e - 03$
30	$-6.3084e - 01$	$1.59e - 05$	$6.6677e - 03$	$1.39e - 03$

Table 2 PCE-based expectation and variance of $S_{20,P}$ for the UDDE (17) varying P computed by using $Q = (P + 4)(P + 5)/2$ Padua integration points

P	E_P	$\|E_P - E_{40}\|/\|E_{40}\|$	V_P	$\|V_P - V_{20}\|/\|V_{40}\|$
10	$-6.3173e-01$	$1.40e-03$	$6.6231e-03$	$8.06e-02$
20	$-6.3074e-01$	$1.69e-04$	$6.6668e-03$	$1.53e-03$
30	$-6.3083e-01$	$2.34e-05$	$6.6703e-03$	$1.01e-03$

Table 3 PCE-based Sobol' indices for the UDDE (17) varying P computed by using $Q = (P + 1)^2$ Gaussian integration points

Sobol' index	$P = 10$	$P = 20$	$P = 30$	$P = 40$
$\mathscr{S}_\nu$	$3.0954e-01$	$3.0351e-01$	$3.0347e-01$	$3.0358e-01$
$\mathscr{S}_\varepsilon$	$2.4775e-01$	$2.4245e-01$	$2.4176e-01$	$2.4172e-01$
$\mathscr{S}_{\nu,\varepsilon}$	$4.4272e-01$	$4.5404e-01$	$4.5477e-01$	$4.5470e-01$
Q	121	441	961	1681

Table 4 PCE-based Sobol' indices for the UDDE (17) varying P computed by using $Q = (P + 4)(P + 5)/2$ Padua integration points

Sobol' index	$P = 10$	$P = 20$	$P = 30$	$P = 40$
$\mathscr{S}_\nu$	$2.8505e-01$	$3.0869e-01$	$3.0342e-01$	$3.0314e-01$
$\mathscr{S}_\varepsilon$	$2.4762e-01$	$2.4309e-01$	$2.4180e-01$	$2.4144e-01$
$\mathscr{S}_{\nu,\varepsilon}$	$4.6733e-01$	$4.4822e-01$	$4.5478e-01$	$4.5542e-01$
Q	105	300	595	990

formulas allow us to compute either the expectation and the variance for the UQ or the Sobol' indices for SA of the stability of system (17) with the needed accuracy to perform the analysis, but lower computational cost w.r.t. Gaussian integration formulas.

5.2 *Thermal Circulation System with Control*

For viscoelasticity measures of materials it is important to maintain a homogeneous temperature at the test point. To this aim a thermal circulation system consisting of a ring pipe and a heater-cooler device has been considered in [41]. The dynamics of the temperature $y(t)$ at time $t \geq 0$, measured at the test point, is described by the following DDE

$$\begin{cases} y_1'(t) &= y_2(t), \\ \rho y_2'(t) &= -y_2(t) + \gamma y_2(t-\tau_1-\tau_2) + \sigma K_1 y_1(t-\tau_1) + \sigma K_2 y_2(t-\tau_1), t \geq 0, \end{cases} \tag{18}$$

where $\tau_1 = 2.4$, $\tau_2 = 2.3$ are two delays, ρ is a coefficient of heat insulation, $K_1 = 1$ and $K_2 = 2$ are the control parameters. To investigate the effect on stability of (18) of the uncertain parameters ρ, σ and γ, we assume that $\xi = (\xi_1, \xi_2, \xi_3)$, with $\xi_i \sim \mathscr{U}[-1, 1], i = 1, 2, 3$, and

$$\begin{aligned} \rho(\xi) = 25\xi_1 + 75 \sim \mathscr{U}[50, 100], \ \sigma(\xi) = 0.2\xi_2 + 1 \sim \mathscr{U}[0.8, 1.2], \\ \gamma(\xi) = 0.4\xi_3 + 0.8 \sim \mathscr{U}[0.4, 1.2], \end{aligned} \tag{19}$$

and we apply the SIG method in Sect. 4 to the resulting UDDE of stochastic dimension $M = 3$

$$\begin{cases} Y_1'(t,\xi) &= Y_2(t,\xi), \\ \rho(\xi) Y_2'(t,\xi) &= -Y_2(t,\xi) + \gamma(\xi) Y_2(t-\tau_1-\tau_2,\xi) + \sigma(\xi) K_1 Y_1(t-\tau_1,\xi) \\ &\quad + \sigma(\xi) K_2 Y_2(t-\tau_1,\xi) \end{cases} \tag{20}$$

In Tables 5 and 6 we show the values $E_P := E[S_{20,p}]$, $V_P := Var[S_{20,P}]$ and their relative errors computed w.r.t the reference values $E_{12} = 1.095776644342109e-01$, $V_{12} = 1.394035291507988e-04$ all rounded to 16 digits. The smoothness of the function $S_{20}(\xi)$ ensures very accurate results for small P. Note that the system is unstable (Tables 5 and 6).

Table 5 PCE-based expectation and variance of $S_{20,P}$ for the UDDE (20) varying P computed by using $Q = (P+1)^3$ Gaussian integration points

P	E_P	$\lvert E_P - E_{12}\rvert/\lvert E_{12}\rvert$	V_P	$\lvert V_P - V_{12}\rvert/\lvert V_{12}\rvert$
3	$1.095776498661167e-01$	$1.33e-07$	$1.393987049873297e-04$	$3.46e-05$
5	$1.095776644252439e-01$	$8.18e-11$	$1.394035252790606e-04$	$2.78e-08$
7	$1.095776644342055e-01$	$4.93e-14$	$1.394035291477865e-04$	$2.16e-11$
9	$1.095776644342110e-01$	$3.80e-16$	$1.394035291508078e-04$	$6.47e-14$

Table 6 PCE-based expectation and variance of $S_{20,P}$ for the UDDE (20) varying P computed by using $Q = (P+1)^3/4$ Padua integration points

P	E_P	$\lvert E_P - E_{12}\rvert/\lvert E_{12}\rvert$	V_P	$\lvert V_P - V_{12}\rvert/\lvert V_{12}\rvert$
3	$1.095776685346089e-01$	$3.74e-08$	$1.394319334773142e-04$	$2.04e-04$
5	$1.095776645648649e-01$	$1.19e-09$	$1.394046985009620e-04$	$8.39e-06$
7	$1.095776644365940e-01$	$2.17e-11$	$1.394037351198232e-04$	$1.48e-06$
9	$1.095776644342536e-01$	$3.89e-13$	$1.394036002600330e-04$	$5.10e-07$

Table 7 PCE-based Sobol' indices of $S_{20,12}$ for the UDDE (20)

Sobol' index	Gaussian formulas	Padua formulas
$\mathscr{S}_\rho$	$7.1654e-01$	$7.1653e-01$
$\mathscr{S}_\sigma$	$2.7581e-01$	$2.7581e-01$
$\mathscr{S}_\gamma$	$5.3898e-03$	$5.3901e-03$
$\mathscr{S}_{\rho,\sigma}$	$2.1489e-03$	$2.1494e-03$
$\mathscr{S}_{\rho,\gamma}$	$1.0597e-04$	$1.0655e-04$
$\mathscr{S}_{\sigma,\gamma}$	$5.2962e-06$	$6.0781e-06$
$\mathscr{S}_{\rho,\sigma,\gamma}$	$2.8152e-07$	$2.8864e-07$
Q	2197	550

5.3 A Model for Bacteriophage Infection

In what follows, we briefly present the delay model of bacteriophage infection formulated and studied in [3] and extended to a stochastic version in [18]. By denoting the density of susceptible bacteria, the density of viruses (phages) and the density of infected bacteria at time $t \geq 0$ respectively by $S(t)$, $V(t)$ and $I(t)$, the following bacteria-phages system with delay has been derived

$$\begin{cases} \frac{dS(t)}{dt} = \alpha S(t)\left(1 - \frac{S(t)+I(t)}{C}\right) - KS(t)V(t) \\ \frac{dV(t)}{dt} = \beta - \mu_p V(t) - KS(t)V(t) + be^{-\mu_i \tau} KS(t-\tilde{\tau})V(t-\tilde{\tau}) \\ \frac{dI(t)}{dt} = -\mu_i I(t) + KS(t)V(t) - e^{-\mu_i \tau} KS(t-\tau)V(t-\tilde{\tau}), \end{cases} \tag{21}$$

where $\alpha > 0$ is the bacteria intrinsic birth rate, $C > 0$ is the bacteria carrying capacity, $\mu_p > 0$ is the rate of spontaneous inactivation of phages, $\mu_i > 0$ is the mortality rate of infected bacteria, $\tilde{\tau} > 0$ is the incubation time for the phages inside the bacteria, $K > 0$ is the contact rate per bacteria with viruses, $\beta \geq 0$ accounts for a constant inflow of phages from the environment and $b \in (1, 100)$ is the virus replication factor, i.e. the number of phages copies releases from an infected bacterium dying by lysis after the period $\tilde{\tau}$. The last differential equation of (21) can be solved w.r.t. I, obtaining the relation

$$I(t) = K \int_{t-\tilde{\tau}}^{t} e^{-\mu_i(t-\theta)} S(\theta)V(\theta)\, d\theta, \quad t \geq 0. \tag{22}$$

Therefore, by assuming that an infected bacterium no longer competes with the susceptible for resources, easy computations allow to write (21) in the following dimensionless form

$$\begin{cases} \frac{ds(t)}{dt} = as(t)(1-s(t)) - s(t)v(t) \\ \frac{dv(t)}{dt} = m_p\frac{C_p}{C} - m_p v(t) - s(t)v(t) + be^{-m_i\tau}s(t-\tau)v(t-\tau), \end{cases} \tag{23}$$

where

$$i(t) = \int_{t-\tau}^{t} e^{-m_i(t-\theta)}s(\theta)v(\theta)\,d\theta,$$

$a = \alpha/KC$, $m_p = \mu_p/KC$, $m_i = \mu_i/KCm$, $\tau = \tilde{\tau}KC$ is the dimensionless incubation time and $C_p = \beta/\mu_p$, with $C_p > C$ is the carrying capacity of the phages. The system (23) admits a trivial equilibrium $E_0 = (0, C_p/C)$ for all $\tau > 0$. However, if $a > C_p/C$ a non trivial positive equilibrium $E_+ = (s^*, v^* = a(1-s^*))$, $0 < s^* < 1$ exists. In particular, we have that

$$s^* = \frac{1}{2a\Delta(\tau)}\left\{a(\Delta(\tau)+m_p) - \sqrt{a^2(\Delta(\tau)+m_p)^2 - 4a\Delta(\tau)m_p\left(a - \frac{C_p}{C}\right)}\right\} \tag{24}$$

if $be^{-m_i\tau} \neq 1$, and

$$s^* = 1 - \frac{C_p}{aC} \tag{25}$$

if $be^{-m_i\tau} = 1$. Finally E_+ coincides with E_0 when $a = C_p/C$. In [18] the author investigates the mean-square stability properties of E_+ under stochastic perturbations. Here we assume that the feasible condition $a > C_p/C$ is satisfied and we study the effect on the stability properties of E_+ induced by $M = 3$ uncertain parameters in the linearized system, namely

$$\begin{cases} \frac{dy_1(t)}{dt} = -as^*y_1(t) - s^*y_2(t) \\ \frac{dy_2(t)}{dt} = -v^*y_1(t) - (m_p+s^*)y_2(t) + bTs^*y_2(t-\tau) + bTv^*y_1(t-\tau), \end{cases} \tag{26}$$

where $T = e^{-m_i\tau}$. In particular we fix $a = 10$, $m_p = 14.925$, $m_i = 0.1$ as in [18], and we choose τ, b and C_p as follows

$$\begin{gathered} \xi = (\xi_1, \xi_2, \xi_3), \quad \xi_i \sim \mathcal{U}[-1,1],\ i = 1,2,3, \\ \tau(\xi_1) = \tfrac{49}{2}\xi_1 + \tfrac{51}{2} \sim \mathcal{U}[1,50],\ b(\xi_2) = \tfrac{97}{2}\xi_2 + \tfrac{101}{2} \sim \mathcal{U}[2,99], \\ C_p(\xi_3) = 4\xi_3 + 5 \sim \mathcal{U}[1,9]. \end{gathered} \tag{27}$$

First we compute the expectation and variance of the P-order numerical spectral abscissa $S_{N,P}(\xi)$ varying P. In this way we experimentally find P which ensures an accuracy of about four decimal digits, which is sufficient for our aims, and we also perform the UQ. In Table 8 we show the values $E_P := E[S_{20,P}]$ and $V_P := var[S_{20,P}]$ computed by using $Q = (P+1)^3$ Gaussian integration points for the computation

Table 8 PCE-based expectation and variance of $S_{20,P}$ for the UDDE (26)–(27) varying P computed by Gaussian integration formulas

P	E_P	$\lvert E_P - E_{20}\rvert/\lvert E_{20}\rvert$	V_P	$\lvert V_P - V_{20}\rvert/\lvert V_{20}\rvert$
5	$-8.1684210e-02$	$1.20e-02$	$1.0402960e-02$	$1.35e-01$
10	$-8.2612167e-02$	$7.80e-04$	$1.1876067e-02$	$1.29e-02$
15	$-8.2675787e-02$	$1.14e-05$	$1.2022156e-02$	$7.07e-04$

Table 9 PCE-based expectation and variance of $S_{20,P}$ for the UDDE (26)–(27) varying P computed by Padua integration formulas.

P	E_P	$\lvert E_P - E_{20}\rvert/\lvert E_{20}\rvert$	V_P	$\lvert V_P - V_{20}\rvert/\lvert V_{20}\rvert$
5	$-8.1279600e-02$	$1.69e-02$	$9.1238625e-03$	$2.41e-01$
10	$-8.2521374e-02$	$1.88e-03$	$1.1646102e-02$	$3.20e-02$
15	$-8.2661855e-02$	$1.80e-04$	$1.1974397e-02$	$4.68e-03$

of the PC coefficients (7) and rounded to 8 digits. The relative errors in Table 8 are computed w.r.t. the reference values

$$E_{20} = -8.267672877780410e-02, \quad V_{20} = 1.203066614563154e-02, \quad (28)$$

and rounded to 3 digits. Moreover by using the $Q = 9261$ values of the numerical spectral abscissa at Gaussian points, we get $\tilde{s} = 6.2017e-01$, and we conclude that E_+ is unstable. In Table 9, we present the results obtained by using the Padua integration formulas with $Q \approx (P+4)^3/4$ (the exact values of Q are given in Table 13). Note that the we lost one decimal digit of accuracy w.r.t. Gaussian integration formulas, but we reduce the computation cost. By using the $Q = 3456$ values of the numerical spectral abscissa at Gaussian points, we also get $\tilde{s} = 6.6615e-01$ rounded to 5-digits.

To emphasise that, contrary to MC-methods, PCE-based methods exploit the regularity of S_N, here we also consider the LS method and the MC method. We recall that MC method provides random estimates with usually limited accuracy, since the convergence rate is $O(n^{-1/2})$, where n is the number of samples. In Tables 10 and 11 we show respectively the results obtained by means of the LS method with R minimization points and MC method with n-samples. All the results give evidence that PCE-based methods are more efficient w.r.t. MC sampling method at least for small stochastic dimension M. Moreover the Padua integration formulas are a valid alternative, since they allow to reach a reasonable balance between accuracy and computation cost. We arrive to the same conclusions by analysing the numerical results concerning the computation of Sobol' indices. In particular in Tables 12 and 13 we show the Sobol' indices of $S_{20,P}$ for different orders P, computed respectively by using Gaussian integration formulas and Padua integration formulas. By comparing for instance the values for $P = 20$, we observe that they coincide up to three decimal

Table 10 PCE-based expectation and variance of $S_{20,P}$ for the UDDE (26)–(27) varying P computed by LS method with R minimization points

P	E_P	$\lvert E_P - E_{20}\rvert/\lvert E_{20}\rvert$	V_P	$\lvert V_P - V_{20}\rvert/\lvert V_{20}\rvert$	R
5	$-6.99995230e-02$	$1.53e-01$	$7.5479075e-03$	$3.73e-01$	76
10	$-8.105586e-02$	$1.96e-02$	$1.0132217e-02$	$1.58e-01$	685
15	$-8.2008553e-02$	$8.08e-03$	$1.1914217e-02$	$9.68e-03$	2455

Table 11 Expectation $E[S_{20}]$ and variance $Var[S_{20}]$ for the UDDE (26)–(27) computed by MC method with n samples

n	$E[S_{20}]$	$Var[S_{20}]$
100	$-8.2802e-02$	$6.7639e-03$
1800	$-8.0662e-02$	$1.3085e-02$
3000	$-8.1403e-02$	$1.1229e-02$
7000	$-8.2247e-02$	$1.2806e-02$
10000	$-8.3203e-02$	$1.2143e-02$
15000	$-8.2520e-02$	$1.1712e-02$

Table 12 PCE-based Sobol' indices of $S_{20,P}$ for the UDDE (26)–(27) varying P computed by using $Q = (P+1)^3$ Gaussian integration points

Sobol' index	$P = 5$	$P = 10$	$P = 15$	$P = 20$
$\mathcal{S}_\tau$	$6.2405e-02$	$4.4067e-02$	$4.4397e-02$	$4.4685e-02$
$\mathcal{S}_b$	$3.0499e-01$	$2.7717e-01$	$2.7435e-01$	$2.7412e-01$
$\mathcal{S}_{C_p}$	$2.4198e-02$	$1.9662e-02$	$1.9332e-02$	$1.9303e-02$
$\mathcal{S}_{\tau,b}$	$5.4642e-01$	$5.7893e-01$	$5.7945e-01$	$5.7916e-01$
$\mathcal{S}_{\tau,C_p}$	$2.3372e-02$	$2.7171e-02$	$2.8028e-02$	$2.8134e-02$
$\mathcal{S}_{b,C_p}$	$4.0984e-03$	$3.7296e-03$	$3.6620e-03$	$3.6466e-03$
$\mathcal{S}_{\tau,b,C_p}$	$3.4523e-02$	$4.9279e-02$	$5.0778e-02$	$5.0955e-02$
Q	216	1331	4096	9261

digits, but the computation cost of the former is three time higher w.r.t. the latter. At last in Table 14 we compute the PCE-based Sobol' indices by LS method: they are less accurate w.r.t. those in Table 13 and have higher computation cost when P increases.

At the end we can back to the sensitivity analysis of (15), underlying again that it can be performed by using small P, namely $P = 10$. The Sobol' indices in show that the replication factor of the phages b and its combination with the incubation period τ are the predominant source of variability, while the other parameters or their combination have the minor effect. In fact the results show that b and the group $\{\tau, b\}$ contribute to the total variance respectively with ≈ 0.27 and ≈ 0.57. Note that the impact of the group $\{b, C_p\}$ is negligible. It is often convenient to summarize the

Table 13 PCE-based Sobol' indices of $S_{20,P}$ for the UDDE (26)–(27) varying P computed by using Q Padua integration points

Sobol' index	$P = 5$	$P = 10$	$P = 15$	$P = 20$
$\mathscr{S}_\tau$	$8.1487e-02$	$4.4853e-02$	$4.4513e-02$	$4.4621e-02$
$\mathscr{S}_b$	$3.2511e-01$	$2.8152e-01$	$2.7515e-01$	$2.7407e-01$
$\mathscr{S}_{C_p}$	$2.7417e-02$	$1.9563e-02$	$1.9369e-02$	$1.9338e-02$
$\mathscr{S}_{\tau,b}$	$5.0300e-01$	$5.7146e-01$	$5.7849e-01$	$5.7907e-01$
$\mathscr{S}_{\tau,C_p}$	$2.3684e-02$	$2.9577e-02$	$2.7809e-02$	$2.8228e-02$
$\mathscr{S}_{b,C_p}$	$5.7566e-03$	$3.9257e-03$	$3.6857e-03$	$3.6548e-03$
$\mathscr{S}_{\tau,b,C_p}$	$3.3551e-02$	$4.9096e-02$	$5.0982e-02$	$5.1024e-02$
Q	189	686	1729	3456

Table 14 PCE-based Sobol' indices of $S_{20,P}$ for the UDDE (26)–(27) varying P computed by LS method with R minimization points

Sobol' index	$P = 5$	$P = 10$	$P = 15$	$P = 20$
$\mathscr{S}_\tau$	$3.0102e-01$	$7.7994e-02$	$4.7932e-02$	$4.4612e-02$
$\mathscr{S}_b$	$2.2867e-01$	$2.9426e-01$	$2.6460e-01$	$2.7480e-01$
$\mathscr{S}_{C_p}$	$3.7794e-02$	$2.8324e-02$	$2.0806e-02$	$1.9208e-02$
$\mathscr{S}_{\tau,b}$	$2.8130e-01$	$5.2330e-01$	$5.4459e-01$	$5.7233e-01$
$\mathscr{S}_{\tau,C_p}$	$8.9502e-02$	$2.6744e-02$	$3.4586e-02$	$2.9818e-02$
$\mathscr{S}_{b,C_p}$	$1.3554e-02$	$7.4237e-03$	$1.1901e-02$	$4.0531e-03$
$\mathscr{S}_{\tau,b,C_p}$	$4.8157e-02$	$4.1955e-02$	$7.5586e-02$	$5.5172e-02$
R	76	685	2455	7236

impact of one parameter using the so-called total sensitivity index, which is the sum of all the sensitivity indexes to which the parameter contributes. We denote $\mathscr{T}_b$, $\mathscr{T}_\tau$, and $\mathscr{T}_{C_p}$, the total sensitivity indexes related to the uncertain b, τ, C_p respectively. The analysis of the following results

$$\begin{aligned}
\mathscr{T}_b &= \mathscr{S}_b + \mathscr{S}_{\tau,b} + \mathscr{S}_{b,C_p} + \mathscr{S}_{\tau,b,C_p} \approx 0.91, \\
\mathscr{T}_\tau &= \mathscr{S}_\tau + \mathscr{S}_{\tau,b} + \mathscr{S}_{b,C_p} + \mathscr{S}_{\tau,b,C_p} \approx 0.70 \\
\mathscr{T}_{C_p} &= \mathscr{S}_{C_p} + \mathscr{S}_{\tau,C_p} + \mathscr{S}_{b,C_p} + \mathscr{S}_{\tau,b,C_p} \approx 0.10
\end{aligned}$$

gives evidence that, among the sources of uncertainties, b has the larger impact, followed by τ, while the uncertainty in the parameter C_p has weaker influence. Further experiments can be found in [59].

6 Conclusions and Future Works

In the work, to study the effect of uncertain parameters on the stability of equilibria, we have proposed a PCE-based approach to analyse the sensitivity of the numerical spectral abscissa of the linearized uncertain system. We have relied to NISP methods to compute the PC coefficients. To reduce the computational cost, we have considered numerical integration formulas based on Padua grids and we have shown experimentally by using some numerical tests, that they represent a valid alternative for UDDEs of stochastic dimension $M = 2, 3$. Indeed even if the convergence rate w.r.t. P can be different, in practice the numerical tests show that the sensitivity analysis can be performed by using small P, namely $P = 5, 10$. In the future, having in mind the paper [13], we aim to investigate the possible extension of Padua integration formulas as well as the recent technique in [48] to greater dimension M. Finally we are also planning to treat stochastic DDEs with uncertain parameters along the same line of the recent paper [39] and to apply the approach to other important quantities of the models (see for instance [30]).

References

1. A. Batkai, S. Piazzera, Semigroup for delay equations. *Research Notes in Mathematics* vol. 10 (A K Peters, Ltd., 2005)
2. A. Bellen, M. Zennaro, *Numerical Methods for Delay Differential Equations* Numerical Mathematics and Scientifing Computing series. (Oxford University Press, 2003)
3. E. Beretta, Y. Kuang, Modeling and analysis of a marine bacteriophage infection with latency period. Nonlinear Anal. Real World Appl. **2**, 35–74 (2001)
4. M. Berveiller, B. Sudret, M. Lemaire, Stochastic finite element: a non-intrusive approach by regression. Rev. Européenne Mécanique Numérique **15**(1–3), 81–92 (2006)
5. D. Breda, Solution operator approximation for characteristic roots of delay differential equations. Appl. Numer. Math. **56**(3–4), 305–317 (2006)
6. D. Breda, S. Maset, R. Vermiglio, Pseudospectral differencing methods for characteristic roots of delay differential equations. SIAM J. Sci. Comput. **27**, 482–495 (2005)
7. D. Breda, S. Maset, R. Vermiglio, Approximation of eigenvalues of evolution operators for linear retarded functional differential equations. SIAM J. Numer. Anal. **50**(3), 1456–1483 (2012)
8. D. Breda, S. Maset, R. Vermiglio, *Stability of Linear Differential Equations-A Numerical Approach with Matlab* (Springer Briefs in Electrical and Computer Engineering, 2015)
9. J.P. Boyd, *Chebyshev and Fourier Spectral Methods* (Dover Books on Mathematics, 2001)
10. L. Bos, S. De Marchi, M. Vianello, Y. Xu, Bivariate Lagrange interpolation at the Padua points: the generating curve approach. J. Approx. Theory **143**(1), 15–25 (2006)
11. L. Bos, S. De Marchi, M. Vianello, Y. Xu, Bivariate Lagrange interpolation at the Padua points: the ideal curve approach. Numer. Math. **108**, 43–57 (2007)
12. L. Bos, S. De Marchi, M. Vianello, Trivariate polynomial approximation on Lissajous curves. IMA J. Numer. Anal. **37**, 519–541 (2017)
13. L. Bos, S. De Marchi, M. Vianello, Polynomial approximation on Lissajous curves in the d-cube. Appl. Numer. Math. **116**, 47–56 (2017)
14. R.H. Cameron, W.T. Martin, The orthogonal development of non-linear functionals in series of fourier-hermite functionals. Ann. Math. Sec. Ser. **48**(2), 385–392 (1947)

15. M. Caliari, S. De Marchi, M. Vianello, Bivariate polynomial interpolation on the square at new nodal sets. Appl. Math. Comput. **165**(20), 261–274 (2005)
16. M. Caliari, S. De Marchi, M. Vianello, Algorithm 886: Padua 2D-Lagrange interpolation at padua points on bivariate domains. ACM Trans. Math. Softw. **35** (2008)
17. M. Caliari, S. De Marchi, M. Vianello, Padua 2DM: fast interpolation and cubature at the Padua points in Matlab/Octave. Numer. Algorithm. **56**, 45–60 (2010)
18. M. Carletti, Mean-square stability of a stochastic model for bacteriophage infection with time delays. Math. Biosc. **210**, 395–414 (2007)
19. G. Chastaing, F. Gamboa, C. Prieur, Generalized Hoeffding-Sobol decomposition for dependent variables—application to sensitivity analysis. Electron. J. Stat. **6**, 2420–2448 (2012)
20. T. Crestaux, O. Le Maître, J. Martinez, Polynomial Chaos expansion for sensitivity analysis. Reliab. Eng. Syst. Saf. **94**, 1161–1172 (2009)
21. M.C. Delfour, State theory of linear hereditary differential systems. J. Math. Anal. Appl. **60**, 8–35 (1977)
22. S. De Marchi, M. Vianello, Y. Xu, New cubature formulae and hyperinterpolation in three variables. BIT Numer. Math. **49**, 55–73 (2009). https://doi.org/10.1007/s10543-009-0210-7
23. O. Diekmann, S.A. van Gils, S.M. Verduyn Lunel, H.O. Walther, Delay equations. Functional, complex and nonlinear analysis. *Applied Mathematical Sciences*, vol. 110 (Springer, New York, 1995)
24. O.G. Ernst, A. Muglera, H-J. Starkloff, E. Ullmann, On the convergence of generalized polynomial chaos expansions. ESAIM: Math. Model. Numer. Anal. **460**(2), 317–339 (2012)
25. K. Engelborghs, T. Luzyanina, D. Roose, Numerical bifurcation analysis of delay differential equations using DDE-BIFTOOL. ACM T. Math. Softw. **28**(1), 1–21 (2002)
26. T. Erneux, Surveys and tutorials in the applied mathematical sciences, in *Applied Delay Differential Equations*. (Springer, New York, 2009)
27. L. Fenzi, Polynomial Chaos Theory: application to the stability of Uncertain Delay Differential Equations. Master Thesis in Mathematics, University of Udine (2015)
28. L. Fenzi, W. Michiels, Robust stability optimization for linear delay systems in a probabilistic framework. Linear Algebr. Appl. **526**, 1–26 (2017)
29. L. Fenzi, W. Michiels, Polynomial (chaos) approximation of maximum eigenvalue functions. *Numerical Algorithms, 82*, 1143–1169 (2019).
30. F. Florian, R. Vermiglio, PC-based sensitivity analysis of the basic reproduction number of population and epidemic models. *Current Trends in Dynamical Systems in Biology and Natural Sciences,21*, 205–222. (2020)
31. R. Ghanem, P. D. Spanos: Stochastic Finite Elements: A Spectral Approach. Springer-Verlag (1991) (Revised edn Dover Publications, 2004)
32. J.K. Hale: Theory of functional differential equations. AMS series **99**. Springer Verlag, New York (1977)
33. T. Insperger, and G. Stépán: Semi-discretization for time-delay systems - Stability and Engineering applications. Applied Mathematical Sciences **178** Springer Verlag, New York (2011)
34. V.B. Kolmanovskii, and A. Myshkis: Applied theory of functional differential equations. No. 85 in Mathematics and Its Applications (Soviet Series). Kluver Academic Press, The Netherlands (1992)
35. V.B. Kolmanovskii, and V.R. Nosov: Stability of functional differential equations. No. 180 in Mathematics in Science and Engineering. Academic Press, London (1986)
36. O. Le Maître, and O. Knio: Spectral Methods for Uncertainty Quantification with Applications to Computational Fluid Dynamics. Springer-Verlag (2010)
37. S. Marelli and B. Sudret: UQLab user manual-Polynomial Chaos Expansions. Report UQLab-V0.9-104, Chair of Risk, Safety & Uncertainty Quantification, ETH Zurich (2015)
38. W. Michiels, S.-I. Niculescu: Stability, control and computation for time-delay systems. An eigenvalue based approach. SIAM: Philadelphia, Series : "Advances in design and control", vol. DC 27, (2014)
39. M. Navarro Jimenez, O. Le Maître,O. Knio: Nonintrusive Polynomial Chaos Expansions for Sensitivity Analysis in Stochastic Differential Equations. SIAM/ASA J. Uncertainty Quantification. **5**, 378–402 (2017)

40. S.-I. Niculescu: Delay effects on stability. A robust control approach. Springer: Heidelberg, Series: LNCIS **269** (2001)
41. S. Pelizzari: Controllo di temperatura di una camera a ricircolazione d'aria per misure di viscoelasticità. Master Thesis in Engineering, University of Udine (1995)
42. J. Shen, T. Tang, Li-Lian Wang, *Spectral Methods, Algorithms, Analysis and Applications*. Springer Series in Computational Mathematics, vol. 4 (Springer, Berlin, Heidelberg, 2011)
43. S. Smolyak, Quadrature and interpolation formulas for tensor products of certain classes of functions. Dokl. Akad. Nauk SSSR **4**, 240–243 (1963)
44. I.M. Sobol, Sensitivity estimates for nonlinear mathematical models. Math. Model. Comput. Exp. **1**, 407–414 (1993)
45. I.M. Sobol, Global sensitivity indices for nonlinear mathematical models and their Monte Carlo estimates. Math. Comput. Simul. **55**, 271–280 (2001)
46. I.M. Sobol, S.S. Kucherenko, Global sensitivity indices for nonlinear mathematical models. Rev. Wilmott Mag. **1**, 56–61 (2005)
47. A. Sommariva, M. Vianello, R. Zanovello, Non tensorial Clenshaw-Curtis cubature. Numer. Algorithm. **49**, 409–427 (2008)
48. A. Sommariva, M. Vianello, Compression of multivariate discrete measures and applications. Numer. Funct. Anal. Optim. **36**, 1198–1223 (2015)
49. B. Sudret, Global sensitivity analysis using polynomial chaos expansions. Reliab. Eng. Syst. Saf. **93**, 964–979 (2008)
50. G. Stépán, *Retarded Dynamical Systems* (Longman, Harlow, UK, 1989)
51. A.W. Van Der Vaart, *Asymptotic Statistics* (Cambridge University Press, Cambridge, 1998)
52. R. Vermiglio, Polynomial Chaos expansions for the stability analysis of uncertain delay differential equations. SIAM/ASA J. Uncertain. Quantif. **5**(1), 278–303 (2017)
53. S. Xiang, An improved error bound on Gauss quadrature. Appl. Math. Lett. **58**, 42–48 (2016)
54. S. Xiang, F. Bornemann, On the convergence rates of Gauss and Clenshaw-Curtis quadrature for functions of limited regularity. SIAM J. Numer. Anal. **50**(5), 2581–2587 (2012)
55. H. Wang, S. Xiang, On the convergence rates of Legendre approximation. Math. Comput. **81**(278), 861–877 (2011)
56. Z. Wu, W. Michiels, Reliably computing all characteristic roots of delay differential equations in a given right half plane using a spectral method. J. Comput. Appl. Math. **236**(9), 2499–2514 (2012)
57. N. Wiener, The homogeneous chaos. Am. J. Math. **60**, 897–936 (1938)
58. D. Xiu, *Numerical Methods for Stochastic Computation* (Princeton University Press, Princeton, NJ, 2010)
59. A. Zamolo, Polynomial Chaos expansion and its application to delay differential equations. Master Thesis in Mathematics, University of Udine (2017)

Lyapunov Stability of a Coupled Ordinary Differential System and a String Equation with Polytopic Uncertainties

Matthieu Barreau, Alexandre Seuret, and Frédéric Gouaisbaut

Abstract This chapter deals with the robust stability analysis of a coupled system made up of an uncertain ordinary differential system and a string equation. The main result states the robust exponential stability of this interconnected system subject to polytopic uncertainties. The Lyapunov theory transforms the stability analysis into the resolution of a set of linear matrix inequalities. They are obtained using projections of the infinite dimensional state onto the orthogonal basis of Legendre polynomials. The special structure of these inequalities is used to derive robust stability results. An example synthesizes the two main contributions of this chapter: an extended stability result and a robustness analysis. The example shows the efficiency of the proposed method.

1 Introduction

Infinite dimensional systems are natural systems which arise when a delay or a partial differential equation appears unavoidably in the modeling phase. In particular, vibrations of structure [9], torsion [8, 11] or heat propagation [3, 21] are examples of phenomena giving rise to infinite dimensional systems. The infinite dimensional behavior is characterized by an infinite dimensional state, belonging to a functional space. This consideration seems rather far from a natural modeling and its utility can be questioned. That is why, as a first step, the infinite dimensional system can be approximated as a high but finite dimensional model. Then, applying classical tools

This work is supported by the ANR project SCIDiS contract number 15-CE23-0014.

M. Barreau (✉) · A. Seuret · F. Gouaisbaut
LAAS - CNRS, Université de Toulouse, CNRS, UPS, Toulouse, France
e-mail: barreau@laas.fr

A. Seuret
e-mail: aseuret@laas.fr

F. Gouaisbaut
e-mail: gouaisb@laas.fr

G. Valmorbida et al. (eds.), *Accounting for Constraints in Delay Systems*,
Advances in Delays and Dynamics 12,
https://doi.org/10.1007/978-3-030-89014-8_9

leads to the design of controllers. Nevertheless, it has been shown in [4, 18], that this truncation brings instability because of the well-known spill-over effect. Indeed, high frequencies can be excited by the controller and as they are not regulated, they can provoke a divergence of the solution.

Then, Distributed Parameters Systems (DPS) can be the most appropriate model for some physical systems since a truncation is not always possible. To overcome this problem, special tools have been developed for DPS systems. Without being exhaustive, we can note the work of Krstic and al. (see for instance [5, 14, 15]) using the backstepping methodology or the use of the semi-group theory [17, 22]. Considering physical modeling, uncertainties of the parameters have to be taken into account to design efficient and robust control laws.

The work conducted here aims at using the second methodology to prove the robust exponential stability of an ordinary differential system coupled with a string equation. One of the main challenge in using semi-group theory consists in finding a scalar product for which the system is dissipative. In [1], we proposed a Lyapunov functional which encompasses the traditional notion of energy as it results from an optimization and is problem-oriented. This framework enabled us to transform the dissipativity proof for the coupled system into the feasibility of Linear Matrix Inequalities (LMI). We propose here to enlarge this study by considering less restrictive LMIs and to the case of uncertainties on the ordinary differential part of the interconnected system. This last part is the main contribution as it is commonly the first extension proposed when using a Lyapunov functional and as it requires a transformation of the LMI to get the robust stability result.

Section 2 presents the problem statement, describing the coupled system, its main characteristics and the slight changes in the framework to deal with polytopic uncertainties. Then, after a brief reminder of some useful lemmas in Sect. 3, an extended exponential stability result is provided in Sect. 4. The last subsection of Sect. 4 is dedicated to the robustness analysis by taking advantage of the convexity of the LMIs with respect to the system's matrices. Finally, in Sect. 5, an example with stability pockets is studied.

Notations: In this paper, Ω is the closed set $[0, 1]$ and $\mathbb{R}^+ = [0, +\infty)$. Then, $(x, t) \mapsto u(x, t)$ is a multi-variable function from $\Omega \times \mathbb{R}^+$ to $\mathbb{R}$. The notation u_t stands for $\frac{\partial u}{\partial t}$. We also use the notations $L^2 = L^2(\Omega; \mathbb{R})$ and for the Sobolov spaces: $\mathscr{H}^n = \{z \in L^2; \forall m \leqslant n, \frac{\partial^m z}{\partial x^m} \in L^2\}$. The dot product in L^2 is defined for $f, g \in L^2$ as $\langle f, g\rangle = \int_\Omega f(x)g(x)dx$. Then the norm associated to this scalar product is $\|f\|^2 = \langle f, f\rangle = \int_\Omega |f(x)|^2 dx$.

For any square matrices A and B, the operations 'He' and 'diag' are defined as follows: $\text{He}(A) = A + A^\top$ and $\text{diag}(A, B) = \left[\begin{smallmatrix} A & 0 \\ 0 & B \end{smallmatrix}\right]$. A positive definite matrix $P \in \mathbb{R}^{n\times n}$ belongs to the set $\mathbb{S}^n_+$ or more simply $P \succ 0$.

2 Problem Statement

The coupled ODE/PDE system under consideration is the following one:

$$\dot{X}(t) = AX(t) + Bu(1,t), \qquad t \geqslant 0, \tag{1a}$$
$$u_{tt}(x,t) = c^2 u_{xx}(x,t), \qquad x \in \Omega, t \geqslant 0, \tag{1b}$$
$$u(0,t) = KX(t), \qquad t \geqslant 0, \tag{1c}$$
$$u_x(1,t) = -c_0 u_t(1,t), \qquad t \geqslant 0, \tag{1d}$$
$$u(x,0) = u^0(x), \; u_t(x,0) = v^0(x), \qquad x \in \Omega, \tag{1e}$$
$$X(0) = X^0, \tag{1f}$$

where Eq. (1a) is an Ordinary Differential Equation (ODE), $X \in \mathbb{R}^n$ with $n \geqslant 1$, A and B are not necessary known matrices of appropriate dimensions. Equation (1b) is a Partial Differential Equation (PDE), and more specifically, a string equation which is a one dimensional wave equation with a speed $c > 0$. The spacial variable x belongs to Ω and t is the time. Equations (1c) and (1d) are the boundary conditions at $x = 0$ and $x = 1$, respectively. The first boundary condition is of Dirichlet kind, so acting on $u(0,t)$, contrary to the second one, which is of Neumann type (i.e. acting on $u_x(1,t)$) and describes a boundary damping condition. This means that, for $c_0 > 0$, the string equation is damped (see [16] for instance). Equations (1e) and (1f) represent the initial conditions of the system, which are said to be compatible with the boundary conditions if they satisfy Eqs. (1c) and (1d).

Remark 1 To ease the reading the variable t will be omitted when it is not needed.

Such systems are used to approximately model the dynamics of vibrations in structures [9], or torsion in a drilling pipe [5, 8, 19]. The ODE (1a) is then the finite dimension controller acting on the top of the pipe. The measure of the angle at the bottom is the input of the ODE. That makes the schematic in Fig. 1.

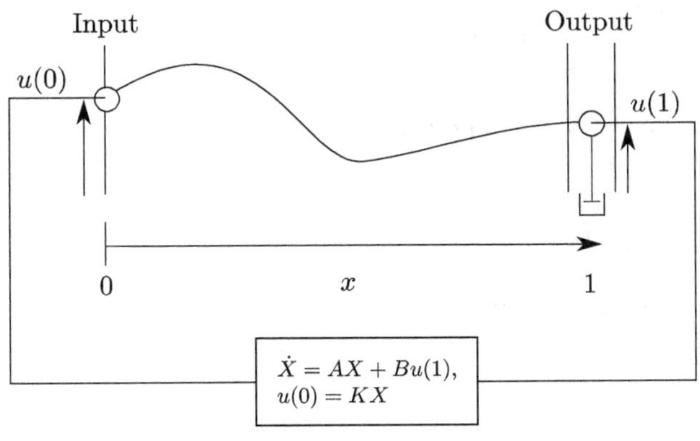

Fig. 1 Scheme of system (1)

This system can be viewed in different manners. First, as it can be noticed in Fig. 1, the PDE can be seen as a communication channel modeled by a string equation. Then, the signal u is the conveyed KX at $x = 0$ there is the input signal while at $x = 1$, this is the output. The communication channel is then stable ($c_0 > 0$) and the ODE is regulated thanks to the string equation. This is the interconnection between a stable PDE plant (i.e. $c_0 > 0$) and a possibly unstable ODE. Then, three scenari come out:

1. If A is Hurwitz, this communication channel is a perturbation and the stability analysis of the coupled ODE/PDE system is nothing more than a robustness analysis subject to this communication channel;
2. If A is not Hurwitz but $A + BK$ is, then the string equation is needed to stabilize the system and is not a perturbation anymore, it helps stabilizing;
3. Finally, if neither A nor $A + BK$ are Hurwitz, the stability of the coupled system relies mostly on the intrinsic properties of the string equation.

This system under the three cases discussed previously has already been studied in [1] and a stability criteria has been derived in terms of LMIs. Several conclusions have been drawn and particularly, c_0 must be positive to get some results using this methodology. This means that the communication channel must be stable. Moreover, the obtained conditions are dependent on a perfect knowledge of all the parameters A, B, K, c and c_0.

To enlarge the result obtained in Theorem 2 of [1], a robustness analysis on the parameters A and B is proposed. The main assumption is that A and B are subject to polytopic uncertainties, that means the following holds:

$$[\ A\ B\ ] \in \mathrm{Co}_{i=1,\ldots,m} \{[\ A_i\ B_i\ ]\}, \tag{2}$$

where $m \in \mathbb{N}$, and the matrices A_i and B_i for $i = 1, \ldots, m$ are known and constant. The notation "Co" means that the matrix $[\ A\ B\ ]$ belongs to a convex set defined by the vertices $[\ A_i\ B_i\ ]$. In other words, there exist weighting scalar functions λ_i, for $i = 1, \ldots, m$ that maps $\mathbb{R}^+$ to $[0,\ 1]$, and such that for all $t \geqslant 0$, $\sum_{i=1}^{m} \lambda_i(t) = 1$ and

$$[\ A\ B\ ] = \sum_{i=1}^{m} \lambda_i(t)\, [\ A_i\ B_i\ ]\,. \tag{3}$$

This kind of uncertainty is commonly used in the LMI formulation because thanks to a convexity argument the robust problem can be easily reformulated in terms of higher complexity LMI. The number of edges to define the polytope is directly related to the complexity of the new LMIs and can then dramatically increase the computational burden.

As the method presented in [1] can also be computationally demanding, after a brief summary of the main lemmas used here, an extension is proposed to decrease the conservatism, keeping the same number of decision variables.

3 Preliminaries

In this preliminary part, we recall two main results used in the sequel.

3.1 Legendre Polynomials and Bessel-Legendre Inequality

To transform integral terms into an LMI formulation, we need to use integral inequalities. Based on the work in [20], the Bessel-Legendre inequality is reminded below:

Lemma 1 (Bessel-Legendre Inequality) *For any orthogonal family $(e_k)_{k\in\mathbb{N}}$ of L^2 with respect to the scalar product $\langle\cdot,\cdot\rangle$ and any function $f \in L^2$, the following inequality holds for $N \in \mathbb{N}$:*

$$\|f\|^2 \geqslant \sum_{k=0}^{N} \langle f, e_k\rangle \|e_k\|^{-2}.$$

This inequality becomes an equality when N goes to infinity. This last property makes us think that as N increases, the conservatism induced by this inequality can be reduced as much as desired.

We choose to apply this result to the orthogonal family of Legendre polynomials $(\mathscr{L}_k)_{k\in\mathbb{N}}$ for several reasons. First, it is a polynomial and consequently, its derivative can be expressed in strictly lower order polynomials. Secondly, the boundary conditions are simple and make calculations easier. The definition of Legendre polynomials is reminded below:

$$\forall k \in \mathbb{N},\ \forall x \in [0,1],\quad \mathscr{L}_k(x) = (-1)^k \sum_{l=0}^{k} (-1)^l \binom{k}{l}\binom{k+l}{l} x^l,$$

with $\binom{k}{l} = \frac{k!}{l!(k-l)!}$. For more properties about this polynomials, the reader can refer to [1, 10, 20].

3.2 Convexity Lemma

This second lemma is useful when it comes to convexify an LMI. This is particularly useful to derive a robustness criterion. This lemma is taken from [6, 7].

Lemma 2 *For some given matrices $\mathscr{M} = \mathscr{M}^\top$, $\mathscr{Y} = \mathscr{Y}^\top$ and $\mathscr{X}$ of appropriate dimensions, the following statements are equivalent:*

1. *The matrix inequality $\mathscr{M} - \mathscr{X}^\top \mathscr{Y} \mathscr{X} \prec 0$ holds.*

2. *There exists a matrix* $\mathscr{X}$ *and a scalar* $\mu > 0$ *such that the following matrix inequality holds*

$$\mathscr{M} + \mathrm{He}(\mathscr{X}^\top \mathscr{Z}) + \mathscr{X}^\top (\mu I + \mathscr{Y})^{-1} \mathscr{X} + \mu \mathscr{Z}^\top \mathscr{Z} \prec 0,$$

together with $\mu I + \mathscr{Y} \succ 0$.

4 From Stability to Robust Stability Analysis

The desired stability property here is the well-known exponential stability for (1) with A, B, K and c known. To define it correctly, we first need to know in which space the solutions of system (1) belong to. This is related to the well-posedness of this system. These results are quite technical and the proof has already been presented in Proposition 1 of [1] but it is reminded here.

Proposition 1 *Let* $\mathscr{H}^m = \mathbb{R}^n \times H^m \times H^{m-1}$ *for* $m > 0$.

If there exists a norm on $\mathscr{H} = \mathscr{H}^1$ *for which the linear operator associated to system (1) is dissipative with* $A + BK$ *non singular, then there exists a unique solution* (X, u, u_t) *of system (1) with initial conditions* $(X^0, u^0, v^0) \in \mathscr{H}^2$ *assumed to be compatible with the boundary conditions. Moreover, the solution has the following regularity property:* $(X, u, u_t) \in C(0, +\infty, \mathscr{H})$.

The solutions belong to space $\mathscr{H}$ and consequently, one can introduce the following natural norm on this space:

$$\forall (X, u, v) \in \mathscr{H}, \quad \|(X, u, v)\|_{\mathscr{H}}^2 = |X|_n^2 + \|u\|^2 + c^2 \|u_x\|^2 + \|v\|^2,$$

where $|\cdot|_n$ is the classical euclidian norm of $\mathbb{R}^n$ and $\|\cdot\|$ refers to the L^2 norm. From this definition, the definition of exponential stability comes to be as follow:

Definition 1 A solution of system (1) belonging to $\mathscr{H}$ with the compatible initial conditions $(X^0, u^0, v^0) \in \mathscr{H}^2$ is exponentially stable if the following exponential estimate holds for $\gamma \geqslant 1$, $\delta > 0$ and $t > 0$:

$$\|(X(t), u(t), u_t(t))\|_{\mathscr{H}}^2 \leqslant \gamma e^{-\delta t} \|(X^0, u^0, v^0)\|_{\mathscr{H}}^2. \tag{4}$$

To prove the existence and uniqueness of the solution along with its exponential stability, the idea is to use a Lyapunov functional. This Lyapunov functional must be equivalent to $\|\cdot\|_{\mathscr{H}}$ and with a strictly negative derivative along the trajectories of the system. This last property ensures the dissipativity of system (1). These two properties are met if an LMI is satisfied. The following subsection gives modified versions of the original LMI condition proposed in [1].

4.1 First Result

An extension of Theorem 2 of [1] to reduce the conservatism is proposed as follows.

Theorem 1 *Consider system (1) with a given speed $c > 0$, a viscous damping $c_0 > 0$ and initial conditions $(X^0, u^0, v^0) \in \mathscr{H}^2$ compatible with the boundary conditions.*

Assume that, for a given integer $N \in \mathbb{N}$, there exist $P_N \in \mathbb{S}_+^{n+2(N+1)}$ and $S, R \in \mathbb{S}_+^2$ such that inequalities

$$\begin{aligned}\Psi_N(A,B) &= \mathrm{He}\left(Z_N^\top(A,B)P_N F_N\right) - c\tilde{R}_N \\ &\quad + c\left(H_N^\top(A,B)(S+R)H_N(A,B) - G_N^\top(A,B)SG_N(A,B)\right) \prec 0, \end{aligned} \tag{5}$$

$$\Phi_N = P_N + \mathrm{diag}(0_n, S, 3S, \dots, (2N+1)S) \succ 0, \tag{6}$$

hold, where

$$\begin{array}{rlrl} F_N &= \left[I_{n+2N+2}\ 0_{n+2N+2,2}\right], & \mathscr{N}_N(A,B) &= \left[A+BK\ \ \tilde{B}\ \ 0_{n,2(N+1)}\right], \\ \tilde{R}_N &= \mathrm{diag}\left(0_n, R, 3R, \cdots, (2N+1)R, 0_2\right), & \tilde{B} &= \frac{1}{2c}B\left[1\ -1\right], \\ Z_N(A,B) &= \left[\mathscr{N}_N^\top(A,B)\ \ c\mathscr{Z}_N^\top(A,B)\right]^\top, && \\ \mathscr{Z}_N(A,B) &= \mathbb{1}_N H_N(A,B) - \bar{\mathbb{1}}_N G_N(A,B) - \left[0_{2N+2,n}\ L_N\ 0_{2N+2,2}\right], && \\ L_N &= \begin{bmatrix} \ell_{0,0}I_2 & \cdots & 0_2 \\ \vdots & \ddots & \vdots \\ \ell_{N,0}I_2 & \cdots & \ell_{N,N}I_2 \end{bmatrix}, & \mathbb{1}_N &= \begin{bmatrix} I_2 \\ \vdots \\ I_2 \end{bmatrix},\ \bar{\mathbb{1}}_N = \begin{bmatrix} I_2 \\ \vdots \\ (-1)^N I_2 \end{bmatrix}, \\ G_N(A,B) &= \left[0_{2,n+2N+2}\ g\right] + \begin{bmatrix} K \\ 0_{1,n} \end{bmatrix}\mathscr{N}_N(A,B), & g &= \begin{bmatrix} 0 & 1 \\ 1+cc_0 & 0 \end{bmatrix}, \\ H_N(A,B) &= \left[0_{2,n+2N+2}\ h\right] + \begin{bmatrix} 0_{1,n} \\ K \end{bmatrix}\mathscr{N}_N(A,B), & h &= \begin{bmatrix} 1-cc_0 & 0 \\ 0 & -1 \end{bmatrix}. \end{array} \tag{7}$$

Then, there exists a unique solution to the coupled infinite dimensional system (1) and it is exponentially stable in the sense of norm $\|\cdot\|_{\mathscr{H}}$, i.e. there exist $\gamma > 1$ and $\delta > 0$ such that energy estimate (4) *holds.*

Remark 2 It is worth mentioning that this theorem introduces a hierarchy in the stability conditions, that means all the systems stable using the previous theorem at an order N_0 are also stable for all higher orders $N \geqslant N_0$.

This theorem comes from the following Lyapunov functional at a given order $N \geqslant 0$:

$$V_N(X_N, u) = X_N^\top P_N X_N + \int_0^1 \chi^\top(x)\,(S + xR)\,\chi(x)dx, \tag{8}$$

where $\chi(x) = \begin{bmatrix} u_t(x)+cu_x(x) \\ u_t(1-x)-cu_x(1-x) \end{bmatrix}$ is a modified Riemann invariant for system (1) and X_N is an extended state made up of the ODE state X and projections of the infinite dimension state χ on a basis of shifted Legendre polynomials on Ω. We denote this projections as follows:

$$\mathfrak{X}_k := \langle \chi, \mathscr{L}_k \rangle = \int_0^1 \chi(x)\mathscr{L}_k(x)dx,$$

where $\mathscr{L}_k$ is the kth Legendre polynomial. Introducing an extended state, we get:

$$X_N = \begin{bmatrix} X^\top & \mathfrak{X}_0^\top & \cdots & \mathfrak{X}_N^\top \end{bmatrix}.$$

Proof This part is inspired by [13, 20] and aims at reducing the conservatism introduced by the positivity condition obtained in [1]. The desired property on the Lyapunov functional is as follow:

$$\begin{aligned} \varepsilon_1 \|(X, u, u_t)\|^2_{\mathscr{H}} \leqslant V_N(X_N, u) &\leqslant \varepsilon_2 \|(X, u, u_t)\|^2_{\mathscr{H}}, \\ \dot{V}_N(X_N, u) &\leqslant -\varepsilon_3 \|(X, u, u_t)\|^2_{\mathscr{H}}, \end{aligned}$$

The existence of ε_2 and ε_3 strictly follows the same lines than as [1]. The difference between Theorem 2 of [1] and this one lies in the existence of ε_1 because of the relaxed condition $\Phi_N \succ 0$ compared to $P_N \succ 0$.

As $S, R \in \mathbb{S}^2_+$ and $\Phi_N \succ 0$, there exists $\varepsilon_1 > 0$ such that the following inequality holds :

$$\begin{aligned} &\Phi_N \succeq \varepsilon_1 \left(\operatorname{diag}\left(I_n + K^\top K, 0_{2(N+1)}\right) + \tfrac{2+c^2}{2c^2} \operatorname{diag}\left(0_n, I_{2(N+1)}\right) \right), \\ &\forall x \in [0, 1], \quad S + xR \succeq S \succeq \varepsilon_1 \tfrac{2+c^2}{2c^2} I_2. \end{aligned}$$

Then, these inequalities and Eq. (8) leads to:

$$\begin{aligned} V_N(X_N, u) \geqslant X_N^\top \Phi_N X_N + \varepsilon_1 \left(|X|_n^2 + \frac{2+c^2}{2c^2} \|\chi\|^2 \right) \\ + \int_0^1 \chi^\top(x)(S - \delta)\chi(x)dx - \sum_{k=0}^{N} (2k+1)\mathfrak{X}_k^\top (S - \delta)\mathfrak{X}_k, \end{aligned}$$

with the shorthand notation $\delta = \varepsilon_1 \frac{2+c^2}{2c^2} I_2$. Using Lemma 1 together with the norm equality $\|\chi\|^2 = 2\left(\|u_t\|^2 + c^2\|u_x\|^2\right)$ implies:

$$V_N(X_N, u) \geqslant \varepsilon_1 \left(|X|_n^2 + \|u_t\|^2 + (2 + c^2)\|u_x\|^2 \right).$$

Finally, using Lemma 1 of [1] brings the conclusion: $V_{N,\alpha}(X_N, u) \geqslant \varepsilon_1 \|(X, u, u_t)\|^2_{\mathscr{H}}$.

Remark 3 Theorem 1 is less restrictive than Theorem 2 of [1] because the condition $P_N \in \mathbb{S}_+^{n+2(N+1)}$ is included in $\Phi_N \succ 0$.

4.2 An Equivalent Formulation

The previous theorem is not affine in A and B and consequently, it cannot be easily extended into a robust formulation. Before stating the robust corollary, the following equivalent formulation is proposed.

Corollary 1 *Consider system (1) with a given speed $c > 0$, a viscous damping $c_0 > 0$, A and B given and known with initial conditions $(X^0, u^0, v^0) \in \mathcal{H}^2$ compatible with the boundary conditions.*

Assume that, for a given integer $N \in \mathbb{N}$, there exist $P_N \in \mathbb{R}^{n+2(N+1)\times n+2(N+1)}$, $S, R \in \mathbb{S}^2_+$, $Y_N \in \mathbb{R}^{2\times(n+2N+6)}$ and $\mu > 0$ such that the following LMIs are satisfied:

$$\begin{cases} \Phi_N = P_N + \mathrm{diag}(0_n, S, 3S, \dots, (2N+1)S) \succ 0, \\ \Theta_N(A, B) \prec 0, \end{cases} \tag{9}$$

with

$$\begin{aligned} \Theta_N(A,B) &= \begin{bmatrix} \mathcal{M}_N(A,B) + \mathrm{He}(Y_N^\top \mathcal{W}_N(A,B)) & Y_N^\top & \mu \mathcal{W}_N^\top(A,B) \\ \star & -\mu I_2 - cS & 0_2 \\ \star & \star & -\mu I_2 \end{bmatrix}, \\ \mathcal{M}_N(A,B) &= \begin{bmatrix} \mathrm{He}\left(Z_N^\top(A,B) P_N F_N\right) - c\tilde{R}_N & H_N^\top(A,B)(S+R) \\ (S+R)H_N(A,B) & -\frac{1}{c}(S+R) \end{bmatrix}, \\ \mathcal{W}_N(A,B) &= \left[G_N(A,B)\ 0_2\right], \end{aligned}$$

where $\star$ denotes the symmetric.

Then, there exists a unique solution to the coupled infinite dimensional system (1) and it is exponentially stable in the sense of norm $\|\cdot\|_{\mathcal{H}}$, i.e. there exist $\gamma > 1$ and $\delta > 0$ such that energy estimate (4) *holds.*

Proof A similar strategy as the one developed in [7] is used here. Starting from Theorem 1, system (1) is exponentially stable if $\Phi_N \succ 0$ and $\Psi_N(A, B) \prec 0$. Ψ_N is not affine in A and B and it is apparently not possible to conclude whether or not it is even convex with respect to A and B. But, applying Schur complement, an equivalent formulation for $\Psi_N(A, B) \prec 0$ is:

$$\begin{bmatrix} \mathrm{He}\left(Z_N^\top(A,B) P_N F_N\right) - c\tilde{R}_N & H_N^\top(A,B)(S+R) \\ (S+R)H_N(A,B) & -\frac{1}{c}(S+R) \end{bmatrix} - c\mathcal{W}_N^\top(A,B) S \mathcal{W}_N(A,B) \prec 0.$$

$\mathcal{M}_N(A, B)$ is affine in A and B but it remains to treat the quadratic term $\mathcal{W}_N(A, B)$ which is still not affine in A and B. Using Lemma 2, the previous statement is equivalent to the existence of $\mu > 0$ and $Y_N \in \mathbb{R}^{2\times(n+2N+6)}$ such that the following hold:

$$\mathcal{M}_N(A,B) + \mathrm{He}(Y_N^\top \mathcal{W}_N(A,B)) + Y_N^\top(\mu I_2 + cS)^{-1} Y_N + \mu \mathcal{W}_N^\top(A,B)\mathcal{W}_N(A,B) \prec 0.$$

The second condition to get the equivalence is $\mu I_2 + cS \succ 0$ which is always guaranteed as $\mu > 0$ and $S \succ 0$.
Using two Schur complements on the first equation, the following equivalence is obtained for a given pair (A, B):

$$\Psi_N(A, B) \prec 0 \quad \Leftrightarrow \quad \Theta_N(A, B) \prec 0. \tag{10}$$

■

Remark 4 Corollary 1 is equivalent to Theorem 1. The two main differences are first in terms of complexity because there are two new variables Y_N and μ and secondly because it is affine in A and B.

4.3 Robustness Analysis

Using the new LMI conditions obtained previously, we can extend Theorem 1 to get the robust exponential stability of uncertain system (1). One of the benefits of using a Lyapunov functional approach relies indeed on the possibility of extending the previous analysis to robustness analysis. The uncertainties on matrices A and B are assumed to be of polytopic type, as described in equation (3). That leads to the following result.

Corollary 2 *Consider uncertain system (1) with a given speed $c > 0$, a viscous damping $c_0 > 0$ and initial conditions $(X^0, u^0, v^0) \in \mathscr{H}^2$ compatible with the boundary conditions. The uncertain pair $[A\ B]$ satisfies equation* (3).
Assume that, for a given integer $N \in \mathbb{N}$, there exist $P_N \in \mathbb{R}^{n+2(N+1) \times n+2(N+1)}$, $S, R \in \mathbb{S}^2_+$, $Y_N \in \mathbb{R}^{2 \times (n+2N+6)}$ and $\mu > 0$ such that the following LMIs are satisfied:

$$\begin{cases} \Phi_N = P_N + \operatorname{diag}(0_n, S, 3S, \ldots, (2N+1)S) \succ 0, \\ \Theta_N(A_i, B_i) \prec 0 \, for\ all\ i \in [1, m], \end{cases} \tag{11}$$

with the notation of Corollary 1.

Then, there exists a unique solution to the coupled uncertain infinite dimensional system (1) and it is exponentially stable in the sense of norm $\| \cdot \|_{\mathscr{H}}$, i.e. there exist $\gamma > 1$ and $\delta > 0$ such that energy estimate (4) *holds.*

Proof The main advantage now is that Θ_N is affine in A and B if P_N, R, S, Y_N and μ are fixed. And consequently, it is convex with respect to A and B at a price of an increased complexity. As it is convex in A and B, if $\Theta_N(A_i, B_i) \prec 0$ for all $i = 1, \ldots, m$, then $\Theta_N(A, B) \prec 0$. Indeed, if equation (3) holds, then we get:

$$\begin{array}{l} Z_N(A, B) = \sum_{i=1}^m \lambda_i Z_N(A_i, B_i), \quad H_N(A, B) = \sum_{i=1}^m \lambda_i H_N(A_i, B_i), \\ G_N(A, B) = \sum_{i=1}^m \lambda_i G_N(A_i, B_i), \quad \text{and consequently,} \\ \mathscr{M}_N(A, B) = \sum_{i=1}^m \lambda_i \mathscr{M}_N(A_i, B_i), \ \mathscr{W}_N(A, B) = \sum_{i=1}^m \lambda_i \mathscr{W}_N(A_i, B_i). \end{array}$$

The previous considerations lead to $\Theta_N(A, B) = \sum_{i=1}^{m} \lambda_i \Theta_N(A_i, B_i)$. As for $i \in [1, m]$, $\Theta_N(A_i, B_i) \prec 0$ then $\Theta_N(A, B) \prec 0$. Corollary 1 can be applied, ensuring the exponential stability for all $[A\ B]$ in $\mathrm{Co}_{i=1,\dots m}\{[A_i\ B_i]\}$ and that ends the proof. ■

Remark 5 Conditions (11) must hold for all $i \in [1, m]$. That means P_N, R, S, $\mathscr{X}_N$ and μ are independent of i, making the condition highly conservative for large m or very different A and B matrices.

Corollary 2 is a robust corollary as A and B may vary inside an uncertainty set but the system keeps its exponential stability behavior. This result can also be applied for some bounded time-varying A and B. The following section will give an example about the robustness property.

5 Examples

For the example section, we will study the following example taken from [13, 23]. This example is a regenerative chatter if the wave is a pure delay (for $cc_0 = 1$). The system is given as follows:

$$A(k) = \begin{bmatrix} 0 & 0 & 1 & 0 \\ 0 & 0 & 0 & 1 \\ -10-k & 10 & 0 & 0 \\ 5 & -15 & 0 & -0.25 \end{bmatrix},\ B(k) = \begin{bmatrix} 0 \\ 0 \\ k \\ 0 \end{bmatrix} \text{ and } K = \begin{bmatrix} 1 \\ 0 \\ 0 \\ 0 \end{bmatrix}^{\top}. \tag{12}$$

For $k = 1$ and different values of $c > 0$ and $c_0 > 0$, Corollary 1 gives Fig. 2. The stability regions are in gray scale. The hatched areas are the exact unstable areas, they are obtained using the methodology described in [2]. As the order increases, more stability pockets are detected. Theorem 1 seems an effective tool for the stability analysis of coupled systems. The chart is different than in [1] because it gives a better illustration of the stability behavior as c ad c_0 vary. Indeed, it displays stability pockets which are known to be of difficult access for Lyapunov-like techniques [12]. Moreover, it is easier to compare with the results obtained with time-delay systems as τ is the delay and for $cc_0 = 1$, system (1) is a time-delay system (see [2] for more information).

For the robust stability analysis, we can use an interval for the k parameter. For instance, if $k \in (1, 2)$, we get the following vertices: $[A(1)\ B(1)]$ and $[A(2)\ B(2)]$. We tested the two scenari $k \in (1, 2)$ and $k \in (0.1, 1.5)$ for some values of c and c_0. The results are displayed in Fig. 3. The black area is stable for Corollary 2 with the desired uncertainties. The white area is unknown and the hashed area is an inner approximation of the unstable area. One can see that there exist areas for which the system is robustly stable to variation of k. To the best of our knowledge, there does not exist an exact robust stability criterion so we cannot compare with the exact

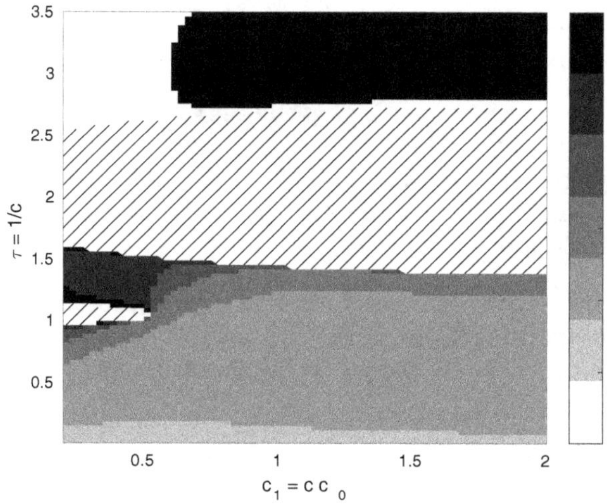

Fig. 2 Stability analysis using Theorem 1 with example (12) for $k = 1$. The hatched area is the exact unstable area. The scale of gray is the stable area using Corollary 1 for $N = 0$ to 5. The white area is the stable area, undetected by the corollary up to the order 5

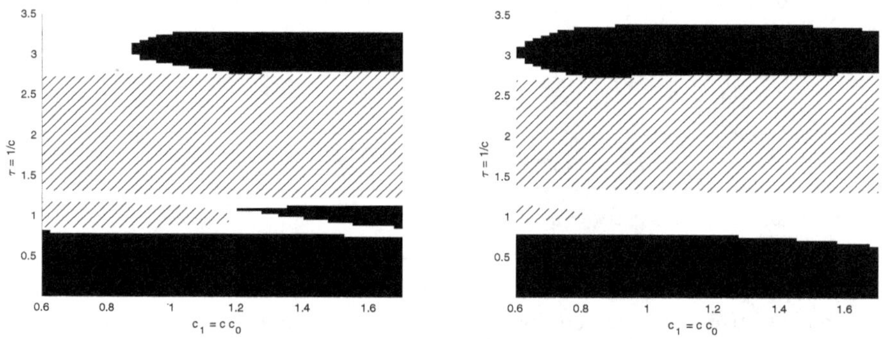

(a) $k \in (1,2)$ (b) $k \in (0.1,1.5)$

Fig. 3 Stability areas for system (1) with matrices defined in (12) with $c \in (0.28, 100)$ and $cc_0 \in (0.6, 1.7)$. The black area is robustly stable in the considered interval while the white area is unknown at an order $N = 7$. The hashed area is an inner-approximation of the unstable area for the system. For (**a**), the hatched area corresponds to the exact unstable area for $k \in \{1, 1.2, 1.5, 2\}$ obtained using a pole location argument. For (**b**), this area is related to the unstable areas for $k \in \{0.1, 0.5, 0.8, 1\}$

unstable area. This inner approximation is obtained as a summation of unstable area for some values of k.

Corollary 2 is able to detect stability pockets and seems to perform with a good accuracy because the stable area is very close to the hashed one. For $c_1 = cc_0 = 1$, an exact robust stability criteria can be achieved using the Control Toolbox of Matlab®

together with the `allmargin` command as said in [20]. The results are displayed below and the stable intervals are written in terms of τ:

Scenario	$k \in (1, 2)$	$k \in (0.1, 1.5)$
`allmargin`	$(0, 0.859) \cup (1.117, 1.264) \cup (2.75, 3.5)$	$(0, 1.328) \cup (2.718, 3.5)$
Cor. 2 ($N = 7$)	$(0, 764) \cup (2.869, 3.254)$	$(0, 764) \cup (2.746, 3.37)$

These two results are very close even if it misses one stability pocket for the first case. So we can say that polytopic uncertainties on A and B are well-addressed using Corollary 2.

6 Conclusion

As a conclusion, the present chapter addressed the problem of robust stability of an interconnected ODE/String system. The robust stability result is obtained considering polytopic uncertainties on the ODE part, i.e. on matrices A and B. The extended theorem proposes to enlarge the stability result by considering a relaxed positivity condition and the robust result takes advantage of the convex form of the LMIs. This corollary has been tested on a well-known example: a regenerative chatter. Comparing with exact stability results, the proposed methodology is indeed very accurate.

References

1. M. Barreau, A. Seuret, F. Gouaisbaut, L. Baudouin, Lyapunov stability analysis of a string equation coupled with an ordinary differential system. IEEE Trans. Autom. Control **63**(11), 3850–3857 (2018)
2. M. Barreau, F. Gouaisbaut, A. Seuret, R. Sipahi, Input/Output stability of a damped string equation coupled with an ordinary differential system. Int. J. Robust Non-linear Control **28**, 6053–6069 (2018)
3. L. Baudouin, A. Seuret, F. Gouaisbaut, M. Dattas, Lyapunov stability analysis of a linear system coupled to a heat equation. *IFAC-PapersOnLine*, vol. 50, Issue 1, July 2017, pp. 11978–11983 (2017)
4. J. Bontsema, R.F. Curtain, A note on spillover and robustness for flexible systems. IEEE Trans. Autom. Control **33**(6), 567–569 (1988)
5. D. Bresch-Pietri, M. Krstic, Output-feedback adaptive control of a wave PDE with boundary anti-damping. Automatica **50**(5), 1407–1415 (2014)
6. C. Briat, Convergence and equivalence results for the Jensens inequality-application to time-delay and sampled-data systems. IEEE Trans. Autom. Control **56**(7), 1660–1665 (2011)
7. C. Briat, A. Seuret, Robust stability of impulsive systems: a functional-based approach, in *4th IFAC conference on Analysis and Design of Hybrid Systems* (ADHS-2012), Eindhoven, Netherlands (2012)
8. N. Challamel, Rock destruction effect on the stability of a drilling structure. J. Sound Vib. **233**(2), 235–254 (2000)

9. G. Chen, J. Zhou, The wave propagation method for the analysis of boundary stabilization in vibrating structures. SIAM J. Appl. Math. **50**(5), 1254–1283 (1990)
10. R. Courant, D. Hilbert, *Methods of Mathematical Physics* (Wiley, 1989)
11. E. Fridman, S. Mondié, B. Saldivar, Bounds on the response of a drilling pipe model. IMA J. Math. Control Inf. **27**(4), 513–526 (2010)
12. E. Fridman, *Introduction to Time-Delay Systems: Analysis and Control* (Springer, 2014)
13. K. Gu, J. Chen, V.L. Kharitonov, *Stability of Time-Delay Systems* (Springer, 2003)
14. M. Krstic and A. Smyshlyaev, *Boundary Control of PDEs: A Course on Backstepping Designs*. Voume 16 of Advances in Design and Control (SIAM, 2008)
15. M. Krstic, *Delay Compensation for Nonlinear, Adaptive, and PDE Systems* (Springer, 2009)
16. J. Lagnese, Decay of solutions of wave equations in a bounded region with boundary dissipation. J. Differ. Equ. **50**(2), 163–182 (2017)
17. Z.H. Luo, B.Z. Guo, Ö. Morgül, *Stability and Stabilization of Infinite Dimensional Systems with Applications* (Springer, 2012)
18. L. Meirovitch, H. Baruh, On the problem of observation spillover in self-adjoint distributed-parameter systems. J. Optim. Theory Appl. **39**(2), 269–291 (1983)
19. M.B. Saldivar, I. Boussaada, H. Mounier, S.I. Niculescu, *Analysis and Control of Oilwell Drilling Vibrations: A Time-Delay Systems Approach* (Springer, 2015)
20. A. Seuret, F. Gouaisbaut, Hierarchy of LMI conditions for the stability analysis of time delay systems. Syst. Control Lett. **81**, 1–7 (2006)
21. S. Tang, C. Xie, State and output feedback boundary control for a coupled PDE-ODE system. Syst. Control Lett. **60**(8), 540–545 (2011)
22. M. Tucsnak, G. Weiss, *Observation and Control for Operator Semigroups* (Springer, 2009)
23. J. Zhang, C.R. Knopse, P. Tsiotras, Stability of time-delay systems: equivalence between Lyapunov and scaled small-gain conditions. IEEE Trans. Autom. Control **46**(3), 482–486 (2001)

Regional Stability of Nonlinear Sampled-Data Controlled Systems Under Actuator Saturation: A Quasi-LPV Approach

Alessandra Helena Kimura Palmeira, Joao Manoel Gomes da Silva Jr., and Jeferson Vieira Flores

Abstract This chapter addresses the stability analysis of sampled-data control for a class of continuous-time nonlinear systems. The proposed approach is based on a local quasi-LPV (Linear Parameter Varying) model for the nonlinear system and the use of a parameter dependent looped-functional to deal with the aperiodic sampling effects. Moreover, the bounds on the control inputs imposed by the physical actuators are also explicitly considered. The effects of the control saturation on the stability are tackled through a parameter dependent generalized sector condition. From these ingredients, conditions in an LMI form are proposed to assess local stability. These conditions are then incorporated in convex optimization problems aiming at obtaining maximized estimates of the region of attraction of the origin or maximizing the inter-sampling time for which the stability is ensured regionally.

Keywords Sampled-data control · Quasi-LPV systems · Nonlinear systems · Input saturation · Linear matrix inequality (LMI)

The authors are also supported by Conselho Nacional de Desenvolvimento Científico e Tecnológico (CNPQ), under grants 306223/2018-0 and 307449/2019-0, and Coordenacão de Aperfeiçoamento de Pessoal de Nível Superior (CAPES)—Finance Code 001, Brazil.

A. H. K. Palmeira · J. M. Gomes da Silva Jr. (✉) · J. V. Flores
School of Engineering, Universidade Federal do Rio Grande do Sul (UFRGS), Porto Alegre, Brazil
e-mail: jmgomes@ufrgs.br

A. H. K. Palmeira
e-mail: kimura.palmeira@ufrgs.br

J. V. Flores
e-mail: jeferson.flores@ufrgs.br

G. Valmorbida et al. (eds.), *Accounting for Constraints in Delay Systems*, Advances in Delays and Dynamics 12,
https://doi.org/10.1007/978-3-030-89014-8_10

1 Introduction

Motivated by the development of networked and embedded control, the interest in sampled-data systems has increased in the last decade [10]. The problem is particularly relevant in the context of nonlinear systems. Differently from the linear ones, exact discretization in general cannot be achieved and discretized models obtained by numerical approximations can lead to unexpected behaviors (such as instability) when used to design sampled-data control laws [16]. Hence, it is of paramount importance to take into account the continuous-time behavior of the plant and the discrete-time update of the control signal. To tackle this problem, many approaches can be found in the literature for sampled-data linear systems. We can cite for instance the ones based on a hybrid system framework [15], where the sampling phenomena is modeled through a impulsive system, and the approaches based on a time-delay system framework, where Lyapunov-Krasovskii [6, 7] and looped-functionals [3, 20] are applied for stability assessment.

On the other hand, differently from linear systems, where the results are stated generically, in the nonlinear case most of the constructive results regard specific classes of systems. In particular, some nonlinear systems can be written in the so-called quasi-LPV (Linear Parameter Varying) form. The quasi-LPV approach is based on the possibility of rewriting the plant in a form where nonlinear terms can be replaced by time-varying parameters that depend on the system state [19]. In this case, techniques of LPV control (see for instance [2, 11] and references therein) can be applied to analysis and synthesis of the closed-looped system. In particular, the controller can be designed as an LPV one, i.e. a time-varying linear control law scheduled by the state dependent parameter. One advantage in this case is that the controller synthesis and the analysis (stability, performance, robustness, etc.) of the closed-loop system can be performed in a framework based on Linear Matrix Inequality (LMI) constraints and convex optimization.

Considering the LPV framework, few papers formally address the sampled-data control problem. In [24], it is assumed that the scheduling parameters do not change in the interval between two samples, which is a strong assumption. A time-delay approach is considered in [18] and the proposed stability conditions are in the form of infinite dimensional LMIs. To tackle this problem, griding techniques are considered. In [22] a model predictive control scheme is considered. In this case, a set of LMIs, obtained from a Lyapunov-Krasovskii approach, must be solved at each sampling instant to determine the control law to be applied. More recently, a parameter dependent looped-functional approach has been proposed in [8]. On the other hand, there is a lack of papers in the literature considering the sampled-data control for quasi-LPV systems. In this context, we can mainly find works dealing with fuzzy Takagi-Sugeno (T-S) models, which can be seen as a particular form of quasi-LPV modeling. For instance, using a Lyapunov-Krasovskii functional approach, LMI conditions to compute a sampled-data T-S fuzzy controller to guarantee exponential stabilization of the closed-loop system are proposed in [27]. In [13], LMI conditions are derived to compute a T-S state feedback or output feedback to asymptotic stabilization and $\mathscr{H}_\infty$ disturbance attenuation for vehicle suspension systems. However, in most cases the quasi-LPV or T-S fuzzy models represent the nonlinear system only regionally [23]. Unfortunately, this fact is ignored in the majority of the papers dealing with this kind of modeling. If a region of validity of the model is not considered, the stability guarantee for the quasi-LPV system in a global context will not hold for the actual nonlinear closed-loop system. In fact, in this case, only the locally stability of the origin may be ensured.

On the other hand, input saturation is an ubiquitous feature in control systems. Due to physical or security limitations the actuators are constrained to provide signals with magnitude inside certain intervals. For linear time-invariant systems, this issue has been addressed considering sampled-data control in [17, 21]. Since the saturation induces a nonlinear closed-loop behavior, conditions are formulated in a local (regional) or global contexts. In the regional case, it is of paramount importance the characterization of stability sets or estimates of the region of attraction [25]. In [22] (mentioned above) and [5], the Lyapunov-Krasovskii approach is used to compute a sampled-data control signal for LPV systems under saturation. In the first paper, a sampled-data model predictive control (MPC) scheme is considered, while the second paper addresses a disturbance attenuation problem. Regarding nonlinear systems, the study of the effects of control saturation on the regional stability is for instance addressed in [9] for Lur'e-type systems and in [4, 26] for rational and polynomial systems. However it should be pointed out that these references do not consider the sampled-data case.

This chapter focus on the problem of stability assessment of nonlinear systems, locally represented by quasi-LPV models, under constrained sampled-data control. It is explicitly assumed that the control signal is updated only at the sampling instants and kept constant between two consecutive samples, while the plant evolves continuously. The stability analysis is based on a local polytopic modeling of the quasi-LPV representation and the use of a parameter dependent looped-functional to take into account the sampling effects. Both periodic and aperiodic sampling cases can be handled by the proposed approach. The possible control saturation imposed by the actuators limits is tackled through a parameter dependent generalized sector condition. From these ingredients, LMI conditions that guarantee the local asymptotic stability of the origin of the closed-loop system are derived. In particular, considering the minimal and maximal bounds on the time between two successive samplings, the conditions allow to characterize estimates of the region of attraction of the origin. Convex optimization problems are then proposed to either obtain a maximized estimate of the region of attraction of the origin or, for a given set of admissible initial conditions, maximize the upper bound on the inter-sampling interval for which the regional stability is ensured.

The chapter is organized as follows. In Sect. 2, the basic assumptions and the problem formulation are presented. Section 3 provides some basic results regarding the local polytopic modeling of quasi-LPV systems, a generalized sector condition to cope with saturation effects in a local framework and the looped functional approach. In Sect. 4, the LMI conditions for the regional stability assessment are derived. These conditions are incorporated in convex optimization problems in Sect. 5. To illustrate the application of the results, a numerical example considering a polynomial system is presented in Sect. 6. The chapter ends with some concluding remarks.

Notation. Throughout the article, the sets $\mathbb{R}^+$, $\mathbb{R}^n$, $\mathbb{R}^{n\times n}$ and $\mathbb{S}^n$ denote respectively the set of positive scalars, n-dimensional vectors, $n \times n$ matrices and symmetric matrices of $\mathbb{R}^{n\times n}$. For a given positive scalar T, define $\mathscr{C}^n_{[0,T]}$ as the set of continuous functions from an interval $[0, T]$ to $\mathbb{R}^n$ and the union set of continuous functions with support in a certain range defined as $\mathbb{K}^n_{[\mathscr{T}_1,\mathscr{T}_2]} = \cup_{T\in[\mathscr{T}_1,\mathscr{T}_2]}\{\mathscr{C}^n_{[0,T]}\}$. $\|\cdot\|$ stands for the Euclidean norm of a vector. The superscript '$'$' stands for matrix transposition. The notation $P > 0$ means that P is positive definite. He$\{A\}$ refers to $A + A'$, symbols I^m and $0^{m\times n}$ represent the identity and the zero matrices of dimension m and $m \times n$. $A_{(i)}$ or $x_{(i)}$ represent the i-th line of the matrix A or i-th element of the vector x. $Co\{\cdot\}$ denotes a convex hull, $Ver(\mathscr{B}_\sigma)$ represents the vertices of polytope $\mathscr{B}_\sigma$. $\mathscr{E}(P, c)$ denotes the ellipsoid set $\mathscr{E}(P, c) = \{x \in \mathbb{R}^n; x'Px \le c\}$, with $P = P' > 0$, $c > 0$. $\partial\mathscr{S}$ defines the boundary of the set $\mathscr{S}$. $\Lambda^n(v)$ is a shortcut for the Kronecker product $v \otimes I^n$.

2 Problem Formulation

Consider the following class of controlled continuous-time nonlinear systems:

$$\dot{x}(t) = f(x(t)) + g(x(t))sat(u(t)), \tag{1}$$

$$u(t) = l(x(t)), \tag{2}$$

where $x \in \mathbb{R}^n$ and $u \in \mathbb{R}^m$ are, respectively, the state and control input vectors, $f(\cdot) : \mathbb{R}^n \to \mathbb{R}^n$, $g(\cdot) : \mathbb{R}^n \to \mathbb{R}^{n\times m}$ and $l(\cdot) : \mathbb{R}^n \to \mathbb{R}^m$ are continuous and (locally) Lipschitz nonlinear functions, with $f(0) = 0$, $l(0) = 0$, i.e. the origin is an equilibrium point of (1)–(2). The control law is a nonlinear state feedback control law represented by $l(x(t))$. The saturation operator, $sat(\cdot) : \mathbb{R}^m \to \mathbb{R}^m$, models the physical limits imposed by the actuator. Each component of the saturation function is classically defined as follows:

$$sat_{(i)}(u) = sign(u_{(i)})min\left\{\left|u_{(i)}\right|, \bar{u}_{(i)}\right\}, \tag{3}$$

for $i = 1, ..., m$, with $\bar{u}_{(i)} > 0$, i.e. $\pm\bar{u}_{(i)}$ are the symmetrical bounds on $i-$th input signal.

To implement the control law, a sampled-data policy is considered. The values of $x(t)$ are assumed to be measured and therefore available for feedback only at the sampling instant t_k, with $k \ge 0$ and t_k being an increasing sequence of positive scalars such that $\bigcup[t_k, t_{k+1}) = [0, +\infty)$. Thus, based on the sampled state, at each sampling time the control signal is updated and its value is kept constant during the sampling interval $[t_k, t_{k+1})$ by means of a zero order holder (ZOH). Hence, the sampled-data implementation of control law (2) is mathematically described as follows:

$$u(t) = l(x(t_k)), \quad \forall t \in [t_k, t_{k+1}), \; \forall k \in \mathbb{N}. \tag{4}$$

The difference between two successive sampling instants is denoted by $T_k = t_{k+1} - t_k$ and it satisfies $0 < \mathscr{T}_1 \leq T_k \leq \mathscr{T}_2$. Note that the particular case where $T_k = \mathscr{T}_1 = \mathscr{T}_2 > 0, \forall k \geq 0$, corresponds to a periodic sampling strategy.

We assume that the nonlinear system (1) can be represented by the following quasi-LPV system

$$\dot{x}(t) = A(\sigma(x(t)))x(t) + B(\sigma(x(t)))sat(u(t)), \tag{5}$$

where $\sigma(x(t))$ is a vector in $\mathbb{R}^N$, $\sigma_{(j)}(x(t))$ are functions of the states, for $j = 1, \ldots, N$. Furthermore, we consider that the control law (4) can also be written in a quasi-LPV form, i.e.:

$$u(t) = l(x(t_k)) = K(\sigma(x(t_k)))x(t_k), \; \forall t \in [t_k, t_{k+1}), \; \forall k \in \mathbb{N}. \tag{6}$$

It is worth noticing that, while the system dynamics in (5) depends continuously on parameter $\sigma(x(t))$, the control law (6) depends on the value of this parameter only at the sampling instants.

Furthermore, we consider the following assumption:

Assumption 1 $A(\sigma)$, $B(\sigma)$ and $K(\sigma)$ depend affinely on σ.

From Assumption 1 and denoting $\sigma(x(t))$ by $\sigma(t)$ for simplicity, $A(\sigma(t))$, $B(\sigma(t))$ and $K(\sigma(t_k))$ can be generically represented as follows:

$$\begin{aligned} A(\sigma(t)) = A_0 + \sum_{j=1}^{N} \sigma_{(j)}(t)A_j, \; B(\sigma(t)) = B_0 + \sum_{j=1}^{N} \sigma_{(j)}(t)B_j, \\ K(\sigma(t_k)) = K_0 + \sum_{j=1}^{N} \sigma_{(j)}(t_k)K_j. \end{aligned} \tag{7}$$

Hence, considering the Assumption 1 and a sampled-data framework, with the matrices defined in (7), the nonlinear system (1) with (4) can be represented by

$$\dot{x}(t) = A(\sigma(t))x(t) + B(\sigma(t))sat(K(\sigma(t_k))x(t_k)), \text{ for } t \in [t_k, t_{k+1}). \tag{8}$$

Assuming that (4) is a stabilizing control law, the set of all initial conditions ($x(0) \in \mathbb{R}^n$) such that $\forall T_k \in [\mathscr{T}_1, \mathscr{T}_2]$, $k \in \mathbb{N}$, the corresponding trajectories of the closed-loop system (1)–(4) converge asymptotically to the origin corresponds to the so-called region of attraction of the origin ($\mathscr{R}_a$) [12]. Due to the difficulty to analytically determine $\mathscr{R}_a$, a problem of interest is to compute an estimate of the $\mathscr{R}_a$ denoted by $\mathscr{X}_0$ such that $\mathscr{X}_0 \subset \mathscr{R}_a \subseteq \mathbb{R}^n$. Then, we can state the following problems:

Problem 1 Given the bounds $\mathscr{T}_1$ and $\mathscr{T}_2$ on the intersampling time T_k, find an estimate $\mathscr{X}_0$ of the region $\mathscr{R}_a$ for the closed-loop system (1)–(4).

Problem 2 Given $\mathscr{T}_1$ and a set of admissible initial conditions $\mathscr{X}_0$, compute the maximum allowable sampling interval $\mathscr{T}_2$ such that $\forall x(0) \in \mathscr{X}_0$ the trajectories of system (1) under the sampled-data control law (4) converge asymptotically to the origin, $\forall T_k \in [\mathscr{T}_1, \mathscr{T}_2]$.

To address Problems 1 and 2, we will use the quasi-LPV model (8), considering that Assumption 1 is satisfied.

3 Preliminaries

In this Section, we present some theoretical tools to be used to obtain conditions to assess the asymptotically stability of the sampled-data quasi-LPV system (8) and, in consequence, the one of the nonlinear system (1)–(4).

3.1 Local Polytopic Modeling

For a set $\mathscr{R}_H \subset \mathbb{R}^n$ containing the origin in its interior, the following assumption regarding the parameter $\sigma(t) = \sigma(x(t))$ in system (8) is considered.

Assumption 2 For $x(t) \in \mathscr{R}_H$, it follows that $\sigma(t) \in \mathscr{B}_\sigma$, with

$$\mathscr{B}_\sigma = \{\sigma \in \mathbb{R}^N;\ \underline{\sigma}_{(j)} \leq \sigma_{(j)} \leq \overline{\sigma}_{(j)},\ j = 1, \cdots, N\}. \tag{9}$$

Set $\mathscr{R}_H$ can be seen as a region of operation for the nonlinear system (1)–(2). In particular, we consider that $\mathscr{R}_H$ is a polyhedral set described as follows:

$$\mathscr{R}_H = \{x \in \mathbb{R}^n;\ \left|H_{(r)}x\right| \leq \eta_{(r)},\ r = 1, \ldots, r_c\}, \tag{10}$$

with $\eta_{(r)} > 0$, $H_{(r)} \in \mathbb{R}^{1\times n}$.

From the bounds on $\sigma(t)$ defined in (9), it follows that $\mathscr{B}_\sigma$ is in fact a convex polytope in $\mathbb{R}^N$ (it is in particular a hypercube) and can be described as the convex hull of its 2^N vertices, i.e.:

$$\mathscr{B}_\sigma = Co\{v_1, v_2, \ldots, v_{2^N}\}.$$

Hence, provided that $x(t) \in \mathscr{R}_H$, $\sigma(t)$ can be obtained as a convex combination of the vertices of $\mathscr{B}_\sigma$, i.e.

$$\sigma(t) = \sum_{g=1}^{2^N} \gamma_g(t) v_g, \text{ with } \gamma_g(t) \geq 0,\ \sum_{g=1}^{2^N} \gamma_g(t) = 1. \tag{11}$$

Defining $\mathbf{A} = [A_1\ A_2\ \cdots\ A_N]$, $\mathbf{B} = [B_1\ B_2\ \cdots\ B_N]$ and $\mathbf{K} = [K_1\ K_2\ \cdots\ K_N]$, assuming that Assumptions 1 and 2 hold and considering the shortcut $\Lambda^n(\nu) = \nu \otimes I^n$, it is therefore possible to represent $A(\sigma(t))$, $B(\sigma(t))$ and $K(\sigma(t_k))$ as follows:

$$\begin{aligned} A(\sigma(t)) &= A_0 + \mathbf{A}\Lambda^n(\sigma(t)) = A_0 + \mathbf{A}(\sigma(t) \otimes I^n) = \sum_{g=1}^{2^N} \gamma_g(t) A(v_g), \\ B(\sigma(t)) &= B_0 + \mathbf{B}\Lambda^m(\sigma(t)) = B_0 + \mathbf{B}(\sigma(t) \otimes I^m) = \sum_{g=1}^{2^N} \gamma_g(t) B(v_g), \\ K(\sigma(t_k)) &= K_0 + \mathbf{K}\Lambda^n(\sigma(t_k)) = K_0 + \mathbf{K}(\sigma(t_k) \otimes I^n) = \sum_{g=1}^{2^N} \gamma_g(t_k) K(v_g), \end{aligned} \tag{12}$$

with

$$A(v_g) = A_0 + \mathbf{A}\Lambda^n(v_g),\ B(v_g) = B_0 + \mathbf{B}\Lambda^m(v_g),\ K(v_g) = K_0 + \mathbf{K}\Lambda^n(v_g). \tag{13}$$

Hence, provided that $x(t) \in \mathscr{R}_H, \forall t \geq 0$, the behavior of closed-looped nonlinear system (1)–(2) can be locally represented by the following polytopic differential inclusion (PDI) for $t \in [t_k,\ t_{k+1}), \forall k \geq 0$:

$$\dot{x}(t) = \left(\sum_{g=1}^{2^N} \gamma_g(t) A(v_g) \right) x(t) + \left(\sum_{g=1}^{2^N} \gamma_g(t) B(v_g) \right) sat \left(\sum_{g=1}^{2^N} \gamma_g(t_k) K(v_g) x(t_k) \right) \tag{14}$$

Set $\mathscr{R}_H$ is therefore the region of validity of the representation (14) for system (1)–(4). In fact, provided that $x(t) \in \mathscr{R}_H$ the dynamics of the actual nonlinear system (1)–(4) is embedded in the ones generated by (14). The idea is therefore to use the PDI (14) to conclude about the stability of (1)–(4). In this case, one should keep in mind that the results are only valid for the trajectories that are kept inside $\mathscr{R}_H$. Hence, if for a certain set $\mathscr{X}_0$ it is guaranteed for (14) that $\forall x(0) \in \mathscr{X}_0$, $x(t) \to 0$ as $t \to \infty$, we can conclude that $\mathscr{X}_0$ is included in the region of attraction of the origin of system (1)–(4) only if $\mathscr{X}_0 \subset \mathscr{R}_H$ and the closed-loop trajectories never leaves $\mathscr{R}_H$. This can be ensured if, in addition, $\mathscr{X}_0$ is a positively invariant set with respect to (14).

3.2 Saturation Handling

Consider the following definition:

$$\psi_k = \psi(u(t_k)) = sat(K(\sigma(t_k))x(t_k)) - K(\sigma(t_k))x(t_k). \tag{15}$$

Note, from the saturation function defined in (3), that the vector valued function $\psi_k : \mathbb{R}^m \to \mathbb{R}^m$ is a dead-zone nonlinearity. From this definition, Eq. (8) can be rewritten as follows:

$$\dot{x}(t) = A(\sigma(t))x(t) + B(\sigma(t))K(\sigma(t_k))x(t_k) + B(\sigma(t))\psi_k, \text{ for } t \in [t_k, t_{k+1}). \tag{16}$$

To deal with the term ψ_k in (16), a parameter dependent version of the generalized sector condition in [25] can be stated.

Lemma 1 *Consider a matrix $S(\sigma(t_k)) \in \mathbb{R}^{m\times n}$. If $x(t_k)$ is such that*

$$\left|(K(\sigma(t_k)) - S(\sigma(t_k)))_{(i)}\, x(t_k)\right| \le \bar{u}_{(i)}, \text{ for } i = 1, \dots, m, \tag{17}$$

then the following inequality is satisfied for any diagonal positive definite matrix $U(\sigma(t_k)) \in \mathbb{R}^{m\times m}$:

$$\psi_k^T U(\sigma(t_k))(\psi_k + S(\sigma(t_k))x(t_k)) \le 0. \tag{18}$$

3.3 Looped-Functional Approach

In order to derive stability conditions for system (1)–(4) from its quasi-LPV representation (8), we will focus on the behavior of the system in the inter-sampling interval $[t_k, t_{k+1})$. As in [3, 20], define $x_k(\tau) = x(t_k + \tau)$ and $\sigma_k(\tau) = \sigma(x(t_k + \tau))$, with $\tau \in [0, T_k)$. Hence, for $\tau \in [0, T_k)$, the closed-loop system (16) can be represented by the following quasi-LPV model

$$\dot{x}_k(\tau) = A(\sigma_k(\tau))x_k(\tau) + B(\sigma_k(\tau))\psi_k + B(\sigma_k(\tau))K(\sigma_k(0))x_k(0). \tag{19}$$

Inspired by the results in [3, 20], the following theorem provides the basis of the looped-functional approach to address Problems 1 and 2. It can also be seen as an extended version of the one proposed in [17].

Theorem 1 *Let a matrix H and vector η that define the set $\mathscr{R}_H$ as in (10) be given. Consider a positive definite function $V : \mathbb{R}^n \to \mathbb{R}^+$ satisfying*

$$\mu_1||x||^2 \le V(x) \le \mu_2||x||^2, \tag{20}$$

with μ_1 and $\mu_2 > 0$, a matrix $S(\sigma_k(0)) \in \mathbb{R}^{m\times n}$ such that for $i = 1, .., m$

$$||\,(K(\sigma_k(0)) - S(\sigma_k(0)))_{(i)}\, x_k(0)||^2 \le \bar{u}_{(i)}^2 V(x_k(0)), \tag{21}$$

and a parameter dependent looped-functional (PDLF)

$$\mathscr{V}_0 : [0, \mathscr{T}_2] \times \mathbb{K}^n_{[\mathscr{T}_1, \mathscr{T}_2]} \times \mathbb{R}^N \times [\mathscr{T}_1, \mathscr{T}_2] \to \mathbb{R},$$

such that

$$\begin{aligned} &\mathscr{V}_0(T_k, x_k, \sigma, T_k) = \mathscr{V}_0(0, x_k, \sigma, T_k) = 0 \\ &\mathscr{V}_0(\tau, x_k, \sigma, T_k) > 0 \end{aligned} \tag{22}$$

$\forall T_k \in [\mathscr{T}_1, \mathscr{T}_2]$, $\forall \tau \in (0, T_k)$, $\forall x_k \in \mathbb{K}^n_{[\mathscr{T}_1, \mathscr{T}_2]}$, $x_k \not\equiv 0$.

Define

$$\mathscr{W}(\tau, x_k, \sigma_k(0), T_k) = V(x_k(\tau)) + \mathscr{V}_0(\tau, x_k, \sigma_k(0), T_k), \tag{23}$$

and let $\dot{\mathscr{W}}(\tau, x_k, \sigma_k(0), T_k)$ be the derivative of $\mathscr{W}(\tau, x_k, \sigma_k(0), T_k)$ with respect to τ. If there exist a positive scalar μ and a positive definite diagonal matrix $U(\sigma_k(0)) \in \mathbb{R}^{m\times m}$, such that the inequalities

$$\dot{\mathscr{W}}(\tau, x_k, \sigma_k(0), T_k) - 2\psi_k' U(\sigma_k(0))(\psi_k + S(\sigma_k(0))x_k(0)) < -\mu \left\| \begin{bmatrix} x_k(\tau) \\ x_k(0) \end{bmatrix} \right\|^2, \tag{24}$$

$$x_k'(\tau) H_{(r)}' H_{(r)} x_k(\tau) \le \eta_{(r)}^2 V(x_k(\tau)), \tag{25}$$

are satisfied along the trajectories of (19), $\forall x_k \in \mathbb{K}^n_{[\mathscr{T}_1, \mathscr{T}_2]}$, $x_k \not\equiv 0$, $\forall \sigma_k(\tau) \in \mathscr{B}_\sigma$, $\forall \tau \in [0, T_k]$, $\forall T_k \in [\mathscr{T}_1, \mathscr{T}_2]$, $\forall k \ge 0$, *then for any initial condition* $x(0) = x_0(0)$ *in the set*

$$\mathscr{L}_V(1) = \left\{ x \in \mathbb{R}^n;\ V(x) \le 1 \right\},$$

it follows that:

(i) $\Delta V(k) = V(x_{k+1}(0)) - V(x_k(0)) < -\mu \mathscr{T}_1 \|x_k(0)\|^2$, $\forall k \ge 0$, *for* $x_k(0) \ne 0$*;*

(ii) the corresponding trajectories of the closed-loop system (1)–(4), *with sampling intervals* $T_k \in [\mathscr{T}_1, \mathscr{T}_2]$, *never leave* $\mathscr{L}_V(1) \subset \mathscr{R}_H$ *and converge asymptotically to the origin.*

Proof Suppose $x(0) = x_0(0) \in \mathscr{L}_V(1)$. If (25) is verified, it follows that $x_0(0) \in \mathscr{L}_V(1) \subset \mathscr{R}_H$ and thus $\sigma_0(0) \in \mathscr{B}_\sigma$. If (21) is satisfied, for $i = 1, ..., m$, it follows that $||(K(\sigma_0(0)) - S(\sigma_0(0)))_{(i)} x_0(0)||^2 \le \bar{u}_{(i)}^2$. This implies, from Lemma 1, that $\psi_0' U(\sigma_0(0))(\psi_0 + S(\sigma_0(0))x_0(0)) \le 0$. Since by definition $\mathscr{V}_0(\tau, x_0, \sigma_0(0), T_0) > 0$ for $\tau \in (0, T_k)$ and $\mathscr{V}_0(0, x_0, \sigma_0(0), T_0) = 0$, if (24) is satisfied, it follows that

$$V(x_0(\tau)) < \mathscr{W}(\tau, x_0, \sigma_0(0), T_0) < \mathscr{W}(0, x_0, \sigma_0(0), T_0) = V(x_0(0)) \le 1,$$

$\forall \tau \in (0, T_0)$. Then, we conclude that $x_0(\tau) \in \mathscr{L}_V(1) \subset \mathscr{R}_H$ and $\sigma(x_0(\tau)) \in \mathscr{B}_\sigma$. Moreover, taking into account (18) and (21), and integrating (24) over any sampling interval it follows that

$$\int_0^{T_k} \dot{\mathscr{W}} d\tau \leq -\mu \int_0^{T_k} \left\| \begin{bmatrix} x_k(\tau) \\ x_k(0) \end{bmatrix} \right\|^2 d\tau. \tag{26}$$

Since from (22) we have $\mathscr{V}_0(0, x_k, \sigma_k(0), T_k) = \mathscr{V}_0(T_k, x_k(0), \sigma_k, T_k)$ and by continuity $x_0(T_0) = x_1(0)$, it follows from (26) that $\Delta V(0) < -\mu \mathscr{T}_1 ||x_0(0)||^2$, $\forall k \geq 0$. Moreover, this implies that $x_1(0) \in \mathscr{L}_V(1)$. Thus repeating the reasoning we can conclude that $\Delta V(k) < -\mu \mathscr{T}_1 ||x_k(0)||^2$, $\forall k \geq 0$ and it follows that $x_k(0) \to 0$ as $k \to \infty$. Furthermore, since $V(x_k(0)) \to 0$, as $k \to \infty$ and $V(x_k(\tau)) < V(x_k(0))$ $\forall \tau \in (0, T_k]$, we conclude that $V(x_k(\tau)) \to 0$ as $k \to \infty$ and thus $x_k(\tau) \to 0$ as $k \to \infty$. Finally, as $\mathscr{L}_V(1)$ is a positively invariant set and, from (25), $\mathscr{L}_V(1) \subset \mathscr{R}_H$, the quasi-LPV system with $\sigma(t) \in \mathscr{B}_\sigma$ indeed represents the behavior of the nonlinear system (1)–(4) and the proof of item (ii) follows. ■

4 Stability Assessment

In this section, from the theoretical results of Theorem 1, we propose conditions in quasi-LMI form to assess the stability of the sampled-data nonlinear system through its quasi-LPV model, under Assumptions 1 and 2. For this we consider a quadratic function

$$V(x) = x'Px, \tag{27}$$

with $P = P' > 0$ and the following PDLF:

$$\mathscr{V}_0(\tau, x_k, \sigma_k(0), T_k) = (T_k - \tau) \Bigg((x_k(\tau) - x_k(0))'[F(\sigma_k(0))(x_k(\tau) - x_k(0)) + 2G(\sigma_k(0))x_k(0)] + \tau \begin{bmatrix} x_k(0) \\ \psi_k \end{bmatrix}' X(\sigma_k(0)) \begin{bmatrix} x_k(0) \\ \psi_k \end{bmatrix} + \int_0^{\tau} \dot{x}_k'(\theta) R \dot{x}_k(\theta) d\theta \Bigg) \tag{28}$$

with

$$R > 0, \tag{29}$$

$$X(\sigma_k(0)) > 0, \ \ \forall \sigma_k(0) \in \mathscr{B}_\sigma, \tag{30}$$

$$\begin{bmatrix} F(\sigma_k(0)) & G(\sigma_k(0))' - F(\sigma_k(0)) \\ * & F(\sigma_k(0)) - \mathrm{He}\{G(\sigma_k(0))\} \end{bmatrix} > 0, \ \ \forall \sigma_k(0) \in \mathscr{B}_\sigma, \tag{31}$$

We assume that $F(\sigma)$ and $G(\sigma)$ have an affine dependence on σ and can be generically described as

$$F(\sigma) = F_0 + \mathbf{F}\Lambda^n(\sigma), \ \ G(\sigma) = G_0 + \mathbf{G}\Lambda^n(\sigma),$$

with $\mathbf{F} = [F_1 \ \ F_2 \ \ \cdots \ \ F_N]$, $\mathbf{G} = [G_1 \ \ G_2 \ \ \cdots \ \ G_N]$, $F_j \in \mathbb{S}^n$, and $G_j \in \mathbb{R}^{n \times n}$, $j = 0, \ldots, N$. By definition, note that $\mathscr{V}_0(T_k, x_k, \sigma_k(0), T_k) = \mathscr{V}_0(0, x_k, \sigma_k(0), T_k) =$

0. Moreover, (29)–(31) imply that $\mathscr{V}_0(\tau, x_k, \sigma_k(0), T_k) > 0$, $\forall \tau \in (0, T_k)$. Hence, $\mathscr{V}_0(\tau, x_k, \sigma_k(0), T_k)$ satisfies condition (22) in Theorem 1.

From the choices of V and $\mathscr{V}_0$ as in (27) and (28), respectively, the following theorem can be stated:

Theorem 2 *Consider a matrix H and vector η that define the set $\mathscr{R}_H$ as in* (10). *If there exist symmetric positive definite matrices $\tilde{P}$ and $\tilde{R} \in \mathbb{S}^n$, symmetric matrices $\tilde{F}_j \in \mathbb{S}^n$ and $\tilde{X}_j \in \mathbb{S}^{n+m}$, matrices $\tilde{G}_j$, $\tilde{Y} \in \mathbb{R}^{n\times n}$, $\tilde{Q}_j \in \mathbb{R}^{(3n+m)\times n}$, $\tilde{S}_j \in \mathbb{R}^{m\times n}$, diagonal matrices $\tilde{U}_j \in \mathbb{S}^m$, $j = 0, 1, \ldots, N$ and a scalar ε satisfying*

$$\tilde{\Pi}_1(v_f, v_g) + T_k\tilde{\Pi}_2(v_f) + T_k\tilde{\Pi}_3(v_f) < 0, \tag{32}$$

$$\begin{bmatrix} \tilde{\Pi}_1(v_f, v_g) - T_k\tilde{\Pi}_3(v_f) & T_k\tilde{Q}(v_f) \\ * & -T_k\tilde{R} \end{bmatrix} < 0, \tag{33}$$

$$\begin{bmatrix} \tilde{P} & \tilde{Y}H_{(r)} \\ * & \eta_{(r)}^2 \end{bmatrix} > 0, \quad r = 1, \ldots, r_c, \tag{34}$$

$$\begin{bmatrix} \tilde{P} & (K(v_f)\tilde{Y} - \tilde{S}(v_f))'_{(i)} \\ * & \bar{u}_{(i)}^2 \end{bmatrix} > 0, \tag{35}$$

$$\begin{bmatrix} \tilde{F}(v_f) & \tilde{G}'(v_f) - \tilde{F}(v_f) \\ \tilde{G}(v_f) - \tilde{F}(v_f) & \tilde{F}(v_f) - \tilde{G}(v_f) - \tilde{G}'(v_f) \end{bmatrix} > 0, \tag{36}$$

$$\tilde{X}(v_f) > 0, \tag{37}$$

$$\tilde{U}(v_f) > 0, \tag{38}$$

$\forall (v_f, v_g) \in Ver(\mathscr{B}_\sigma) \times Ver(\mathscr{B}_\sigma)$, $T_k \in \{\mathscr{T}_1, \mathscr{T}_2\}$ *and* $i = 1, \ldots, m$, *with*

$$\begin{aligned} \tilde{\Pi}_1(v_f, v_g) &= \mathrm{He}\{M_1'\tilde{P}M_3 - M_{12}'\tilde{G}(v_f)M_2 - \tilde{Q}(v_f)M_{12} - M_4'\tilde{S}(v_f)M_2 \\ &\quad + (\varepsilon M_1' + M_3')(A(v_g)\tilde{Y}M_1 + B(v_g)K(v_f)\tilde{Y}M_2 - \tilde{Y}M_3 \\ &\quad + B(v_g)\tilde{U}(v_f)M_4)\} - 2M_4'\tilde{U}(v_f)M_4 - M_{12}'\tilde{F}(v_f)M_{12}, \\ \tilde{\Pi}_2(v_f) &= M_3'\tilde{R}M_3 + \mathrm{He}\left\{M_3'(\tilde{F}(v_f)M_{12} + \tilde{G}(v_f)M_2)\right\}, \\ \tilde{\Pi}_3(v_f) &= M_{24}'\tilde{X}(v_f)M_{24}, \end{aligned} \tag{39}$$

matrices $A(v_g)$, $B(v_g)$ and $K(v_f)$ defined as in (13),

$$\begin{aligned} \tilde{Q}(v_f) &= \tilde{Q}_0 + \tilde{\boldsymbol{Q}}\Lambda^n(v_f), \quad \tilde{F}(v_f) = \tilde{F}_0 + \tilde{\boldsymbol{F}}\Lambda^n(v_f), \quad \tilde{G}(v_f) = \tilde{G}_0 + \tilde{\boldsymbol{G}}\Lambda^n(v_f), \\ \tilde{X}(v_f) &= \tilde{X}_0 + \tilde{\boldsymbol{X}}\Lambda^{n+m}(v_f), \quad \tilde{U}(v_f) = \tilde{U}_0 + \tilde{\boldsymbol{U}}\Lambda^m(v_f), \quad \tilde{S}(v_f) = \tilde{S}_0 + \tilde{\boldsymbol{S}}\Lambda^n(v_f), \end{aligned} \tag{40}$$

and the auxiliary matrices

$$\begin{aligned} M_1 &= \begin{bmatrix} I^n & 0^n & 0^n & 0^{n\times m} \end{bmatrix}, \; M_2 = \begin{bmatrix} 0^n & I^n & 0^n & 0^{n\times m} \end{bmatrix}, \; M_3 = \begin{bmatrix} 0^n & 0^n & I^n & 0^{n\times m} \end{bmatrix}, \\ M_4 &= \begin{bmatrix} 0^{m\times n} & 0^{m\times n} & 0^{m\times n} & I^m \end{bmatrix}, \; M_{12} = M_1 - M_2, \; M_{24} = [M_2' \; M_4']', \end{aligned} \tag{41}$$

then for any initial condition $x(0) \in \mathcal{E}(P, 1) = \{x \in \mathbb{R}^n;\ x'Px \leq 1\}$, where $P = \tilde{Y}^{-1'}\tilde{P}\tilde{Y}^{-1}$ the corresponding trajectories of the sampled-data nonlinear closed-loop system (1)–(4)*, represented by the quasi-LPV system* (8) *under Assumptions 1 and 2, converge asymptotically to the origin, $\forall T_k \in [\mathcal{T}_1, \mathcal{T}_2]$.*

Proof Consider the function V and the PDLF $\mathcal{V}_0$ defined in (27) and (28), respectively. Considering $\mathcal{W}(\tau, x_k, \sigma_k(0), T_k)$ as defined in (23) and differentiating it with respect to τ gives

$$\begin{aligned}\dot{\mathcal{W}} &= 2\dot{x}_k'(\tau)Px_k(\tau) - (x_k(\tau) - x_k(0))'[F(\sigma_k(0))(x_k(\tau) - x_k(0)) \\ &\quad + 2G(\sigma_k(0))x_k(0)] + (T_k - \tau)\left[\dot{x}_k'(\tau)2G(\sigma_k(0))x_k(0)\right. \\ &\quad \left. + 2\dot{x}_k'(\tau)\ F(\sigma_k(0))(x_k(\tau) - x_k(0)) + \dot{x}_k'(\tau)R\dot{x}_k(\tau)\right] \\ &\quad + (T_k - 2\tau)\begin{bmatrix} x_k(0) \\ \psi_k \end{bmatrix}' X(\sigma_k(0))\begin{bmatrix} x_k(0) \\ \psi_k \end{bmatrix} - \textstyle\int_0^\tau \dot{x}_k'(\theta)R\dot{x}_k(\theta)d\theta.\end{aligned} \tag{42}$$

Define $\chi(\tau) = [x_k'(\tau)\ \ x_k'(0)\ \ \dot{x}_k'(\tau)\ \ \psi_k']'$ and consider a matrix $Q(\sigma_k(0)) = Q_0 + \mathbf{Q}\Lambda^n(\sigma_k(0)) = Q_0 + \mathbf{Q}(\sigma_k(0) \otimes I^n)$, where $\mathbf{Q} = [Q_1, \ldots, Q_N]$, with $Q_j \in \mathbb{R}^{(3n+m)\times n}$, for $j = 0, 1, \ldots, N$. Since $R > 0$, we have that [1]:

$$\begin{aligned}\int_0^\tau \dot{x}_k'(\theta)R\dot{x}_k(\theta)d\theta &\geq 2\chi'(\tau)Q(\sigma_k(0))(x_k(\tau) - x_k(0)) \\ &\quad - \tau\chi'(\tau)Q(\sigma_k(0))R^{-1}Q'(\sigma_k(0))\chi(\tau).\end{aligned} \tag{43}$$

Hence, based on auxiliary matrices $M_1,\ M_2,\ M_3,\ M_4,\ M_{24}$ and M_{12} defined in (41), it follows from (42) and (43) that

$$\begin{aligned}&\dot{\mathcal{W}} - 2\psi_k'U(\sigma_k(0))(\psi_k + S(\sigma_k(0))x_k(0)) \leq \chi'(\tau)[\hat{\Pi}_1(\sigma_k(0)) \\ &\quad + (T_k - \tau)\Pi_2(\sigma_k(0)) + (T_k - 2\tau)\Pi_3(\sigma_k(0)) + \tau Q(\sigma_k(0))R^{-1}Q'(\sigma_k(0))]\chi(\tau),\end{aligned} \tag{44}$$

with

$$\begin{aligned}\hat{\Pi}_1(\sigma_k(0)) &= \text{He}\{M_3'PM_1 - Q(\sigma_k(0))M_{12} - M_{12}'G(\sigma_k(0))M_2 \\ &\quad - M_4'U(\sigma_k(0))S(\sigma_k(0))M_2\} - M_{12}'F(\sigma_k(0))M_{12} - 2M_4'U(\sigma_k(0))M_4, \\ \Pi_2(\sigma_k(0)) &= \text{He}\left\{M_3'F(\sigma_k(0))M_{12} + M_3'G(\sigma_k(0))M_2\right\} + M_3'RM_3, \\ \Pi_3(\sigma_k(0)) &= M_{24}'X(\sigma_k(0))M_{24}.\end{aligned}$$

From (19), for any matrices Y_1 and Y_2 of appropriate dimensions, we have the equality

$$\begin{aligned}&(Y_1x_k(\tau) + Y_2\dot{x}_k(\tau))'[A(\sigma_k(\tau))x_k(\tau) + (B(\sigma_k(\tau))K(\sigma_k(0)))x_k(0) + B(\sigma_k(\tau))\psi_k \\ &\quad - \dot{x}_k(\tau)] = 0.\end{aligned}$$

Define $\Pi_1(\sigma_k(0), \sigma_k(\tau)) = \hat{\Pi}_1(\sigma_k(0)) + \text{He}\{\Theta(\sigma_k(0), \sigma_k(\tau))\}$, with

$$\Theta(\sigma_k(0), \sigma_k(\tau)) = (Y_1 M_1 + Y_2 M_3)'[A(\sigma_k(\tau))M_1 + B(\sigma_k(\tau))K(\sigma_k(0))M_2 - M_3 + B(\sigma_k(\tau))M_4].$$

Suppose now that Y_2 and $U(\sigma_k(0))$ are nonsingular $\forall \sigma_k(0) \in \mathscr{B}_\sigma$ and consider $Y_1 = \varepsilon Y_2$, $\tilde{Y} = Y_2^{-1}$ and $\tilde{U}(\sigma_k(0)) = U(\sigma_k(0))^{-1} = \tilde{U}_0 + \tilde{\mathbf{U}}\Lambda^m(\sigma_k(0)) = \tilde{U}_0 + \tilde{\mathbf{U}}(\sigma_k(0) \otimes I^m)$ with $\tilde{\mathbf{U}} = [\tilde{U}_1 \dots \tilde{U}_N]$. Thus, defining $\tilde{\chi}(\tau) = \Xi^{-1}\chi(\tau)$, with $\Xi = diag\{\tilde{Y}, \tilde{Y}, \tilde{Y}, \tilde{U}(\sigma_k(0))\}$, we can rewrite (44) as follows:

$$\begin{aligned}\dot{\mathscr{W}} - 2\psi_k' U(\sigma_k(0))(\psi_k + S(\sigma_k(0))x_k(0)) \le \tilde{\chi}'(\tau)\Xi'[\Pi_1(\sigma_k(0), \sigma_k(\tau)) \\ + (T_k - \tau)\Pi_2(\sigma_k(0)) + (T_k - 2\tau)\Pi_3(\sigma_k(0)) \\ + \tau Q(\sigma_k(0), \sigma_k(\tau))\tilde{R}^{-1}Q'(\sigma_k(0), \sigma_k(\tau))]\Xi\tilde{\chi}(\tau).\end{aligned}$$

Hence, by applying Schur's complement, a sufficient condition to verify $\dot{\mathscr{W}} - 2\psi_k' U(\sigma_k(0))(\psi_k + S(\sigma_k(0))x_k(0)) \le -\mu \left\| \begin{bmatrix} x_k(\tau) \\ x_k(0) \end{bmatrix} \right\|^2$, along the trajectories of (19), $\forall x_k \in \mathbb{K}^n_{[\mathscr{T}_1, \mathscr{T}_2]}, x_k \not\equiv 0, \forall \sigma_k(\tau) \in \mathscr{B}_\sigma, \forall \tau \in [0, T_k], T_k \in [\mathscr{T}_1, \mathscr{T}_2], \forall k \ge 0$, for some $\mu > 0$, is given by

$$\Psi(\tau, \sigma_k(\tau), \sigma_k(0), T_k) = \begin{bmatrix} \tilde{\Pi}(\tau, \sigma_k(\tau), \sigma_k(0), T_k) & \tau \Xi' Q(\sigma_k(0))\tilde{Y} \\ * & -\tau \tilde{R} \end{bmatrix} < 0, \tag{45}$$

$\forall \sigma_k(\tau) \in \mathscr{B}_\sigma$, $\forall \sigma_k(0) \in \mathscr{B}_\sigma$, $\forall \tau \in [0, T_k)$ and $T_k \in [\mathscr{T}_1, \mathscr{T}_2]$, with $\tilde{\Pi}(\tau, \sigma_k(\tau), \sigma_k(0), T_k) = \Xi'\Pi_1(\sigma_k(0), \sigma_k(\tau))\Xi + (T_k - \tau)\Xi'\Pi_2(\sigma_k(0))\Xi + (T_k - 2\tau)\Xi'\Pi_3(\sigma_k(0))\Xi$.

Note that (45) is affine on $\sigma_k(0)$ and $\sigma_k(\tau)$. In this case, by convexity, a necessary and sufficient condition to satisfy (45) consists in satisfying it at any combination of the vertices of $\mathscr{B}_\sigma$, i.e. we must verify

$$\Psi(\tau, v_f, v_g, T_k) < 0, \;\; \forall (v_f, v_g) \in Ver(\mathscr{B}_\sigma) \times Ver(\mathscr{B}_\sigma). \tag{46}$$

$\forall \tau \in [0, T_k)$ and $\forall T_k \in [\mathscr{T}_1, \mathscr{T}_2]$, with $\tilde{\Pi}_1(v_f, v_g)$, $\tilde{\Pi}_2(v_f)$ and $\tilde{\Pi}_3(v_f)$ defined as in (39), where the change of variables $\tilde{P} = \tilde{Y}'P\tilde{Y}$, $\tilde{F}_j = \tilde{Y}'F_j\tilde{Y}$, $\tilde{G}_j = \tilde{Y}'G_j\tilde{Y}$, $\tilde{Q}_j = \Xi'Q_j\tilde{Y}$, $\tilde{S}_j = S_j\tilde{Y}$, $\tilde{X}_j = \begin{bmatrix} \tilde{Y} & 0 \\ 0 & \tilde{U}_j \end{bmatrix}' X_j \begin{bmatrix} \tilde{Y} & 0 \\ 0 & \tilde{U}_j \end{bmatrix}$, for $j = 0, \cdots, N$, have been considered.

Noting now that since (46) is also affine with respect to the variable $\tau \in [0, T_k)$, from convexity arguments, it suffices to ensure that this relation is verified for $\tau = 0$ and $\tau = T_k$. In other words, it is satisfied $\forall \tau \in [0, T_k)$ iff

$$\tilde{\Pi}_1(v_f, v_g) + T_k\tilde{\Pi}_2(v_f) + T_k\tilde{\Pi}_3(v_f) < 0, \tag{47}$$

$$\begin{bmatrix} \tilde{\Pi}_1(v_f, v_g) - T_k\tilde{\Pi}_3(v_f) & T_k\tilde{Q}(v_f) \\ * & -T_k\tilde{R} \end{bmatrix} < 0. \tag{48}$$

Furthermore since $T_k \in [\mathscr{T}_1, \mathscr{T}_2]$, applying the same convexity argument, it follows that (32) and (33) verified for $T_k = \mathscr{T}_1$ and $T_k = \mathscr{T}_2$ (i.e. for $T_k \in \{\mathscr{T}_1, \mathscr{T}_2\}$) are equivalent to (47) and (48), respectively. Hence (32) and (33) ensure that $\dot{\mathscr{W}} - 2\psi_k' U(\sigma_k(0))(\psi_k + S(\sigma_k(0))x_k(0)) \le -\mu \left\| \begin{bmatrix} x_k(\tau) \\ x_k(0) \end{bmatrix} \right\|^2$, provided Ξ is nonsingular. Moreover, also by convexity arguments, if (36)–(37) are verified $\forall v_f \in \mathscr{B}_\sigma$ it ensures that $\mathscr{V}_0(\tau, x_k, \sigma_k(0), T_k)$ satisfies (22), provided $\sigma_k(0) \in \mathscr{B}_\sigma$. Since the block-diagonal term in position (3,3) in (32) is given by $-\tilde{Y} - \tilde{Y}' + \tilde{R} < 0$, it follows that $\tilde{Y}$ is nonsingular. Note also that, from (38), we have $\tilde{U}(\sigma(0)) > 0$ and thus we can conclude that Ξ is nonsingular. On the other hand, by right and left-multiplying (35) and (34) by $diag\{\tilde{Y}^{-1}, I\}$, using convexity arguments and applying Schur's complement, it follows that (21) and (25) are satisfied. Thus, the simultaneous satisfaction of conditions (32)–(38) ensures that conditions of Theorem 1 are satisfied, which guarantees that $\mathscr{L}_V(1) = \mathscr{E}(P, 1) \subset \mathscr{R}_H$ is included in the basin of attraction of the closed-loop nonlinear system (1)–(4), $\forall T_k \in [\mathscr{T}_1, \mathscr{T}_2]$. ■

Remark 1 In Theorem 2 there are $2^{2N+1} + 2^{2N+1} + 2^N r_c + 2^N m + 3(2^N)$ matrix inequalities to be verified. The numerical complexity can be reduced, at the expense of a conservatism increasing, considering for instance $\mathscr{V}_0$ parameter independent. This can be straightforwardly done by considering $X(\sigma_k(0)) = X_0$, $F(\sigma_k(0)) = F_0$ and $G(\sigma_k(0)) = G_0$ in the conditions.

5 Optimization Problems

5.1 *Maximization of the Estimate of the Region of Attraction*

Note that the set $\mathscr{L}_V(1) = \mathscr{E}(P, 1) \subset \mathscr{R}_H$ is, by definition, included in $\mathscr{R}_a$ and can be used to estimate it. Then, from Problem 1 (given sampling interval bounds $\mathscr{T}_1$ and $\mathscr{T}_2$) the goal is to maximize the set $\mathscr{E}(P, 1)$ considering some size criterion, such as the maximization of its minor axis. This can be accomplished by the solution of the following optimization problem:

$$\begin{array}{l} \min \ \delta \\ \text{subject to} \\ (32) - (38), \\ \begin{bmatrix} \delta I^n & I^n \\ I^n & \tilde{Y}' + \tilde{Y} - \tilde{P} \end{bmatrix} > 0 \\ \forall (v_f, v_g) \in Ver(\mathscr{B}_\sigma) \times Ver(\mathscr{B}_\sigma), \ \text{for } T_k \in \{\mathscr{T}_1, \mathscr{T}_2\}. \end{array} \tag{49}$$

The last matrix inequality in (49) ensures that the maximal eigenvalue of P is smaller than δ (see [21]). Thus the minimization of δ ensures the maximization of the minor axis of $\mathscr{E}(P, 1)$. Due to the scalar ε in $\tilde{\Pi}_1(\nu_f, \nu_g)$, it follows that (32)–(33) are not LMIs in the decision variables. However, (49) can be solved from the solution of LMI problems over a grid in ε.

5.2 Maximization of the Sampling Interval

To address Problem 2, we can consider a given region of admissible initial conditions $\mathscr{X}_0 = \mathscr{E}(X_0, 1) = \{x \in \mathbb{R}^n;\ x'X_0x \leq 1\}$, with $X_0 = X_0' > 0$. Then, the aim is to maximize the value of $\mathscr{T}_2$ for which the closed-loop stability is ensured for $T_k \in [\mathscr{T}_1, \mathscr{T}_2]$ with $\mathscr{E}(X_0, 1) \subset \mathscr{R}_a$. This can be accomplished from the solution of the following optimization problem:

$$\begin{array}{l} \max\ \mathscr{T}_2 \\ \text{subject to} \\ (32) - (38), \\ \begin{bmatrix} X_0 & I^n \\ I^n & \tilde{Y}' + \tilde{Y} - \tilde{P} \end{bmatrix} > 0 \\ \forall (\nu_f, \nu_g) \in Ver(\mathscr{B}_\sigma) \times Ver(\mathscr{B}_\sigma),\ \text{for } T_k \in \{\mathscr{T}_1, \mathscr{T}_2\}. \end{array} \tag{50}$$

Note that the last inequality in (50) ensures that $P < X_0$ (see [21]) and then $\mathscr{E}(X_0, 1) \subset \mathscr{E}(P, 1)$.

Hence, considering $\mathscr{T}_1$ given, the optimization problem (50) can be solved by iteratively increasing $\mathscr{T}_2$ and testing the feasibility of the LMIs over a grid in ε.

Remark 2 The conditions of Theorem 2 depend on the choices of η and H that define set $\mathscr{R}_H$. Increasing the size of set $\mathscr{R}_H$ enable to reduce the restriction on eigenvalues of P and to potentially obtain a larger region $\mathscr{E}(P, 1)$. However it also increases the set $\mathscr{B}_\sigma$ of admissible parameters, which can lead to the non feasibility of the conditions. In order to obtain $\mathscr{E}(P, 1)$ as large as possible, we can iteratively increase $\mathscr{R}_H$ until the conditions became unfeasible. This can be done by fixing the shape of $\mathscr{R}_H$ through the choice of H and $\tilde{\eta}$ and considering $\eta = \rho\tilde{\eta}$, with $\rho > 0$ being a scaling factor.

Remark 3 It should be pointed out that, in general, the quasi-LPV representation for a nonlinear system is not unique, i.e., the parameters depending on the state can be chosen in more than one way. Of course, different quasi-LPV representations for the same nonlinear system will lead to different results for problems (49) and (50). However, in particular, the union of estimates obtained with different representations will be a guaranteed set of regional stability for the actual nonlinear system, i.e. it will be included in its basin of attraction.

6 Numerical Example

Consider the nonlinear system borrowed from [4]:

$$\dot{x}_1(t) = x_2(t),$$
$$\dot{x}_2(t) = (1 + x_1^2(t))x_1(t) + (2 + 8x_2^2(t))x_2(t) + sat(u(t)),$$

with $\bar{u} = 1$. Defining $\sigma(x(t)) = [x_1^2(t) \;\; x_2^2(t)]'$, this nonlinear system can be represented by the quasi-LPV form (5) with matrices

$$A_0 = \begin{bmatrix} 0 & 1 \\ 1 & 2 \end{bmatrix}, \; A_1 = \begin{bmatrix} 0 & 0 \\ 1 & 0 \end{bmatrix}, \; A_2 = \begin{bmatrix} 0 & 0 \\ 0 & 8 \end{bmatrix}, \; B_0 = \begin{bmatrix} 0 \\ 1 \end{bmatrix}, \; B_1 = B_2 = 0.$$

Consider a local stabilizing control law in the form (6), with the following matrices:

$$K_0 = \begin{bmatrix} -1.9285 & -5.2143 \end{bmatrix}, K_1 = \begin{bmatrix} -0.0014 & -0.0039 \end{bmatrix}, K_2 = \begin{bmatrix} -0.0032 & -0.0099 \end{bmatrix}.$$

We assume that the system operates in a region such that $|x_1| \leq 0.6$ and $|x_2| \leq 0.4$, i.e. $x \in \mathscr{R}_H$ as defined in (10) with $H = I^2$ and $\eta = [0.6 \;\; 0.4]'$. In this case it follows that $\sigma_{(1)} \in [0, \; 0.36]$ and $\sigma_{(2)} \in [0, \; 0.16]$.

Let us consider first the periodic sampling case, i.e. $T_k = \mathscr{T}_1 = \mathscr{T}_2 = \mathscr{T} \; \forall k \in \mathbb{N}$. Applying the optimization problem (50) aiming at maximizing $\mathscr{T}$ and considering $X_0 = 100 \times I^2$, the maximum feasible sampling period for which the stability can be ensured for the set $\mathscr{E}(X_0, 1)$ is equal to 0.285s.

We consider now optimization problem (49) for $\mathscr{T}_1 = 0.010$s and different values of $\mathscr{T}_2$. The optimal values obtained for δ in function of $\mathscr{T}_2$ are depicted in Fig. 1. It can be observed that as $\mathscr{T}_2$ increases, the value of δ increases, which implies that the size of set $\mathscr{E}(P, 1)$ is smaller, which indicates that the actual the region of attraction of the origin is also smaller. Observe that the case of periodic sampling (i.e. $\mathscr{T}_1 = \mathscr{T}_2$) leads to a larger set when compared to the aperiodic cases.

For $T_k \in [0.010, \; 0.200]$s, the optimization problem (49) leads to $\delta = 19.4718$ and an estimate $\mathscr{E}(P, 1)$ of the region of attraction defined by the matrix

$$P = \begin{bmatrix} 7.9787 & 5.1168 \\ 5.1168 & 16.5544 \end{bmatrix}.$$

For this case, Fig. 2 shows the trajectories of the closed-loop system for some initial conditions for T_k randomly generated, such that $T_k \in [0.01, \; 0.2]$s. The initial conditions leading to convergent and divergent trajectories are marked with '$*$' and 'o', respectively. As expected, the trajectories starting in $\mathscr{L}_V(1) = \mathscr{E}(P, 1)$ converge to the origin. Note that there are divergent trajectories starting outside $\mathscr{E}(P, 1)$, but not far from its boundary. The sets $\mathscr{R}_H$ and $\mathscr{E}(P, 1)$ are shown in dashed line and solid line, respectively.

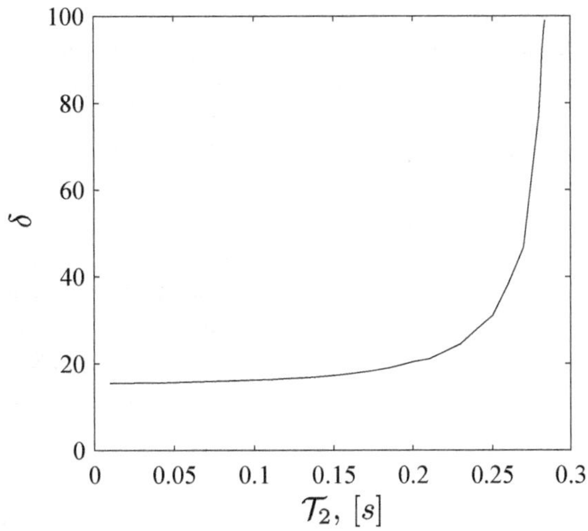

Fig. 1 Obtained values of δ for different values of $\mathscr{T}_2$, for $\mathscr{T}_1 = 0.010$s

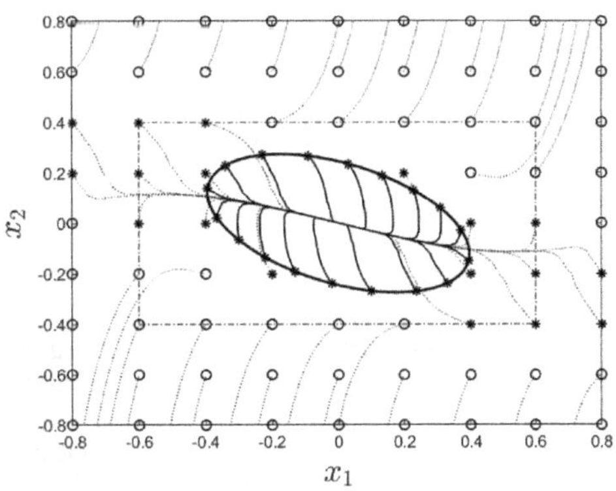

Fig. 2 The sets $\mathscr{R}_H$ and $\mathscr{L}_V(1) = \mathscr{E}(P, 1)$ in dashed and in solid lines, respectively. For $T_k \in [0.010, 0.200]$s, some convergent (starting at points '*') and divergent trajectories (starting at points 'o')

7 Conclusion

This chapter has addressed the stability analysis problem of constrained sampled-data controlled nonlinear systems considering quasi-LPV models. The control signal has been assumed to be constant between two successive sampling instants and the continuous behavior of the nonlinear plant has been explicitly considered (i.e., no discretization has been performed). Based on a parameter dependent looped-functional to take into account the sampling effects and a parameter dependent generalized sector condition to tackle control saturation, conditions in quasi LMI-form (i.e. they are LMIs provided that a scalar is fixed) have been derived to assess the regional stability of the nonlinear closed-loop system. Then, two convex optimization problems have been formulated. The first one allows to compute an estimate of the region of attraction of the origin, which is a level set of a quadratic function. Since the matrices defining the level set are free variables, the optimization problem provides the best estimate considering this class of function. The second problem regards the determination of the maximal bound on the intersampling time for which is possible to guarantee the regional stability with respect to a pre-specified set of admissible initial conditions.

As a broad class of nonlinear systems can be cast in a quasi-LPV framework, the proposed approach provides an effective and numerically tractable conditions to evaluate the stability of sampled-data controlled nonlinear systems. In particular the practical amplitude constraints imposed by the actuators, i.e. the control saturation, can be properly taken into account by the approach.

References

1. C. Briat, Convergence and equivalence results for the Jensen's inequality-application to time-delay and sampled-data systems. IEEE Trans. Autom. Control **56**(7), 1660–1665 (2011)
2. C. Briat, Linear parameter-varying and time-delay systems, in *Analysis, Observation, Filtering & Control* (Springer, 2015)
3. C. Briat, A. Seuret, A looped-functional approach for robust stability analysis of linear impulsive systems. Syst. Control Lett. **61**(10), 980–988 (2012)
4. D.F. Coutinho, J.M. Gomes da Silva Jr., Computing estimates of the region of attraction for rational control systems with saturating actuators. IET Control Theory Appl. **4**(3), 315–325 (2010)
5. H. Du, N. Zhang, B. Samali, F. Naghdy, Robust sampled-data control of structures subject to parameter uncertainties and actuator saturation. Eng. Struct. **36**, 39–48 (2012)
6. E. Fridman, A refined input delay approach to sampled-data control. Automatica **46**(2), 421–427 (2010)
7. E. Fridman, A. Seuret, J.P. Richard, Robust sampled-data stabilization of linear systems: an input delay approach. Automatica **40**(8), 1141–1446 (2004)
8. J.M. Gomes da Silva Jr., A.H.K. Palmeira, J.V. Flores, $\mathscr{L}_2$-disturbance attenuation for LPV systems under sampled-data control. Int. J. Robust Nonlinear Control **28**, 5019–5032 (2018)
9. W.M. Haddad, V. Kapila, V.S. Chellaboina, Guaranteed domains of attraction for multivariable luré systems via open lyapunov surfaces. Int. J. Robust Nonlinear Control **7**(10), 935–949 (1997)
10. J.P. Hespanha, P. Naghshtabrizi, Y. Xu, A survey of recent results in networked control systems. Proc.-IEEE **95**(1), 138 (2007)
11. C. Hoffmann, H. Werner, A survey of linear parameter-varying control applications validated by experiments or high-fidelity simulations. IEEE Trans. Control Syst. Technol. **23**(2), 416–433 (2015)
12. H. Khalil, *Nonlinear Systems*, 3rd edn. (Prentice-Hall, NJ, 2002)
13. H. Li, X. Jing, H.K. Lam, P. Shi, Fuzzy sampled-data control for uncertain vehicle suspension systems. IEEE Trans. Cybern. **44**(7), 1111–1126 (2014)
14. A. Megretski, A. Rantzer, System analysis via integral quadratic constraints. IEEE Trans. Autom. Control **42**(6), 819–830 (1997)
15. P. Naghshtabrizi, J. Hespanha, A. Teel, Exponential stability of impulsive systems with application to uncertain sampled-data systems. Syst. Control Lett. **57**(5), 378–385 (2008)
16. D. Nešić, A.R. Teel, P.V. Kokotović, Sufficient conditions for stabilization of sampled-data nonlinear systems via discrete-time approximations. Syst. Control Lett. **38**(4–5), 259–270 (1999)

17. A.H.K. Palmeira, J.M. Gomes da Silva Jr., S. Tarbouriech, I.M. Ghiggi, Sampled-data control under magnitude and rate saturating actuators. Int. J. Robust Nonlinear Control **26**(15), 3232–3252 (2016)
18. A. Ramezanifar, J. Mohammadpour, K. Grigoriadis, Sampled-data control of LPV systems using input delay approach, in *51st IEEE Conference on Decision and Control* (IEEE, Maui, Hawaii, USA, 2012), pp. 6303–6308
19. W.J. Rugh, J.S. Shamma, Research on gain scheduling. Automatica **36**(10), 1401–1425 (2000)
20. A. Seuret, A novel stability analysis of linear systems under asynchronous samplings. Automatica **48**(1), 177–182 (2012)
21. A. Seuret, J.M. Gomes da Silva Jr., Taking into account period variations and actuator saturation in sampled-data systems. Syst. Control Lett. **61**(12), 1286–1293 (2012)
22. T. Shi, H. Su, Sampled-data MPC for LPV systems with input saturation. IET Control Theory Appl. **8**(17), 1781–1788 (2014)
23. L.F. Silva, V.J. Leite, E.B. Castelan, M. Klug, Local stabilization of time-delay nonlinear discrete-time systems using T-S models and convex optimization. Math. Probl. Eng. (2014)
24. K. Tan, K. Grigoriadis, F. Wu, Output-feedback control of LPV sampled-data systems. Int. J. Control **75**(4), 252–264 (2002)
25. S. Tarbouriech, G. Garcia, J.M. Gomes da Silva Jr., I. Queinnec, *Stability and Stabilization of Linear Systems with Saturating Actuators* (Springer, 2011)
26. G. Valmorbida, S. Tarbouriech, G. Garcia, Design of polynomial control laws for polynomial systems subject to actuator saturation. IEEE Trans. Autom. Control **58**(7), 1758–1770 (2013)
27. Z.G. Wu, P. Shi, H. Su, J. Chu, Sampled-data fuzzy control of chaotic systems based on a T-S fuzzy model. IEEE Trans. Fuzzy Syst. **22**(1), 153–163 (2014)

A Method for Computation and Analysis of Partial Synchronization Manifolds of Delay Coupled Systems

Libo Su, Wim Michiels, Erik Steur, and Henk Nijmeijer

Abstract In this chapter, we present a systematic method for computation and analysis of partial synchronization manifolds of networks with linear diffusive couplings which is implemented as a MATLAB function. For the computation part, the method is developed based on a recently proposed algorithm for characterizing partial synchronization manifolds, with additional applicability to networks with different systems at the nodes. For the analysis part, it performs the decomposition of network dynamics into synchronization error dynamics and dynamics on manifolds in a systematic way, which is important in the context of stability analysis of partial synchronization manifolds.

This research was supported by the project C14/17/072 of the KU Leuven Research Council, by the project G0A5317N of the Research Foundation-Flanders (FWO—Vlaanderen), and by the project UCoCoS, funded by the European Unions Horizon 2020 research and innovation programme under the Marie Sklodowska-Curie Grant Agreement No 675080.

L. Su · W. Michiels (✉)
Department of Computer Science, Leuven, KU, Belgium
e-mail: wim.michiels@cs.kuleuven.be

L. Su
e-mail: libo.su@cs.kuleuven.be

E. Steur
Delft Center for Systems and Control, Delft University of Technology, Delft, The Netherlands
e-mail: e.steur@tudelft.nl

H. Nijmeijer
Department of Mechanical Engineering, Eindhoven University of Technology, Eindhoven, The Netherlands
e-mail: h.nijmeijer@tue.nl

G. Valmorbida et al. (eds.), *Accounting for Constraints in Delay Systems*,
Advances in Delays and Dynamics 12,
https://doi.org/10.1007/978-3-030-89014-8_11

1 Introduction

During recent decades, attention has been attracted to synchronization of interconnected dynamical systems. This phenomenon has been widely observed by the scientific community. Examples can be found from nature to engineering: Fireflies flash synchronously [1]; Geese fly in flock during migration [2]; Vehicles move in platoon [3]; etc. The most noticeable form of synchronous behavior is *full synchronization*, which refers to the situation where all systems of a network behave identically [4]. Many researchers have proposed conditions for full synchronization of networks, see, *e.g.* , [5–7].

However, networks may show a form of incomplete synchronization, or *partial synchronization*, which refers to the phenomenon that only some but not all the systems in the networks synchronize [4]. Partial synchronization has been observed in a variety of complex systems, for example collective behaviors of neurons in parts of human brain, [8]. It has also attracted considerable attention recently, see, *e.g.* , [4, 9–12], and the references therein. In practice, there may exist time delays between, and even inside the systems in networks. Interestingly, these time delays may affect the existence and stability of partial synchronization. Some studies have been conducted on partial synchronization of networked systems with time-delay couplings [13–17]. For example, in [13], a master stability function (see [6]) based approach is used to characterize cluster (partial) and group synchronization and also determine their stability of networks with time delay couplings, where group synchronization is a specific type of partial synchronization where the nodes in different clusters host different local dynamics, see detailed definition in [13]. In [14], a decomposition method of network dynamics around cluster states is proposed to derive delayed modal equations about cluster states for stability analysis of partial synchronization in networks with identical systems. In [15], an LMI-based condition for partial synchronization is derived, and it is also shown that partial synchronization is delay-independent for certain coupling strength larger than some threshold provided that the proposed condition is satisfied.

In this work, we focus on partial synchronization in networks of systems interacting via linear time-delay couplings. To describe the patterns of partial synchronization, the concept of partial synchronization manifolds is commonly used, which are linear invariant subspaces of the state space of the systems in networks, [16]. The existence of a partial synchronization manifold, by itself, does not imply that in experiments synchronous behavior within each cluster of the corresponding partition occurs. The latter also requires the manifold to be stable. In [4], such manifolds are identified by exploiting the symmetry under permutation of the networks with identical systems coupled through diffusion. In [15], a so-called equitable partition is defined to identify partial synchronization manifolds. More equivalent existence conditions for partial synchronization manifolds for network of identical systems interacting via linear diffusive time-delay couplings have been presented in [16]. Using one of the criteria, which utilizes the block structure of the reordered adjacency matrix of a network, an algorithm for computing partial synchronization manifolds has been developed in [16]. In [18], an improved algorithm is developed which is more computationally efficient when the couplings are invasive, *i.e.*, they do not vanish when full synchronization emerges. Most work on partial synchronization considers networks consisting of identical systems. In some research, systems with different dynamics are also considered, see, *e.g.* , [13, 19]. As an extension of [16, 18], in this work, we first extend the method proposed in [16] for computing partial synchronization manifolds to allow for networks with (some but not all) different systems at the nodes. Furthermore, we complete this method by deriving the synchronization error dynamics and the dynamics on the manifolds, which serve as a foundation for stability analysis in future work.

The structure of this chapter is as follows. Section 2 shows how to characterize and compute partial synchronization manifolds, summarizing the results from [16, 18]. Section 3 shows the synchronization errors dynamics and dynamics on the partial synchronization manifolds. Section 4 presents numerical examples. Section 5 provides the conclusions.

2 Characterization and Computation of Partial Synchronization Manifolds

In this section, we introduce some relevant concepts on partial synchronization manifolds from [16], and also extend the existence condition for partial synchronization manifolds and software in [16, 18] to allow for networks with different systems at the nodes. As an extension of [16, 18], we follow the settings and conventions from them throughout this section.

2.1 Characterization

We consider a network consisting of N systems whose coupling is represented by a directed graph $\mathscr{G} = (\mathscr{V}, \mathscr{E}, A)$.

- $\mathscr{V}$ is a finite set of nodes with cardinality $|\mathscr{V}| = N$;
- $\mathscr{E} \subset \mathscr{V} \times \mathscr{V}$ is the *ordered* set of edges, where the edge (i, j) points from node j to node i;
- $A = \left[a_{i,j}\right] \in \mathbb{R}^{N\times N}$ is the weighted adjacency matrix, where $a_{i,j} > 0$ represents the weight of edge (i, j) when $(i, j) \in \mathscr{E}$, and $a_{i,j} = 0$ when $(i, j) \notin \mathscr{E}$.

Note that the graphs considered here are simple and strongly connected. A graph is simple if it has no self-loops or multiple edges, where self-loops are edges joining a node to itself and multiple edges are two or more edges connecting an ordered pair of nodes [20]. A graph is strongly connected if, for any two nodes i, j in it, there exist a directed path from i to j and a directed path from j to i, [21].

Every node hosts a dynamical system of the following form

$$\begin{cases} \dot{x}_i(t) = f_i(x_i(t)) + B_i u_i(t) \\ y_i(t) = C_i x_i(t) \end{cases} \tag{1}$$

where $i \in \mathscr{V}$, state $x_i(t) \in \mathbb{R}^n$, function $f_i : \mathbb{R}^n \to \mathbb{R}^n$, input(s) $u_i(t) \in \mathbb{R}^m$, output(s) $y_i(t) \in \mathbb{R}^m$, input matrices $B_i \in \mathbb{R}^{n\times m}$ and output matrices $C_i \in \mathbb{R}^{m\times n}$, $i = 1, \ldots, N$.

The systems (1) interact via either one of the following two types of diffusive couplings:

$$u_i(t) = k \sum_{j\in\mathscr{N}_i} a_{i,j}[y_j(t-\tau) - y_i(t)] \tag{2}$$

or

$$u_i(t) = k \sum_{j\in\mathscr{N}_i} a_{i,j}[y_j(t-\tau) - y_i(t-\tau)], \tag{3}$$

where, $\mathscr{N}_i$ is the neighboring set of node i, *i.e.*, $\mathscr{N} := \{j \in \mathscr{V} \mid (i, j) \in \mathscr{E}\}$, $a_{i,j}$ are the entries of the weighted adjacency matrix A, τ is the time-delay, and k is the coupling strength. Coupling (2) is called *invasive* coupling since the coupling does not vanish when all the nodes synchronize; while coupling (3) is called *non-invasive* coupling since the coupling vanishes when all the nodes synchronize, [22].

For the coupled systems (1), (2) or (1), (3), a solution starting at time t_0 is a partial synchronization solution if there exist nodes $i, j \in \mathscr{V}$ with $i \neq j$ such that

$$x_i(t) = x_j(t), \quad \forall t \geq t_0 - \tau. \tag{4}$$

Here $x_i(\cdot)$ is the component of solution corresponding to node i in the network.

In order to describe the clustering of nodes in a graph $\mathscr{G}$, the concept of partition is used. A partition $\mathscr{P}$ is a set of nonempty, disjoint subsets of $\mathscr{V}$ satisfying the condition that $\mathscr{V}$ is the union of these subsets. These subsets are refereed as *parts* of the partition. In this way, a part represents a cluster of the nodes. The number of parts is denoted by κ. A partition is parameterized by a permutation matrix Π. While, the permutation matrix Π is associated with an equivalent relation $\sim$, which is such that if $i \sim j$ if the ijth entry of Π is equal to 1. Note that $\kappa = \dim\ker(I_N - \Pi)$.

Following the notations of [23], we denote the space of continuous functions that map the interval $[0, \tau] \subset \mathbb{R}$ into $\mathbb{R}^{Nn}$ by $\mathscr{C}([-\tau, 0], \mathbb{R}^{Nn})$ and the states of the coupled systems by $x_t \in \mathscr{C}([-\tau, 0], \mathbb{R}^{Nn})$ with

$$x_t(\theta) := \begin{pmatrix} x_{t,1}(\theta) \\ x_{t,2}(\theta) \\ \vdots \\ x_{t,N}(\theta) \end{pmatrix} = \begin{pmatrix} x_1(t+\theta) \\ x_2(t+\theta) \\ \vdots \\ x_N(t+\theta) \end{pmatrix}, \quad \theta \in [-\tau, 0].$$

Then, with a permutation matrix Π, the condition of form (4) can be expressed as

$$x_t(\theta) = (\Pi \otimes I_n)x_t(\theta), \quad \forall \theta \in [-\tau, 0], \quad \forall t \geq t_0. \tag{5}$$

The set of partially synchronous states induced by the permutation matrix Π is denoted by $\mathscr{M}(\Pi)$, *i.e.*,

$$\mathscr{M}(\Pi) := \{\phi \in \mathscr{C}([-\tau, 0], \mathbb{R}^{Nn}) \mid \phi(\theta) = \mathrm{col}(\phi_1(\theta), \phi_2(\theta), \dots, \phi_N(\theta)),$$
$$\phi_i(\theta) \in \mathbb{R}^n, \ i = 1, \dots, N, \ \phi(\theta) = (\Pi \otimes I_n)\phi(\theta) \, \forall \theta \in [-\tau, 0]\}.$$

Note that $\mathscr{M}(\Pi)$ is a linear subspace of $\mathscr{C}([-\tau, 0], \mathbb{R}^{Nn})$.

We call $\mathscr{M}(\Pi)$ a partial synchronization manifold if it is a forward invariant set for the coupled systems (1), (2) or (1), (3), *i.e.*, once the states are partially synchronous according to $\mathscr{M}(\Pi)$, they remain so as time evolves. This brings us to the following definition.

Definition 1 [16] The set $\mathscr{M}(\Pi)$ with permutation matrix Π for which $1 < \kappa < N$ is a partial synchronization manifold for the coupled systems (1), (2), or (1), (3), if it holds that $x_t \in \mathscr{M}(\Pi), \forall t \geq t_0$ whenever $x_{t_0} \in \mathscr{M}(\Pi)$.

Partial synchronization manifolds are in fact linear invariant subspaces of state space of the network dynamics. Once the states starts inside these subspaces, they will remain inside these subspaces with the evolution of the dynamics.

A partition $\mathscr{P}$ associated with the permutation matrix Π is called *viable* if the corresponding set $\mathscr{M}(\Pi)$ is a partial synchronization manifold.
Given a partition $\mathscr{P}$, the nodes can be relabelled by clusters, *i.e.*, , the first κ_1 nodes belong to cluster 1, the second κ_2 nodes belong to cluster 2, and so on,

$$\{\{1, \dots, \kappa_1\}, \{\kappa_1 + 1, \dots, \kappa_1 + \kappa_2\}, \dots\},$$

where κ_i, $i = 1, \dots, \kappa$ are the sizes of parts of $\mathscr{P}$. Mathematically, this can be done by using another permutation matrix R, which is referred as the *reordering matrix.* A $N \times N$ permutation matrix R is a reordered matrix if it satisfies

$$R^\top \Pi R = \begin{pmatrix} \Pi_C(\kappa_1) & & & 0 \\ & \Pi_C(\kappa_2) & & \\ & & \ddots & \\ 0 & & & \Pi_C(\kappa_\kappa) \end{pmatrix}, \quad \sum_{\ell=1}^{\kappa} k_\ell = N, \tag{6}$$

i.e., $R^\top \Pi R$ is a block diagonal matrix with κ blocks $\Pi_C(\kappa_\ell)$, each of which is a $\kappa_\ell \times \kappa_\ell$-dimensional cyclic permutation matrix.

Using R, the *reordered adjacency matrix* can be constructed as follows.

$$R^\top A R = \left(\begin{array}{c|c|c|c} A_{11} & A_{12} & \cdots & A_{1\kappa} \\ \hline A_{21} & A_{22} & \cdots & A_{2\kappa} \\ \hline \vdots & \ddots & \ddots & \vdots \\ \hline A_{\kappa 1} & A_{\kappa 2} & \cdots & A_{\kappa\kappa} \end{array}\right), \quad A_{ij} \in \mathbb{R}^{k_i \times k_j}. \tag{7}$$

Now, we can introduce the existence condition for partial synchronization manifolds. To allow for systems to have different dynamics ($f_i(\cdot) \neq f_j(\cdot)$, $B_i \neq B_j$ and $C_i \neq C_j$, for some $i, j \in \mathscr{V}$), we extend Theorems 3 and 4 in [16].

Theorem 1 *Assume that the systems (1) are left-invertible (the system input-output map is injective), then the following statements are equivalent:*

1) *$\mathscr{M}(\Pi)$ is a partial synchronization manifold for (1) and (2), respectively (1) and (3);*
2) *all blocks, respectively all off-diagonal blocks, of the reordered adjacency matrix (7), partitioned in blocks of size $k_i \times k_j$ have constant row-sums, and $\mathscr{F} = (\Pi \otimes I_n)\mathscr{F}$, $\mathscr{B} = (\Pi \otimes I_n)\mathscr{B}$, $\mathscr{C} = (\Pi \otimes I_n)\mathscr{C}$, with*

$$\mathscr{F} = \begin{bmatrix} f_1(\cdot) \\ f_2(\cdot) \\ \vdots \\ f_N(\cdot) \end{bmatrix}, \mathscr{B} = \begin{bmatrix} B_1 \\ B_2 \\ \vdots \\ B_N \end{bmatrix}, \mathscr{C} = \begin{bmatrix} C_1^\top \\ C_2^\top \\ \vdots \\ C_N^\top \end{bmatrix}.$$

Here, the conditions $\mathscr{F} = (\Pi \otimes I_n)\mathscr{F}$, $\mathscr{B} = (\Pi \otimes I_n)\mathscr{B}$ and $\mathscr{C} = (\Pi \otimes I_n)\mathscr{C}$ indicate that all the nodes in the same cluster host systems with the same dynamics, but the nodes in different clusters can host systems with different dynamics. Hence, this theorem can be also applied to the situation where some (but not all) systems in the networks are different, thus marking it as an extension of Theorems 3 and 4 of [16] where only networks of identical systems are considered.

From Theorem 1, it can be found that only the topology (adjacency matrix) and, to a very limited extent, the types of dynamics of the systems play a role on the existence of partial synchronization manifolds, while the other parameters such as delay τ and coupling strength k have no impact. The intuition behind Theorem 1 is the following. The row conditions on the adjacency matrix guarantee that every system inside the same cluster receives same (amount of) interactions from other systems, and the restriction on the dynamics types of the systems ensure that every system in the same cluster have the same input-output dynamics. Under those conditions, the solutions will be able to stay on the $\mathcal{M}(\Pi)$ if they start on $\mathcal{M}(\Pi)$.

It should be noted that the existence of partial synchronization manifold is only necessary for the networks to be partially synchronized. In order to observe partial synchronization of a network, the attractiveness (stability) of a partial synchronization manifold is also required. It will be shown in the subsequent sections the coupling parameters k and τ can affect the stability of a partial synchronization manifold.

2.2 Computation

Here, we briefly introduce the algorithm for computing the partial synchronization manifolds for networks with identical/non-identical dynamical systems at the nodes. It is an extension of the algorithms in [16, 18] which only work for networks with identical systems.

The algorithm consists of two steps:

- generate possible partitions;
- check the viability of a partition (*i.e.*, check whether a partition is related to a partial synchronization manifold).

For the first step, a vector $A_R \in \mathbb{N}^N$ indicating (in)equalities of row-sums of the adjacency matrix is exploited in [18] to limit the number of partitions to be generated for networks with invasive couplings, since for any two nodes to be in the same cluster of a viable partition, their corresponding rows in the adjacency matrix should have the same row sum. Here, we use a similar vector $A_F \in \mathbb{N}^N = [F_i], i = 1, 2, \ldots, N$ representing the types of dynamical systems in each node. $F_i = F_j$ if and only if $f_i(\cdot) = f_j(\cdot)$, $B_i = B_j$ and $C_i = C_j$ for any $i, j \in \{1, 2, \ldots, N\}$.

For networks with invasive couplings (2), we can use the Algorithm 1 to generate the partitions. Here, each partition is presented by a N-digit code $d_1 \cdots d_i \cdots d_N$, which is defined in such a way that the nodes i, j belong to the same cluster if $d_i = d_j$. For example, for a network of 6 nodes $\{1, 2, 3, 4, 5, 6\}$, the code 001201 represents the partition $\{\{1, 2, 5\}, \{3, 6\}, \{4\}\}$.

For networks with non-invasive couplings (3), the procedure is similar with the exception that there is no A_R, *i.e.*, when adding the sequential system i, $D_i = \{d_j \mid F_i = F_j,\ j = 1, \ldots, i-1\} \cup \{\max(d_1, d_2, \ldots, d_{i-1}) + 1\}$. With the partitions generated, the viability of these partitions can be checked using the row-sum test from Theorem 1.

Algorithm 1: Generate partitions (invasive coupling case), [18]

Input: $A_R = [R_1 \cdots R_N]^\top$, $A_F = [F_1 \cdots F_N]^\top$
Output: set of partition codes $\{d_1 \cdots d_N\}$ for systems $\{1, \ldots, N\}$
let $D_1 = \{0\}$ be the set of partition codes for system $\{1\}$ (d_1=0);
for i=2,...,N **do**
  set the set of partition codes for systems $\{1, \ldots, i\}$ empty;
  foreach *element in the set of partition codes for systems* $\{1, \ldots, i-1\}$ **do**
    retrieve $d_1, \ldots, d_{i-1}$;
    compute the set $D_i =$
    $\underbrace{\{d_j \mid R_j = R_i \,\&\, F_i = F_j,\ j = 1, \ldots, i-1\}}_{\text{System i allowed in the same clsuter of other systems due to same row sums and same dynamics}} \cup \underbrace{\{\max(d_1, \ldots, d_{i-1}) + 1\}}_{\text{System i in a new clsuter}}$;
    append every element of D_i to the current code $d_1 \cdots d_{i-1}$ and add all corresponding codes $d_1 \cdots d_i$ in the set of partition codes for systems $\{1, \ldots, i\}$;
  end
end

3 Dynamics Decomposition of Partially Synchronized Network

In this section, we present a systematic approach to split the dynamics of the network into two parts: (1) dynamics of synchronization error; (2) dynamics of solutions lying on the partial synchronization manifolds. The asymptotic stability of the synchronization error dynamics reveals the stability (attractiveness) of the partial synchronization manifolds. That is, when the synchronization error dynamics are stable, the partial synchronization manifolds will be able to attract nearby solutions, and partially synchronous solutions will be robust against small perturbations. So the synchronization error dynamics are important for stability analysis of partial synchronization manifolds, while, the dynamics on the manifolds can be used to study the behaviours of the systems after synchronization.

For simplicity, we assume the systems are ordered in a preliminary phase according to a viable partition $\mathscr{P}$ associated with Π in κ clusters. The state variables are also *re-indexed* such that $x_{i,j}$ denote the j-th state in the i-th cluster, that is,

$$\begin{array}{ll} x_{1,1},\ x_{1,2}, \ldots, x_{1,\kappa_1} & \text{cluster } 1, \\ x_{2,1},\ x_{2,2}, \ldots, x_{2,\kappa_2} & \text{cluster } 2, \\ \vdots & \vdots \\ x_{\kappa,1},\ x_{\kappa,2}, \ldots, x_{\kappa,\kappa_\kappa} & \text{cluster } \kappa, \end{array}$$

where κ_i is the number of nodes in cluster i and $\kappa_1 + \cdots + \kappa_\kappa = N$.

In order to split the dynamics on the manifold from the error dynamics, we define another new set of coordinates

$$z = Hx = \begin{bmatrix} x_{1,1} \\ x_{2,1} \\ \vdots \\ x_{\kappa,1} \\ E_1 \\ E_2 \\ \vdots \\ E_\kappa \end{bmatrix}, \text{ with } E_i = \begin{bmatrix} e_{i,2} \\ \vdots \\ e_{i,\kappa_i} \end{bmatrix} = \begin{bmatrix} x_{i,2} - x_{i,1} \\ \vdots \\ x_{i,\kappa_i} - x_{i,1} \end{bmatrix}, i = 1, \ldots, \kappa. \quad (8)$$

Here, $H \in \mathbb{R}^{nN \times nN}$ is an invertible matrix describing the transformation of the coordinates, and E_i denote the synchronization errors within cluster i. We refer the systems corresponding to $x_{1,1}, x_{2,1}, \ldots, x_{\kappa,1}$ as the reference systems of each cluster.

Since Π is a viable partition, the row sum of all (off-diagonal) blocks in (7) are constant for networks with invasive (non-invasive) couplings. Let $R_{i,j}$ be the value of the row sums of the (i, j)-th block of matrix A, with $i, j \in \{1, \ldots, \kappa\}$ and, in case of non-invasive coupling, $i \neq j$. Additionally, for $i = j$, we define $R_{i,j} = 0$ in case of non-invasive coupling. Also, the dynamical systems at the nodes in each cluster are the same. In this case, let us denote the dynamics of the nodes in cluster i by g_i, $\tilde{B}_i$, $\tilde{C}_i$, $i = 1, 2, \ldots, \kappa$. That is, $g_1 = f_1 = f_2 = \cdots = f_{\kappa_1}$, $g_2 = f_{\kappa_1+1} = f_{\kappa_1+1} = \cdots = f_{\kappa_1+\kappa_2}$, $\tilde{B}_1 = B_1 = B_2 = \cdots = B_{\kappa_1}$, $\tilde{C}_1 = C_1 = C_2 = \cdots = C_{\kappa_1}$, and so on.

3.1 Synchronization Error Dynamics

Network with Invasive Coupling
Let

$$R_i = \sum_{j=1}^{\kappa} R_{i,j}, \quad i = 1, \ldots, \kappa$$

denote the row sums of the adjacency matrix, corresponding to each cluster. Then the error dynamics can be expressed as

$$\begin{bmatrix} \dot{e}_{1,2}(t) \\ \vdots \\ \dot{e}_{1,\kappa_1}(t) \\ \vdots \\ \dot{e}_{\kappa,2}(t) \\ \vdots \\ \dot{e}_{\kappa,\kappa_\kappa}(t) \end{bmatrix} = \begin{bmatrix} g_1(x_{1,2}(t)) - g_1(x_{1,1}(t)) \\ \vdots \\ g_1(x_{1,\kappa_1}(t)) - g_1(x_{1,1}(t)) \\ \vdots \\ g_\kappa(x_{\kappa,2}(t)) - g_\kappa(x_{\kappa,1}(t)) \\ \vdots \\ g_\kappa(x_{\kappa,\kappa_\kappa}(t)) - g_\kappa(x_{\kappa,1}(t)) \end{bmatrix} - \begin{bmatrix} k\tilde{B}_1\tilde{C}_1 R_1 e_{1,2}(t) \\ \vdots \\ k\tilde{B}_1\tilde{C}_1 R_1 e_{1,\kappa_1}(t) \\ \vdots \\ k\tilde{B}_\kappa\tilde{C}_\kappa R_\kappa e_{\kappa,2}(t) \\ \vdots \\ k\tilde{B}_\kappa\tilde{C}_\kappa R_\kappa e_{\kappa,\kappa_\kappa}(t) \end{bmatrix}, \tag{9}$$

$$+ k\tilde{\mathscr{B}}(A_{\text{red}} \otimes I_m)\tilde{\mathscr{C}} \begin{bmatrix} e_{1,2}(t-\tau) \\ \vdots \\ e_{1,\kappa_1}(t-\tau) \\ \vdots \\ e_{\kappa,2}(t-\tau) \\ \vdots \\ \dot{e}_{\kappa,\kappa_\kappa}(t-\tau) \end{bmatrix}$$

with

$$\tilde{\mathscr{B}} = \text{diag}(I_{\kappa_1-1} \otimes \tilde{B}_1, \ldots, I_{\kappa_\kappa-1} \otimes \tilde{B}_\kappa),$$

$$\tilde{\mathscr{C}} = \text{diag}(I_{\kappa_1-1} \otimes \tilde{C}_1, \ldots, I_{\kappa_\kappa-1} \otimes \tilde{C}_\kappa),$$

and

$$A_{\text{red}} = T_1 A T_1^T - T_2 A T_1^T, \tag{10}$$

where $T_1, T_2 \in \mathbb{R}^{(N-\kappa)\times N}$ are defined as

$$T_1 = \text{diag}(T_{11}, \ldots, T_{1\kappa}), T_2 = \text{diag}(T_{21}, \ldots, T_{2\kappa}),$$

with

$$T_{1i} = \begin{bmatrix} 0 & 1 & \cdots & & 0 \\ \vdots & & \ddots & & \vdots \\ & & & 1 & 0 \\ 0 & & \cdots & 0 & 1 \end{bmatrix}, T_{2i} = \begin{bmatrix} 1 & 0 & \cdots & 0 \\ 1 & 0 & \cdots & 0 \\ \vdots & \vdots & & \vdots \\ 1 & 0 & \cdots & 0 \end{bmatrix} \in \mathbb{R}^{(\kappa_i-1)\times\kappa_i}, i = 1, \ldots, \kappa.$$

The linearized error dynamics are described by

$$\begin{bmatrix} \delta \dot{E}_1(t) \\ \vdots \\ \delta \dot{E}_\kappa(t) \end{bmatrix} =$$

$$\begin{bmatrix} I_{\kappa_1-1} \otimes (\frac{\partial g_1}{\partial x}(x_{1,1}(t)) - kR_1\tilde{B}_1\tilde{C}_1) & & 0 \\ & \ddots & \\ 0 & & I_{\kappa_\kappa-1} \otimes (\frac{\partial g_\kappa}{\partial x}(x_{\kappa,1}(t)) - kR_\kappa\tilde{B}_\kappa\tilde{C}_\kappa) \end{bmatrix}$$

$$\begin{bmatrix} \delta E_1(t) \\ \vdots \\ \delta E_\kappa(t) \end{bmatrix} + k\tilde{\mathscr{B}}(A_{\text{red}} \otimes I_m)\tilde{\mathscr{C}} \begin{bmatrix} \delta E_1(t-\tau) \\ \vdots \\ \delta E_\kappa(t-\tau) \end{bmatrix}. \tag{11}$$

The first term in (10) corresponds to matrix A where the rows and columns, corresponding to the reference systems (described by $x_{1,1}, \ldots, x_{\kappa,1}$) are taken out. The second term has the following form, which corresponds to matrix A where the columns corresponding to the reference systems are taken out, and the rows corresponding to the reference systems are repeated for $(\kappa_i - 1)$ times in the rows corresponding to cluster i.

$$T_2 A T_1^T = \left[\begin{array}{cc|cc|c} a_{1,2} \cdots & a_{1,\kappa_1} & a_{1,\kappa_1+2} \cdots & a_{1,\kappa_1+\kappa_2} & \\ \vdots & \vdots & \vdots & \vdots & \cdots \\ a_{1,2} \cdots & a_{1,\kappa_1} & a_{1,\kappa_1+2} \cdots & a_{1,\kappa_1+\kappa_2} & \\ \hline & \vdots & & & \vdots \\ \hline a_{N-\kappa_\kappa+1,2} \cdots & a_{N-\kappa_\kappa+1,\kappa_1} & a_{N-\kappa_\kappa+1,\kappa_1+2} \cdots & a_{N-\kappa_\kappa+1,\kappa_1+\kappa_2} & \\ \vdots & \vdots & \vdots & \vdots & \cdots \\ a_{N-\kappa_\kappa+1,2} \cdots & a_{N-\kappa_\kappa+1,\kappa_1} & a_{N-\kappa_\kappa+1,\kappa_1+2} \cdots & a_{N-\kappa_\kappa+1,\kappa_1+\kappa_2} & \end{array}\right].$$

Note that if $B_i = B$ and $C_i = C$ for all $i = 1, \ldots, N$, the term $k\tilde{\mathscr{B}}(A_{\text{red}} \otimes I_m)\tilde{\mathscr{C}}$ reduces to $kA_{\text{red}} \otimes BC$.

The following result concerns the spectrum of A_{red}.

Theorem 2 *Let $\sigma_\Pi(A)$ be the part of the spectrum of A, corresponding to the invariant subspace* $\ker(I - \Pi)$, *i.e., all eigenvalues of A whose corresponding eigenspaces belong to* $\ker(I - \Pi)$. *Then we have*

$$\sigma(A_{\text{red}}) = \sigma(A) \setminus \sigma_{\Pi}(A). \tag{12}$$

Proof When the set $M(\Pi)$ is a partial synchronization manifold, $\ker(I - \Pi)$ is a right invariant subspace of A, see [16] for details. That is,

$$\ker(I - \Pi) = \Big\{ \sum_{i=1}^{\kappa} c_i v_i | c_i \in \mathbb{R} \Big\},$$

where v_i are the (generalized) eigenvectors of A, corresponding to $\sigma_{\Pi}(A)$.

Using H defined in (8), we can construct matrix $\bar{A}$ as follows

$$\bar{A} = HAH^{-1} = \begin{bmatrix} A_{11} & A_{12} \\ 0 & A_{\text{red}} \end{bmatrix}, \tag{13}$$

where $A_{11} \in \mathbb{R}^{\kappa \times \kappa}$, $A_{12} \in \mathbb{R}^{\kappa \times (N-\kappa)}$ and $0 \in \mathbb{R}^{(N-\kappa)\times\kappa}$.

Based on the structure of $\bar{A}$, we have

$$\sigma(A) = \sigma(\bar{A}) = \sigma(A_{11}) \cup \sigma(A_{\text{red}}).$$

As $v_i \in \ker(I - \Pi)$, *i.e.*, $v_i = \Pi v_i$, the vectors $w_i = Hv_i$ will have the following form

$$w_i = \begin{bmatrix} \bar{w}_i \\ \mathbf{0} \end{bmatrix} \tag{14}$$

where $\bar{w}_i \in \mathbb{R}^{\kappa}$ and $\mathbf{0} \in \mathbb{R}^{N-\kappa}$, $i = 1, 2, \ldots, \kappa$.

If for some $i \in \{1, 2, \ldots, \kappa\}$, v_i is an eigenvector of A, w_i is an eigenvector of $\bar{A}$, which indicates

$$\bar{A} \begin{bmatrix} \bar{w}_i \\ \mathbf{0} \end{bmatrix} = \begin{bmatrix} A_{11} & A_{12} \\ 0 & A_{red} \end{bmatrix} \begin{bmatrix} \bar{w}_i \\ \mathbf{0} \end{bmatrix} = \lambda_i \begin{bmatrix} \bar{w}_i \\ \mathbf{0} \end{bmatrix}.$$

Therefore,

$$A_{11}\bar{w}_i = \lambda_i \bar{w}_i.$$

A similar argument holds for the generalized eigenvectors of A. It follows that

$$\sigma(A_{11}) = \sigma_{\Pi}(A).$$

Hence, we can conclude (12), completing the proof. □ ■

Network with Non-invasive Coupling

When defining a series of vectors $\tilde{R}_i \in \mathbb{R}^{\kappa_i - 1}$ with their elements being the row sums of the rows in A corresponding to $x_{i,2}, \ldots, x_{i,\kappa_i}$ for $i = 1, \ldots, \kappa$, the error dynamics of networks with non-invasive couplings can be expressed as

$$\begin{bmatrix} \dot{e}_{1,2}(t) \\ \vdots \\ \dot{e}_{1,\kappa_1}(t) \\ \vdots \\ \dot{e}_{\kappa,2}(t) \\ \vdots \\ \dot{e}_{\kappa,\kappa_\kappa}(t) \end{bmatrix} = \begin{bmatrix} g_1(x_{1,2}(t)) - g_1(x_{1,1}(t)) \\ \vdots \\ g_1(x_{1,\kappa_1}(t)) - g_1(x_{1,1}(t)) \\ \vdots \\ g_\kappa(x_{\kappa,2}(t)) - g_\kappa(x_{\kappa,1}(t)) \\ \vdots \\ g_\kappa(x_{\kappa,\kappa_\kappa}(t)) - g_\kappa(x_{\kappa,1}(t)) \end{bmatrix} - k\tilde{\mathscr{B}}(L_{\text{red}} \otimes I_m)\tilde{\mathscr{C}} \begin{bmatrix} e_{1,2}(t-\tau) \\ \vdots \\ e_{1,\kappa_1}(t-\tau) \\ \vdots \\ e_{\kappa,2}(t-\tau) \\ \vdots \\ \dot{e}_{\kappa,\kappa_\kappa}(t-\tau) \end{bmatrix}, \tag{15}$$

with

$$L_{\text{red}} = \operatorname{diag}\left(\tilde{R}_1, \ldots, \tilde{R}_\kappa\right) - A_{\text{red}}. \tag{16}$$

The linearized error dynamics is given by

$$\begin{bmatrix} \delta\dot{E}_1(t) \\ \vdots \\ \delta\dot{E}_\kappa(t) \end{bmatrix} = \begin{bmatrix} I_{\kappa_1 - 1} \otimes \frac{\partial g_1}{\partial x}(x_{1,1}(t)) & & 0 \\ & \ddots & \\ 0 & & I_{\kappa_\kappa - 1} \otimes \frac{\partial g_\kappa}{\partial x}(x_{\kappa,1}(t)) \end{bmatrix} \begin{bmatrix} \delta E_1(t) \\ \vdots \\ \delta E_\kappa(t) \end{bmatrix} - k\tilde{\mathscr{B}}(L_{\text{red}} \otimes I_m)\tilde{\mathscr{C}} \begin{bmatrix} \delta E_1(t-\tau) \\ \vdots \\ \delta E_\kappa(t-\tau) \end{bmatrix}. \tag{17}$$

Again, if $B_i = B$ and $C_i = C$ for all $i = 1, \ldots, N$, the term $k\tilde{\mathscr{B}}(L_{\text{red}} \otimes I_m)\tilde{\mathscr{C}}$ reduces to $kL_{\text{red}} \otimes BC$.

The following result concerns the spectrum of L_{red}.

Theorem 3 *Let $\sigma_\Pi(L)$ be the part of the spectrum of L, corresponding to the invariant subspace* $\ker(I - \Pi)$, *i.e., all eigenvalues of L whose corresponding eigenspaces belong to* $\ker(I - \Pi)$. *Then we have*

$$\sigma(L_{\text{red}}) = \sigma(L) \setminus \sigma_\Pi(L). \tag{18}$$

Proof Replace A, A_{11}, A_{12} and A_{red} with L, L_{11}, and L_{red} in the proof of Theorem 2. □ ■

3.2 Dynamics on the Synchronization Manifolds

Network with Invasive Coupling

For a network with invasive coupling, the dynamics on the synchronization manifolds can be described by a network of κ coupled systems:

$$\dot{x}_{i,1}(t) = g_i(x_{i,1}(t)) + k\tilde{B}_i \sum_{j=1}^{\kappa} R_{i,j} \left(\tilde{C}_j x_{j,1}(t-\tau) - \tilde{C}_i x_{i,1}(t) \right), \tag{19}$$

where $i = 1, \ldots, \kappa$.

This reduced network can be represented by graph $\tilde{\mathscr{G}} = (\tilde{\mathscr{V}}, \tilde{\mathscr{E}}, A_{\text{syn}})$, where $|\tilde{\mathscr{V}}| = \kappa$ and $A_{\text{syn}} = \left[R_{i,j}\right]_{i,j=1}^{\kappa} \in \mathbb{R}^{\kappa \times \kappa}$.

Note that the "reduced" adjacency matrix $\left[R_{i,j}\right]_{i,j=1}^{\kappa}$ has in general diagonal elements different from zero, implying the presence of self-loops in (19).

Network with Non-invasive Coupling

For a network with non-invasive coupling, the dynamics on the synchronization manifolds can be described by a network of κ coupled systems:

$$\dot{x}_{i,1}(t) = f_i(x_{i,1}(t)) + k\tilde{B}_i \sum_{j=1,\ j\neq i}^{\kappa} R_{i,j} \left(\tilde{C}_j x_{j,1}(t-\tau) - \tilde{C}_i x_{i,1}(t-\tau) \right), \tag{20}$$

where $i = 1, \ldots, \kappa$.

This reduced network can be represented by graph $\tilde{\mathscr{G}} = (\tilde{\mathscr{V}}, \tilde{\mathscr{E}}, A_{\text{syn}})$, where $|\tilde{\mathscr{V}}| = \kappa$ and $A_{\text{syn}} = \left[R_{i,j}\right]_{i,j=1}^{\kappa} \in \mathbb{R}^{\kappa \times \kappa}$. Note that $R_{ij} = 0$ if $i = j$.

It is important to note that this coupling type may affect the dynamics on the synchronization manifold. Hence, strictly speaking, the terminology "non-invasive" only applies to fully synchronous solutions.

Note: For a stable partial synchronization manifold, to observe the corresponding partial synchrony in practice, it also requires that $x_{i,1}(t) \neq x_{j,1}(t), \forall t \geq t_0,$ for every $i \neq j \in \{1, \ldots, \kappa\}$, which means the reduced network has no stable partial synchronization manifold.

3.3 Software

The algorithm proposed in Sect. 2 and the results we presented in this section are implemented in the software such that it is able to:

- compute partial synchronization manifolds given a network (adjacency matrix A and coupling type, dynamical system types of each nodes)
- compute the matrices describing the synchronization error dynamics (A_{red} or L_{red}) and the dynamics on the manifolds (A_{syn}) given a viable partition.

The second part can be achieved by using (10) or (16), and $A_{\text{syn}} = [R_{ij}]$ easily.

The input of the software includes the adjacency matrix A of a network. If A contains non-integer elements, a pre-defined tolerance is required for performing the row sum test. If A only contains integer elements, the calculation in the software can be performed in exact arithmetic. If the adjacency matrix A is decomposable in the way defined in [16], (*i.e.*, $A = \omega_1 A_1 + \ldots + \omega_r A_r$, where $\omega_i, i = 1, \ldots, r$ are rationally independent and $A_i, i = 1, \ldots, r$ are adjacency matrices with non-negative integer entries), and if the input $\{A_1, A_2, \ldots, A_r\}$ is provided, the calculations can also be done in exact arithmetic.

The software is implemented as a MATLAB function and can be downloaded from the following link: http://twr.cs.kuleuven.be/research/software/delay-control/manifolds/.

4 Example

We present a didactic example of a network with 4 interconnected systems to show the analysis process presented in the preceding sections. The systems are of the following form:

$$\begin{aligned} \dot{x}_i(t) &= A_0 x_i(t) + B u_i(t), \\ y_i(t) &= C x_i(t), \end{aligned} \tag{21}$$

with

$$A_0 = \begin{bmatrix} -\varepsilon & 1 \\ -\Omega^2 & -\varepsilon \end{bmatrix}, B = \begin{bmatrix} 0 \\ 1 \end{bmatrix}, C = \begin{bmatrix} 1 & 0 \end{bmatrix}, \Omega, \varepsilon \in \mathbb{R}_{\geq 0} \text{ and } i = 1, 2, 3, 4.$$

Here u_i describes the couplings between the systems:

$$u_i = k \sum_{j \in N_i} a_{i,j} [y_j(t-\tau) - y_i(t-\tau)], \tag{22}$$

where $a_{i,j}$ are the elements of the adjacency matrix A defined as below

$$A = \begin{bmatrix} 0 & 0.1 & 2 & 1 \\ 0.1 & 0 & 1 & 2 \\ 4 & 2 & 0 & 0.3 \\ 3 & 3 & 0.3 & 0 \end{bmatrix}, \tag{23}$$

and where the coupling strength k and the time delay τ are considered as parameters.

Since the network only contains 4 nodes which host the same linear system, it is possible to characterize and analyze the partial manifolds of the network analytically.

Combining the systems of all the nodes, we have

$$\begin{bmatrix} \dot{x}_1(t) \\ \dot{x}_2(t) \\ \dot{x}_3(t) \\ \dot{x}_4(t) \end{bmatrix} = I \otimes A_0 \begin{bmatrix} x_1(t) \\ x_2(t) \\ x_3(t) \\ x_4(t) \end{bmatrix} - kL \otimes BC \begin{bmatrix} x_1(t-\tau) \\ x_2(t-\tau) \\ x_3(t-\tau) \\ x_4(t-\tau) \end{bmatrix}. \tag{24}$$

Here L is the weighted Laplacian matrix, defined by

$$L := D - A = \begin{bmatrix} 3.1 & -0.1 & -2 & -1 \\ -0.1 & -3.1 & -1 & -2 \\ -4 & -2 & -6.3 & -0.3 \\ -3 & -3 & -0.3 & 6.3 \end{bmatrix}, \tag{25}$$

where

$$D = \begin{bmatrix} \sum_{j=1}^{N} a_{1j} & & 0 \\ & \ddots & \\ 0 & & \sum_{j=1}^{N} a_{Nj} \end{bmatrix} = \begin{bmatrix} 3.1 & 0 & 0 & 0 \\ 0 & 3.2 & 0 & 0 \\ 0 & 0 & 6.3 & 0 \\ 0 & 0 & 0 & 6.3 \end{bmatrix}.$$

The eigenvalues and eigenvectors of L are

$$\sigma(L) = \begin{bmatrix} 0 \\ 2.93 \\ 9.00 \\ 6.87 \end{bmatrix}, V = \begin{bmatrix} 0.50 & 0.56 & -0.32 & 0.15 \\ 0.50 & -0.79 & -0.32 & -0.22 \\ 0.50 & 0.18 & 0.63 & -0.63 \\ 0.50 & -0.19 & 0.63 & 0.73 \end{bmatrix}. \tag{26}$$

First, we explain how to compute the stability regions of (21)–(22) in the (k, τ) parameter space. The characteristic function of (24) is given by

$$\gamma(\lambda; k, \tau) = \det(I \otimes (\lambda I - A_0) + L \otimes kBCe^{-\lambda\tau}). \tag{27}$$

As in [24], this function can be factorized as

$$\gamma(\lambda; k, \tau) = \prod_{i=1}^{4} \gamma_i(\lambda; k, \tau), \tag{28}$$

with

$$\gamma_i(\lambda; k, \tau) = \det(\lambda I - A_0 + kBC\lambda_i(L)e^{-\lambda\tau}), i = 1, 2, 3, 4. \tag{29}$$

Denoting $\tilde{k} = \lambda_i(L)k$, each corresponding characteristic equation is of the following form

$$(\lambda + \varepsilon)^2 + \Omega^2 + \tilde{k}e^{-\lambda\tau} = 0. \tag{30}$$

To analyze the stability of (24) at the equilibrium, substitute $\lambda = j\omega, \omega > 0$, yields:

$$-\omega^2 + \varepsilon^2 + 2j\varepsilon\omega + \Omega^2 = -\tilde{k}e^{-j\omega\tau}. \tag{31}$$

Equating the module and phase of the left side with the right side of (31), we have

$$(-\omega^2 + \varepsilon^2 + \Omega^2)^2 + (2\varepsilon\omega)^2 = \tilde{k}^2, \tag{32}$$

$$\arctan\left(\frac{2\varepsilon\omega}{-\omega^2 + \varepsilon^2 + \Omega^2}\right) + 2\pi l = -\omega\tau, l \in \mathbb{Z}. \tag{33}$$

From these two equations, we can express $\tilde{k}$ and τ as function of ω, *i.e.*, $\tilde{k} := \tilde{k}(\omega)$ and $\tau := \tau(\omega)$. Therefore, by sweeping ω, we can depict the crossing curves of the time delay system numerically in the parameter space $(\tau, \tilde{k})$. Figure 1 shows the stability crossing curves of the system with $\Omega = 1$ and $\varepsilon = 0.02$. These values are used in the remainder of the example. The stability region is marked with the symbol "S". Note that for $\omega = 0$, we have $\tilde{k} = -(\varepsilon^2 + \Omega^2)$.

Next, we analyze the partial manifolds of this system using the procedure and software we presented in the preceding section. Using the software developed in this work, we can easily identify a viable partition $\mathscr{P} = \{\{1, 2\}, \{3, 4\}\}$ and derive the separated dynamics.

The synchronization error dynamics are described by

$$\begin{bmatrix} \dot{E}_1(t) \\ \dot{E}_2(t) \end{bmatrix} = \begin{bmatrix} A_0 & 0 \\ 0 & A_0 \end{bmatrix} \begin{bmatrix} E_1(t) \\ E_2(t) \end{bmatrix} - kL_{red} \otimes BC \begin{bmatrix} E_1(t-\tau) \\ E_2(t-\tau) \end{bmatrix}, \tag{34}$$

with

$$L_{\text{red}} = \begin{bmatrix} 3.2 & -1 \\ -1 & 6.6 \end{bmatrix}, \sigma(L_{\text{red}}) = \begin{bmatrix} 2.93 \\ 6.87 \end{bmatrix}. \tag{35}$$

We can see that $\sigma(L_{\text{red}}) \subset \sigma(L)$.

By comparing (24) and (34), we observe they are of the same form (but different dimensions). Therefore, the factorized characteristic function is

$$\prod_{i=1}^{2} \gamma_i(\lambda; k, \tau) = \prod_{i=1}^{2} \det(\lambda I - A_0 + kBC\lambda_i(L_{\text{red}})e^{-\lambda\tau}), \tag{36}$$

which is part of (29).

The dynamics on the manifold are given by

$$\begin{bmatrix} \dot{x}_1 \\ \dot{x}_3 \end{bmatrix} = \begin{bmatrix} A_0 & 0 \\ 0 & A_0 \end{bmatrix} \begin{bmatrix} x_1(t) \\ x_3(t) \end{bmatrix} + \begin{bmatrix} kBCR_{12}(x_3(t-\tau) - x_1(t-\tau)) \\ kBCR_{21}(x_1(t-\tau) - x_3(t-\tau)) \end{bmatrix}, \tag{37}$$

i.e.,

$$\begin{bmatrix} \dot{x}_1 \\ \dot{x}_3 \end{bmatrix} = \begin{bmatrix} A_0 & 0 \\ 0 & A_0 \end{bmatrix} \begin{bmatrix} x_1(t) \\ x_3(t) \end{bmatrix} - kL_{\text{syn}} \otimes BC \begin{bmatrix} x_1(t-\tau) \\ x_3(t-\tau) \end{bmatrix}, \tag{38}$$

where

$$L_{\text{syn}} = \begin{bmatrix} R_{12} & -R_{12} \\ -R_{21} & R_{21} \end{bmatrix} = \begin{bmatrix} 3 & -3 \\ -6 & 6 \end{bmatrix}, \sigma(L_{\text{syn}}) = \begin{bmatrix} 0 \\ 9 \end{bmatrix}. \tag{39}$$

We can see that $\sigma(L_{\text{red}}) = \sigma(L) \setminus \sigma(L_{\text{syn}})$, which verifies Theorem 3.

Again, the factorized characteristic function of (38) shown below is part of (29):

$$\prod_{i=1}^{2} \gamma_i(\lambda; k, \tau) = \prod_{i=1}^{2} \det(\lambda I - A_0 + kBC\lambda_i(L_{\text{syn}})e^{-\lambda\tau}). \tag{40}$$

A comparison between (26) on one hand, and (35), (39) on the other hand, yields additional information that the eigenvalues $\lambda_1(L) = 0$ and $\lambda_3(L) = 9$ are related to the dynamics on the manifolds, and that the eigenvalues $\lambda_2(L) = 2.93$ and $\lambda_4(L) = 6.87$ are related to the synchronization errors dynamics. Note that the decomposition can already be observed from the patterns of the eigenvectors in (26), see also Theorem 2 in [25]. However, when the network become complicated (*e.g.* increased size), analysis by hand may become extremely difficult. In this case, the procedure we proposed will show its advantage, as it is automatic and more systematic.

Now, let us check the stability crossing curve of the time delay system described by (31). From Fig. 1, we can see that the parameters k and τ impact the stability of the partial synchronization manifolds. To illustrate the impact, we show two following cases:

- Case 1 with $\tau = 5.0$ and $k = 0.05$;
- Case 2 with $\tau = 5.0$ and $k = 0.08$.

For Case 1, we have $[\tilde{k}_1\, \tilde{k}_2\, \tilde{k}_3\, \tilde{k}_4] = [0\, 0.15\, 0.45\, 0.34]$. The parameter pairs $(\tau, \tilde{k}_i)$ are shown in Fig. 1. All the points are inside the stability region. Therefore, all the dynamics are stable.

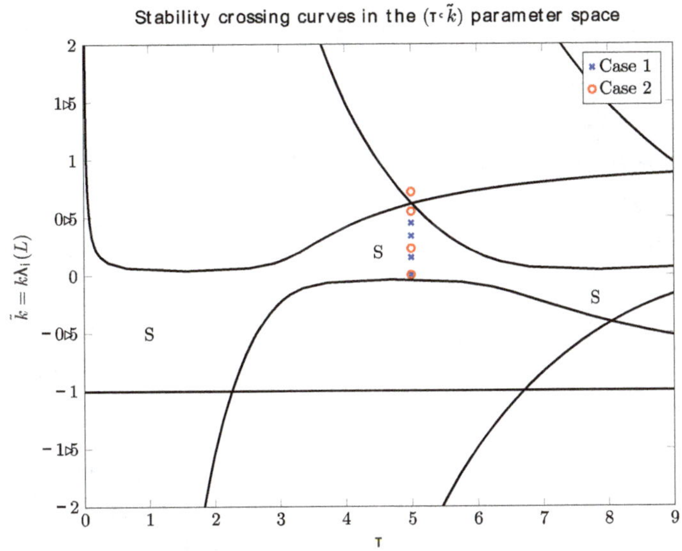

Fig. 1 Stability regions of (21)–(22) with $\Omega = 1$ and $\varepsilon = 0.02$

For Case 2, we have $[\tilde{k}_1\ \tilde{k}_2\ \tilde{k}_3\ \tilde{k}_4] = [0\ 0.23\ 0.75\ 0.55]$. The parameter pairs $(\tau, \tilde{k}_i)$ are shown in Fig. 1. As we can see from it, the pair $(\tau, \tilde{k}_3)$ is outside of the stability region while the other three pairs are inside the stability region. From the analysis above, we know $\tilde{k}_3$ is related to the dynamics on the manifold. Therefore, the partial synchronization manifold is stable, while the dynamics on the manifold is unstable. Figure 2 shows the simulation results of the network. Note that the differences between y_1 and y_2, respectively, between y_3 and y_4 are invisible after some time.

Finally, we also add a nonlinear term in each system.

$$\begin{aligned} \dot{x}_i(t) &= A_0 x_i(t) + B u_i(t) + h(x_i(t)), \\ y_i(t) &= C x_i(t), \end{aligned} \tag{41}$$

where

$$h(x_i(t)) = \begin{bmatrix} -v_i(t)(v_i^2(t) + w_i^2(t)) \\ -w_i(t)(v_i^2(t) + w_i^2(t)) \end{bmatrix}, \tag{42}$$

and

$$x_i(t) = \begin{bmatrix} v_i(t) \\ w_i(t) \end{bmatrix},\ v_i, w_i \in \mathbb{R}.$$

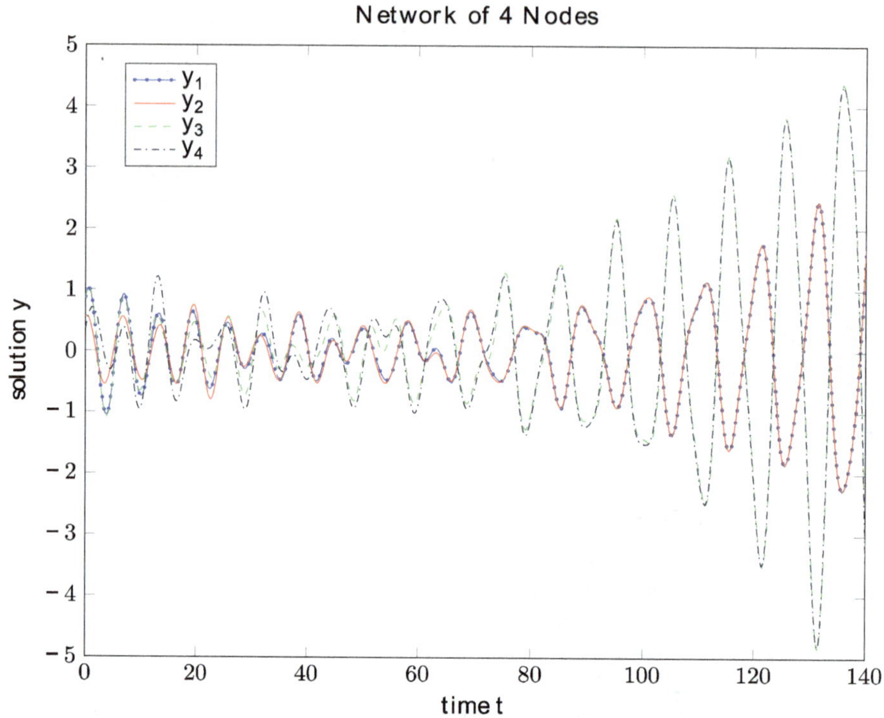

Fig. 2 Simulation results of the network with $\tau = 5.00$ and $k = 0.08$

The effect induced by the nonlinear term is that after the stability of the synchronized equilibrium is lost by increasing k, the partially synchronous solutions converge to a limit cycle. The latter is shown in Fig. 3.

It is worth mentioning that matrix (25) reveals that the coupling of the nodes within clusters is weak, and that the coupling of the nodes between clusters is strong. In this sense, it might be surprising that, when k is increased, the dynamics on the manifold associated with $\mathscr{P} = \{\{1, 2\}, \{3, 4\}\}$ become unstable first, instead of breaking the synchrony within the clusters. However, this can be explained by the presence of the time delay and the sensitivity of high gain feedback with respect to it.

5 Conclusion

In this chapter, we present a method to compute and systematically analyze the partial synchronization manifolds of delay coupled systems, which is realized as a MATLAB software. First, we extend the method from [16] and [18] to allow the systems in the networks to be different. Secondly, we split the dynamics of the network into two parts: (1) synchronization error dynamics; (2) dynamics on manifolds. The stability

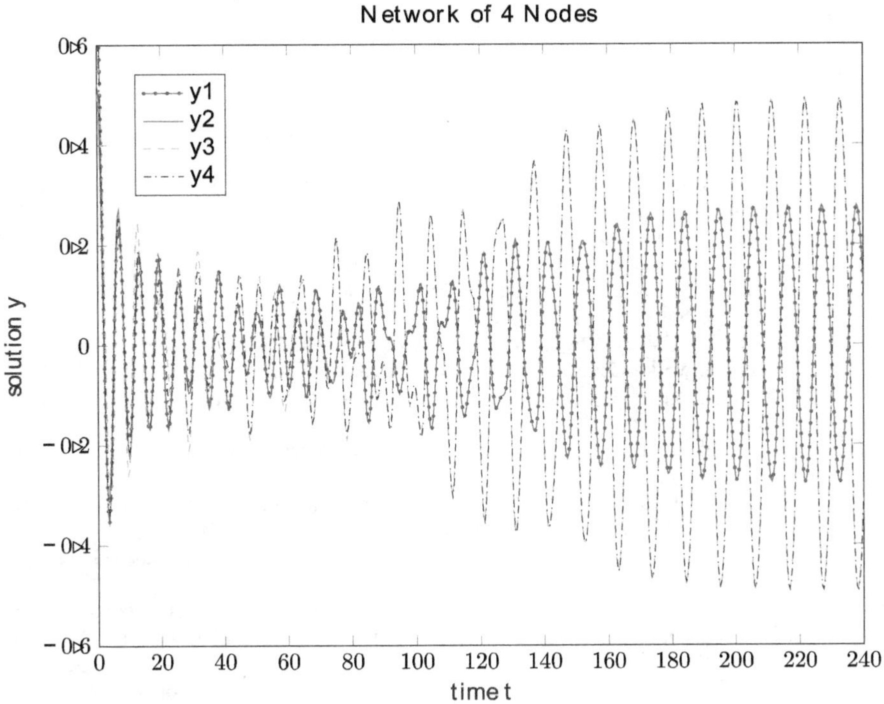

Fig. 3 Simulation of the network ($\tau = 5.00$, $k = 0.08$) with the nonlinear term

of the synchronization error dynamics is a direct indicator of the stability of partial synchronization manifolds. We also specify the relations between the spectrum of the separated dynamics and that of the original networks. To demonstrate the procedure we propose for analysis of partial synchronization manifolds, we present a didactic example. Using the procedure along with the software, the dynamics of the network can be systematically separated, which can be used for stability analysis of the manifolds.

References

1. J. Buck, E. Buck, Synchronous fireflies. Sci. Am. **234**(74–9), 82–5 (1976)
2. F.L. Lewis, H. Zhang, K. Hengster-Movric, A. Das, *Introduction to Synchronization in Nature and Physics and Cooperative Control for Multi-agent Systems on Graphs* (Springer London, London, 2014), pp. 1–21
3. J. Ploeg, D.P. Shukla, N. van de Wouw, H. Nijmeijer, Controller synthesis for string stability of vehicle platoons. IEEE Trans. Intell. Transp. Syst. **15**(2), 854–865 (2014)
4. A. Pogromsky, G. Santoboni, H. Nijmeijer, Partial synchronization: from symmetry towards stability. Physica D: Nonlinear Phenomena **172**(1), 65–87 (2002)

5. H. Fujisaka, T. Yamada, Stability theory of synchronized motion in coupled-oscillator systems. Prog. Theor. Phys. **69**(1), 32–47 (1983)
6. L.M. Pecora, T.L. Carroll, Master stability functions for synchronized coupled systems. Phys. Rev. Lett. **80**, 2109–2112 (1998)
7. A. Pogromsky, H. Nijmeijer, Cooperative oscillatory behavior of mutually coupled dynamical systems. IEEE Trans. Circuits Syst. I: Fund. Theory Appl. **48**(2), 152–162 (2001)
8. C.M. Gray, Synchronous oscillations in neuronal systems: mechanisms and functions. J. Comput. Neurosci. **1**, 11–38 (1994)
9. I. Stewart, M. Golubitsky, M. Pivato, Symmetry groupoids and patterns of synchrony in coupled cell networks. SIAM J. Appl. Dyn. Syst. **2**(4), 609–646 (2003)
10. M. Golubitsky, I. Stewart, A. Török, Patterns of synchrony in coupled cell networks with multiple arrows. SIAM J. Appl. Dyn. Syst. **4**(1), 78–100 (2005)
11. V.N. Belykh, G.V. Osipov, V.S. Petrov, J.A.K. Suykens, J. Vandewalle, Cluster synchronization in oscillatory networks. Chaos: Interdiscip. J. Nonlinear Sci. **18**(3), 037106 (2008)
12. A.Y. Pogromsky, A partial synchronization theorem. Chaos: Interdiscip. J. Nonlinear Sci. **18**(3), 037107 (2008)
13. T. Dahms, J. Lehnert, E. Schöll, Cluster and group synchronization in delay-coupled networks. Phys. Rev. E **86**, 016202 (2012)
14. G. Orosz, Decomposition of nonlinear delayed networks around cluster states with applications to neurodynamics. SIAM J. Appl. Dyn. Syst. **13**(4), 1353–1386 (2014)
15. K. Ryono, T. Oguchi, Partial synchronization in networks of nonlinear systems with transmission delay couplings. IFAC-PapersOnLine **48**(18), 77–82 (2015)
16. E. Steur, H.U. Ünal, C. van Leeuwen, W. Michiels, Characterization and computation of partial synchronization manifolds for diffusive delay-coupled systems. SIAM J. Appl. Dyn. Syst. **15**(4), 1874–1915 (2016)
17. K. Ryono, T. Oguchi, Delay-independent synchronization and network topology of systems with transmission delay couplings. SICE J. Control Meas. Syst. Integr. **10**(3), 198–203 (2017)
18. L. Su, W. Michiels, E. Steur, H. Nijmeijer, Computing partial synchronization manifolds of delay-coupled systems, in *Proceedings of the 9th European Nonlinear Dynamics Conference* (2017)
19. T. Oguchi, M. Suzuki, D. Yanagi, Delay-independent partial synchronization in networks of non-identical nonlinear systems with transmission delay coupling, in *Proceedings of the 9th European Nonlinear Dynamics Conference* (2017)
20. A. Gibbons, *Algorithmic Graph Theory* (Cambridge University Press, 1985)
21. B. Bollobas, *Modern Graph Theory* (Springer-Verlag GmbH, 1998)
22. E. Steur, C. van Leeuwen, W. Michiels, Partial synchronization manifolds for linearly time-delay coupled systems, in *Proceedings of the 21st International Symposium on Mathematical Theory of Networks and Systems, MTNS 2014* (University of Groningen, 2014), pp. 867–847
23. V. Kolmanovskii, A. Myshkis, *Introduction to the Theory and Applications of Functional Differential Equations* (Springer, Netherlands, 2013)
24. W. Michiels, H. Nijmeijer, Synchronization of delay-coupled nonlinear oscillators: An approach based on the stability analysis of synchronized equilibria. Chaos: Interdiscip. J. Nonlinear Sci. **19**(3), 033110 (2009)
25. H.U. Ünal, W. Michiels, Prediction of partial synchronization in delay-coupled nonlinear oscillators, with application to Hindmarsh-Rose neurons. Nonlinearity **26**(12), 3101 (2013)

Analysis of Linear Partial Differential Equations Using Convex Optimization

Aditya Gahlawat and Giorgio Valmorbida

Abstract We propose a framework for the exponential stability analysis for a large class of linear Partial Differential Equations (PDEs). The class of PDEs studied are parameterized by polynomial data. The class contains parabolic, first and second order hyperbolic, partial integro-differential equations, and systems coupled in-domain and on the boundaries. The proposed method is numerically tractable since we formulate the analysis test as a convex optimization problem. We use polynomial positive semi-definite matrices, the fundamental theorem of calculus, and the Green's theorem to reduce the analysis problem to the construction and verification of integral inequalities on Hilbert spaces.

1 Introduction

An important consideration when studying spatio-temporal processes is the determination of stability of the underlying processes. This serves various purposes such as ensuring that the safety-critical constraints of the systems are respected, or, determining whether feedback systems have a stabilizing effect. A vast set of spatio-temporal processes are accurately modeled by Partial Differential Equations (PDEs), a few examples of which can be found in [4, 23, 30]. Therefore, stability analysis of PDEs is a crucial component of control theory for infinite dimensional systems. The relative maturity of the theory for systems governed by Ordinary Differential Equations (ODEs) and the infinite dimensional nature of the state-space of PDEs ([6, 10]) has led to analysis methods based on finite-dimensional approximation being a

A. Gahlawat
Department of Mechanical, Materials and Aerospace Engineering (MMAE), Illinois Institute of Technology, Chicago, IL 60616, USA
e-mail: agahlawa@hawk.iit.edu

G. Valmorbida (✉)
Laboratoire des Signaux et Systèmes, CentraleSupélec, CNRS, Univ. Paris-Sud, Université Paris-Saclay, 3 Rue Joliot Curie, Gif-sur-Yvette 91192, France
e-mail: giorgio.valmorbida@centralesupelec.fr

G. Valmorbida et al. (eds.), *Accounting for Constraints in Delay Systems*, Advances in Delays and Dynamics 12,
https://doi.org/10.1007/978-3-030-89014-8_12

popular approach. A well-known example is the finite-element approximation based method [17]. In such methods, first the PDE is approximated by a set of ODEs. Then, finite-dimensional tools are applied to determine the stability of the ODE model while proving the convergence of the system of ODEs to the PDE as the resolution of the approximation is increased. We refer to such methods as *indirect methods* [9, 14]. Another indirect method for certain PDEs, for example, for the hyperbolic PDEs of the type considered in [15], is to reduce them to delay systems and then to apply tools pertinent to the reduced systems [16]. Conversely, *direct methods* are the ones wherein the infinite-dimensional nature of the state-space is considered without the application of any form of approximation. A well known direct method for the controller synthesis for PDEs is the backstepping method [18, 19, 21, 25, 26] which relies on the construction of an invertible state transformation mapping the system to be controlled to a provably stable system.

In this chapter we use the direct method known as as the Lyapunov's second method [7, 22] to establish the exponential stability of a large class of linear PDEs. The application of the Lyapunov's method for the controller synthesis for semilinear parabolic and hyperbolic PDEs can be found in [11]. Further examples can also be found in [2, 5]. The application of the Lyapunov's method requires the choice of the type of Lyapunov Functional (LF). A simple choice of an LF type may lead to solvable Lyapunov inequalities, however it can also lead to inherent conservatism as illustrated in [12]. Conversely, due to the infinite dimensional nature of the systems considered, choosing more general LF candidates may lead to Lyapunov inequalities for which analytical or numerical solutions may be difficult to obtain. In this chapter we propose a class of LF candidates along with a framework for the solution of the related Lyapunov inequalities. The LF candidates we consider are an extension of the candidates considered in [28] and a generalization of the LF candidates considered in [12]. For the PDE systems studied, we choose LF candidates which are a sum of weighted norm on chosen Hilbert spaces and integrals with quadratic kernels defined on two dimensional spatial domains. The resultant Lyapunov inequalities appear in the form of integral inequalities on Hilbert spaces. Using Sum-of-Squares (SOS) polynomials [1, Sect. 2.1], the Fundamental Theorem of Calculus (FTC), and the Green's theorem we reduce the problem of verification of the Lyapunov integral inequalities to the problem of Linear Matrix Inequalities (LMIs) subject to affine constraints, which is a Semi-Definite Programming (SDP) problem [3, 29], a type of convex optimization problem, and hence numerically tractable. We have previously applied the similar methodology on a reduced class of PDEs in [13].

1.1 Notation

We denote by $\overline{\Omega}$ the set $\{(x, y) \in \mathbb{R}^2 : 0 \leq y \leq x \leq 1\}$, $\underline{\Omega} = \{(x, y) \in \mathbb{R}^2 : 0 \leq x \leq y \leq 1\}$ and $\Delta = [0, 1] \times [0, 1]$. In the following definitions $\alpha, \beta \in \mathbb{N}$. For $w : [0, 1] \to \mathbb{R}^\beta$, $w \in \mathscr{C}^\alpha([0, 1])$, we denote by $\partial_x^i w(x)$ the i-th derivative of w and

$$w_\alpha(x) = \left[\, w(x)^T,\ \partial_x w(x)^T,\ \cdots,\ \partial_x^\alpha w(x)^T \,\right]^T, \quad w_\alpha^b = \left[\, w_{\alpha-1}(1)^T,\ w_{\alpha-1}(0)^T \,\right]^T,$$
$$\bar{w}_\alpha(x) = \left[\, w_\alpha(x)^T,\ \left(w_\alpha^b\right)^T \,\right]^T.$$

Thus, $w_\alpha : [0, 1] \to \mathbb{R}^{\beta(\alpha+1)}$, $w_\alpha^b \in \mathbb{R}^{2\beta\alpha}$ and $\bar{w}_\alpha : [0, 1] \to \mathbb{R}^{\beta(3\alpha+1)}$. We denote by $N_\partial, N \in \mathbb{N}^{\beta\alpha\times\beta(\alpha+1)}$ and $N_0, N_1 \in \mathbb{N}^{\beta\alpha\times 2\beta\alpha}$ the matrices satisfying $\partial_x w_{\alpha-1}(x) = N_\partial w_\alpha(x)$, $w_{\alpha-1}(x) = N w_\alpha(x)$, $w_{\alpha-1}(0) = N_0 w_\alpha^b$, and $w_{\alpha-1}(1) = N_1 w_\alpha^b$.

We denote

$$\mathscr{H}^\alpha\left([0, 1]; \mathbb{R}^\beta\right) = \Big\{ w : [0, 1] \to \mathbb{R}^\beta \ :\ w, \partial_x w, \ldots, \partial_x^{\alpha-1} w \text{ are absolutely continuous on } [0, 1] \text{ and } \int_0^1 \left(\partial_x^\alpha w(x)\right)^T \left(\partial_x^\alpha w(x)\right) dx < \infty \Big\}.$$

We also write $\mathscr{L}_2\left([0, 1]; \mathbb{R}^\beta\right) = \mathscr{H}^0\left([0, 1]; \mathbb{R}^\beta\right)$. The space $\mathscr{H}^\alpha\left([0, 1]; \mathbb{R}^\beta\right)$ is endowed with the standard inner product and norm $\|w\|_{\mathscr{H}^\alpha} = \sqrt{\langle w_\alpha, w_\alpha\rangle_{\mathscr{H}^\alpha}} = \sqrt{\int_0^1 w_\alpha(x)^T w_\alpha(x) dx}$, and the space $\mathscr{L}_2\left([0, 1]; \mathbb{R}^\beta\right)$ has the norm and inner product $\|\cdot\|_{\mathscr{L}_2} = \|\cdot\|_{\mathscr{H}^0}$ and $\langle\cdot, \cdot\rangle_{\mathscr{L}_2} = \langle\cdot, \cdot\rangle_{\mathscr{H}^0}$, respectively.

For any $m, n \in \mathbb{N}$, we denote by $0_{m,n}$ the matrix of zeros of dimensions m-by-n and 0_m when $m = n$. Similarly, we denote by I_n the identity matrix of dimensions n-by-n. The set of symmetric matrices of dimension n-by-n is donated by $\mathbb{S}^n$. For any square matrix Q, we denote $He(Q) = \frac{1}{2}\left(Q + Q^T\right)$.

We denote by $\mathscr{S}^n[x]$ and $\mathscr{S}^n[(x, y)]$ the sets of symmetric polynomial matrices of size n-by-n in variables x, and x and y, respectively. Here, $x \in \mathbb{R}$ and $y \in \mathbb{R}$. Similarly, we denote by $\mathscr{R}^{m\times n}[x]$ and $\mathscr{R}^{m\times n}[(x, y)]$ the sets of real polynomial matrices of size m-by-n in variables x and y. We denote by $\Sigma^n[x]$ the set of Sum-of-Squares (SOS) polynomials which, for all $x \in [0, 1]$, belong to $\mathbb{S}^n$. Note that, by definition, an SOS polynomial is non-negative for all $x \in \mathbb{R}$ [3, Chap. 3]. We say that a polynomial $S \in \mathscr{S}^n[x]$ is positive semi-definite on $[0, 1]$ if $S(x) \succeq 0$, for all $x \in [0, 1]$. Given $\alpha, \beta, d \in \mathbb{N}$, we define

$$Y_{q(\alpha,\beta,d)}(x, y) = I_{\beta(\alpha+1)} \otimes z(x, y) \in \mathbb{R}^{q(\alpha,\beta,d)\times\beta(\alpha+1)}, \tag{1}$$

where $q(\alpha, \beta, d) = \frac{1}{2}\beta(\alpha + 1)(d + 2)(d + 1)$, $z(x, y) \in \mathbb{R}^{\frac{1}{2}(d+2)(d+1)}$ is the vector of monomials in the standard polynomial basis in x and y up to degree d and $\otimes$ denotes the Kronecker product. For example, for $d = 2$, we have

$$z(x, y) = \left[\, 1\ x\ y\ x^2\ xy\ y^2 \,\right]^T.$$

For any bivariate function $K : \overline{\Omega} \to \mathbb{R}^{n\times n}$, we define the linear map

$$\Gamma[K] = \begin{cases} K(x, y), & x \ge y \\ K(y, x)^T, & y > x \end{cases},$$

thus satisfying, for any $v : [0, 1] \to \mathbb{R}^n$,

$$\int_{\Delta} v(x)^T \Gamma [K] v(y) d\Delta$$
$$= \int_0^1 \int_0^x v(x)^T K(x, y) v(y) dy dx + \int_0^1 \int_x^1 v(x)^T K(y, x)^T v(y) dy dx.$$

2 Problem Statement

We consider the following class of linear Partial Differential Equations (PDEs) with polynomial coefficients

$$\partial_t w(x, t) = A_1(x) w_\alpha(x, t) + \int_0^x A_2(x, y) w_\alpha(y, t) dy + \int_x^1 A_3(x, y) w_\alpha(y, t) dy, \tag{2a}$$

$$w(\cdot, t) \in \mathscr{B}, \quad \forall t \geq 0, \tag{2b}$$

where $A_1 \in \mathscr{R}^{\beta \times \beta(\alpha+1)}[x]$, $A_2, A_3 \in \mathscr{R}^{\beta \times \beta(\alpha+1)}[(x, y)]$, and the boundary conditions define the set

$$\mathscr{B} = \left\{ v \in \mathscr{H}^\alpha \left([0, 1]; \mathbb{R}^\beta \right) : \int_0^1 \left[F_1(x) \; F_2 \right] \bar{v}_\alpha(x) dx = 0_{\beta\alpha, 1} \right\}, \tag{2c}$$

where $F_1 \in \mathscr{R}^{\beta\alpha \times \beta(\alpha+1)}[x]$ and $F_2 \in \mathbb{R}^{\beta\alpha \times 2\beta\alpha}$.

We assume that the PDE (2) is well posed, that is, the PDE admits a unique solution which belongs to $\mathscr{H}^\alpha \left([0, 1]; \mathbb{R}^\beta \right)$ for all $t > 0$.

We will establish the stability of (2) by constructing Lyapunov Functional (LF) certificates of exponential stability. In particular, we consider the following class of LF candidates

$$\mathcal{V}(w) := \frac{1}{2} \int_0^1 w(x)^T T_b(x) w(x) dx + \frac{1}{2} \int_{\Delta} w(x)^T \Gamma \left[\bar{T} \right] w(y) d\Delta, \tag{3}$$

where $T_b \in \mathscr{S}^\beta[x]$ and $\bar{T} \in \mathscr{R}^{\beta \times \beta}[(x, y)]$ are polynomial functions to be determined.

In order to formulate the conditions for the exponential stability of (2), we first establish the time derivative of (3) along the trajectories of (2). We define

$$\mathcal{V}_d(w) := -\partial_t \mathcal{V}(w) - 2\delta \mathcal{V}(w), \tag{4}$$

where δ is a positive scalar to be determined. As we show in Appendix 6, $\mathcal{V}_d(w)$ can be written as

$$\mathcal{V}_d(w) = \int_0^1 \bar{w}_\alpha(x)^T U_b(x) \bar{w}_\alpha(x) dx + \int_\Delta w_\alpha(x)^T \Gamma\left[\bar{U}\right] w_\alpha(y) d\Delta, \tag{5}$$

where

$$\begin{aligned}
U_b(x) =& - He\left(\begin{bmatrix} T_b(x)A_1(x) & 0_{\beta,2\beta\alpha} \\ 0_{3\beta\alpha,\beta(\alpha+1)} & 0_{3\beta\alpha,2\beta\alpha} \end{bmatrix} + \begin{bmatrix} \delta T_b(x) & 0_{\beta,3\beta\alpha} \\ 0_{3\beta\alpha,\beta} & 0_{3\beta\alpha} \end{bmatrix}\right), \\
\bar{U}(x, y) =& -\frac{1}{2}\left(\begin{bmatrix} \bar{T}(x, y)^T A_1(x) \\ 0_{\beta\alpha,\beta(\alpha+1)} \end{bmatrix}^T + \begin{bmatrix} \bar{T}(x, y) A_1(y) \\ 0_{\beta\alpha,\beta(\alpha+1)} \end{bmatrix}\right) \\
& -\frac{1}{2}\left(\begin{bmatrix} T_b(x) A_2(x, y) \\ 0_{\beta\alpha,\beta(\alpha+1)} \end{bmatrix} + \begin{bmatrix} T_b(y) A_3(y, x) \\ 0_{\beta\alpha,\beta(\alpha+1)} \end{bmatrix}^T\right) \\
& -\frac{1}{2}\left(\underline{U}(x, y) + \begin{bmatrix} 2\delta\bar{T}(x, y) & 0_{\beta,\beta\alpha} \\ 0_{\beta\alpha,\beta} & 0_{\beta\alpha} \end{bmatrix}\right),
\end{aligned}$$

and

$$\begin{aligned}
&\underline{U}(x, y) = \\
&\int_0^y \left(\begin{bmatrix} \bar{T}(x, z) \\ 0_{\beta\alpha,\beta} \end{bmatrix} A_3(z, y) + \left(\begin{bmatrix} \bar{T}(y, z) \\ 0_{\beta\alpha,\beta} \end{bmatrix} A_3(z, x)\right)^T\right) dz \\
&+ \int_y^x \left(\begin{bmatrix} \bar{T}(x, z) \\ 0_{\beta\alpha,\beta} \end{bmatrix} A_2(z, y) + \left(\begin{bmatrix} \bar{T}(z, y)^T \\ 0_{\beta\alpha,\beta} \end{bmatrix} A_3(z, x)\right)^T\right) dz \\
&+ \int_x^1 \left(\begin{bmatrix} \bar{T}(z, x)^T \\ 0_{\beta\alpha,\beta} \end{bmatrix} A_2(z, y) + \left(\begin{bmatrix} \bar{T}(z, y)^T \\ 0_{\beta\alpha,\beta} \end{bmatrix} A_2(z, x)\right)^T\right) dz.
\end{aligned}$$

We now state the conditions for the exponential stability of (2) in the following theorem.

Theorem 1 *Given the PDE (2), suppose there exist positive scalars μ_1, μ_2, δ and polynomials $T_b \in \mathcal{S}^\beta[x]$ and $\bar{T} \in \mathcal{R}^{\beta\times\beta}[(x, y)]$ such that*

$$\mu_1 \|w\|^2_{\mathcal{L}_2} \le \mathcal{V}(w) \le \mu_2 \|w\|^2_{\mathcal{L}_2}, \quad \forall w \in \mathcal{L}_2\left([0, 1]; \mathbb{R}^\beta\right), \tag{6}$$

$$\mathcal{V}_d(w) \ge 0, \quad \forall w \in \mathcal{B} \subset \mathcal{H}^\alpha\left([0, 1]; \mathbb{R}^\beta\right), \tag{7}$$

where $\mathcal{V}(w)$ and $\mathcal{V}_d(w)$ are defined in (3) and (5), respectively. Then, there exists a positive scalar κ such that the solutions of (2) satisfy

$$\|w(\cdot, t)\|_{\mathcal{L}_2} \le \kappa e^{-\delta t} \|w(\cdot, 0)\|_{\mathcal{L}_2}, \quad \forall t \ge 0. \tag{8}$$

Proof Let us choose the LF candidate $\mathcal{V}(w)$. Since (7) holds for all $w \in \mathcal{B}$, for all solutions of (2) we have that

$$\mathcal{V}_d(w) = -\partial_t \mathcal{V}(w) - 2\delta \mathcal{V}(w) \geq 0, \quad \forall t \geq 0.$$

Integrating this expression in time produces

$$\mathcal{V}(w) \leq e^{-2\delta t} \mathcal{V}(w(0)),$$

and thus, using (6) produces

$$\mu_1 \|w(\cdot, t)\|_{\mathcal{L}_2}^2 \leq e^{-2\delta t} \mu_2 \|w(\cdot, 0)\|_{\mathcal{L}_2}^2.$$

Then, we conclude that (8) holds with $\kappa = \sqrt{\mu_2/\mu_1}$. ■

We have reduced the problem of assessing the exponential stability to the problem of verification of the integral inequalities in (6) and (7). Verification of (6) is straightforward. We generalize the results in [12, 28] to verify

$$\mu_1 \|w\|_{\mathcal{L}_2}^2 \leq \mathcal{V}(w) \leq \mu_2 \|w\|_{\mathcal{L}_2}^2, \quad \forall w \in \mathcal{L}_2\left([0, 1]; \mathbb{R}^\beta\right),$$

using a Positive Semi-Definite (PSD) matrix valued function, which ensures that the integral inequality holds.

Since (7) needs to be verified on the subset $\mathcal{B}$, and not on the whole space $\mathcal{H}^\alpha\left([0, 1]; \mathbb{R}^\beta\right)$, we cannot use the same approach as for the verification of (7) since doing so would introduce conservatism. Therefore, our approach for the verification of (7) is to construct *slack integrals* which we define to be integral expressions $\mathcal{S}(w)$ which have the same structure as $\mathcal{V}_d(w)$ and satisfy

$$\mathcal{S}(w) = 0, \quad \forall w \in \mathcal{B}.$$

The use of slack integrals will then allow us to verify (7) using the generalization of the results in [12, 28].

3 Integral Inequalities on Hilbert Spaces

In this section we will construct two integral inequalities which will assist in the verification of the integral inequalities in (6) and (7). The main property of such inequalities is that they are parameterized by non-negative polynomial matrices, a property which we will exploit in order to formulate a numerically tractable method to verify the conditions of Theorem 1.

We begin with the construction of non-negative integral inequalities on $\mathcal{H}^\alpha\left([0, 1]; \mathbb{R}^\beta\right)$. For any $w \in \mathcal{H}^\alpha\left([0, 1]; \mathbb{R}^\beta\right)$, let us define

$$\mathcal{R}(w) = \int_0^1 \bar{w}_\alpha(x)^T R_b(x) \bar{w}_\alpha(x) dx + \int_\Delta w_\alpha(x)^T \Gamma\left[\bar{R}\right] w_\alpha(y) d\Delta, \tag{9}$$

where $R_b \in \mathscr{S}^{\beta(3\alpha+1)}[x]$ and

$$\begin{aligned}\bar{R}(x, y) =& R_{12}(x)Y_{q(\alpha,\beta,d)}(x, y) + Y_{q(\alpha,\beta,d)}(y, x)^T R_{13}(y)^T \\ &+ \int_0^y Y_{q(\alpha,\beta,d)}(z, x)^T R_{33} Y_{q(\alpha,\beta,d)}(z, y)dz \\ &+ \int_y^x Y_{q(\alpha,\beta,d)}(z, x)^T R_{23}^T Y_{q(\alpha,\beta,d)}(z, y)dz \\ &+ \int_x^1 Y_{q(\alpha,\beta,d)}(z, x)^T R_{22} Y_{q(\alpha,\beta,d)}(z, y)dz, \end{aligned} \tag{10}$$

for some (polynomial) matrices $R_{12}, R_{13} \in \mathscr{R}^{\beta(\alpha+1)\times q(\alpha,\beta,d)}[x]$, $R_{22}, R_{33} \in \mathbb{S}^{q(\alpha,\beta,d)}$ and $R_{23} \in \mathbb{R}^{q(\alpha,\beta,d)\times q(\alpha,\beta,d)}$, which also define

$$R(x) = \left[\begin{array}{c|c|c} R_b(x) & \begin{bmatrix} R_{12}(x) \\ 0_{2\beta\alpha,q(\alpha,\beta,d)} \end{bmatrix} & \begin{bmatrix} R_{13}(x) \\ 0_{2\beta\alpha,q(\alpha,\beta,d)} \end{bmatrix} \\ \hline \begin{bmatrix} R_{12}(x) \\ 0_{2\beta\alpha,q(\alpha,\beta,d)} \end{bmatrix}^T & R_{22} & R_{23} \\ \hline \begin{bmatrix} R_{13}(x) \\ 0_{2\beta\alpha,q(\alpha,\beta,d)} \end{bmatrix}^T & R_{23}^T & R_{33} \end{array}\right] \in \mathscr{S}^{\beta(3\alpha+1)+2q(\alpha,\beta,d)}[x]. \tag{11}$$

The following theorem states the conditions on the polynomial matrix $R(x)$ whose submatrices define R_b and $\bar{R}$ such that (9) is non-negative on $\mathscr{H}^\alpha\left([0, 1]; \mathbb{R}^\beta\right)$.

Theorem 2 *Given (polynomial) matrices which define $\mathcal{R}(w)$ in (9) and the matrix $R(x)$ in (11), if*

$$R(x) \succeq 0, \quad \forall x \in [0, 1], \tag{12}$$

then

$$\mathcal{R}(w) \geq 0, \quad \forall w \in \mathscr{H}^\alpha\left([0, 1]; \mathbb{R}^\beta\right). \tag{13}$$

The proof of this theorem is provided in Appendix 6.

We now construct strictly positive integral inequalities on $\mathscr{L}_2\left([0, 1]; \mathbb{R}^\beta\right)$. The following corollary corresponds to $\alpha = 0$ in Theorem 2.

Given $w \in \mathscr{L}_2\left([0, 1]; \mathbb{R}^\beta\right)$ and $d \in \mathbb{N}$, let us define

$$\mathcal{T}(w) = \int_0^1 w(x)^T T_b(x)w(x)dx + \int_\Delta w(x)^T \Gamma\left[\bar{T}\right] w(y)d\Delta, \tag{14}$$

where $T_b \in \mathscr{S}^\beta[x]$ and

$$\begin{aligned}\bar{T}(x,y) = & T_{12}(x)Y_{q(0,\beta,d)}(x,y) + Y_{q(0,\beta,d)}(y,x)^T T_{13}(y)^T \\ & + \int_0^y Y_{q(0,\beta,d)}(z,x)^T T_{33} Y_{q(0,\beta,d)}(z,y)dz \\ & + \int_y^x Y_{q(0,\beta,d)}(z,x)^T T_{23}^T Y_{q(0,\beta,d)}(z,y)dz \\ & + \int_x^1 Y_{q(0,\beta,d)}(z,x)^T T_{22} Y_{q(0,\beta,d)}(z,x)dz. \end{aligned} \tag{15}$$

for any $T_{12}, T_{13} \in \mathscr{R}^{\beta \times q(0,\beta,d)}[x]$, $T_{22}, T_{33} \in \mathbb{S}^{q(0,\beta,d)}$ and $T_{23} \in \mathbb{R}^{q(0,\beta,d) \times q(0,\beta,d)}$, which also define

$$T(x) = \begin{bmatrix} T_b(x) & T_{12}(x) & T_{13}(x) \\ T_{12}(x)^T & T_{22} & T_{23} \\ T_{13}(x)^T & T_{23}^T & T_{33} \end{bmatrix} \in \mathscr{S}^{\beta+2q(0,\beta,d)}[x]. \tag{16}$$

Corollary 1 *Given (polynomial) matrices which define $\mathcal{T}(w)$ in (14) and $T(x)$ in (16). If there exists a positive scalar ϵ such that*

$$T(x) - \begin{bmatrix} \epsilon I_\beta & 0_{\beta,2q(0,\beta,d)} \\ 0_{2q(0,\beta,d),\beta} & 0_{2q(0,\beta,d)} \end{bmatrix} \succeq 0, \quad \forall x \in [0,1], \tag{17}$$

then,

$$\mathcal{T}(w) \geq \epsilon \|w\|^2_{\mathscr{L}_2}, \quad \forall w \in \mathscr{L}_2\left([0,1];\mathbb{R}^\beta\right). \tag{18}$$

See Appendix 6 for the proof.

4 Slack Integrals

In this section we construct slack integrals which will allow us to verify the inequality in (7) on the set $\mathscr{B} \subset \mathscr{H}^\alpha\left([0,1];\mathbb{R}^\beta\right)$. Recall that we defined slack integrals $\mathcal{S}(w)$ as integral expressions that have the same structure as $\mathcal{V}_d$ in (5) and which satisfy

$$\mathcal{S}(w) = 0, \quad \forall w \in \mathscr{B}.$$

To construct slack integrals we will use the quadratic forms of the Fundamental Theorem of Calculus (FTC) and the Green's theorem which are presented below.

Lemma 1 (FTC quadratic form) *For any $K_1 \in \mathscr{R}^{\beta\alpha \times \beta\alpha}[x]$ and $K_2 \in \mathscr{R}^{\beta\alpha \times 2\beta\alpha}[x]$ the following identity holds*

$$\int_0^1 \bar{w}_\alpha(x)^T He\left(K_b(x)\right)\bar{w}_\alpha(x)dx = 0, \quad \forall w \in \mathscr{H}^\alpha\left([0,1];\mathbb{R}^\beta\right), \tag{19}$$

where

$$\begin{aligned} K_b &= \begin{bmatrix} K_{b1}(x) & K_{b2}(x) \\ 0_{2\beta\alpha,\beta(\alpha+1)} & K_{b3} \end{bmatrix}, \\ K_{b1}(x) &= N^T \partial_x K_1(x) N + N^T K_1(x) N_\partial + N_\partial^T K_1(x) N, \\ K_{b2}(x) &= N_\partial^T K_2(x) + N^T \partial_x K_2(x), \\ K_{b3} &= N_0^T K_1(0) N_0 - N_1^T K_1(1) N_1 + N_0 K_2(0) - N_1 K_2(1). \end{aligned}$$

Proof The identity (19) is produced by expanding

$$\int_0^1 \frac{d}{dx} g(x) dx - (g(1) - g(0)) = 0,$$

with

$$g(x) = w_{\alpha-1}(x)^T K_1(x) w_{\alpha-1}(x) + w_{\alpha-1}(x)^T K_2(x) w_\alpha^b.$$

■

Lemma 2 (Green's theorem quadratic form) *For any $H_1, H_2 \in \mathscr{R}^{\beta\alpha\times\beta\alpha}[(x, y)]$, the following identity holds*

$$\int_0^1 \bar{w}_\alpha(x)^T He\left(H_b(x)\right) \bar{w}_\alpha(x) dx + \int_\Delta w_\alpha(x)^T \Gamma\left[\bar{H}\right] w_\alpha(y) d\Delta = 0,$$

for all $w \in \mathscr{H}^\alpha\left([0, 1]; \mathbb{R}^\beta\right)$, where

$$\begin{aligned} H_b(x) &= \begin{bmatrix} H_{b1}(x) & H_{b2}(x) \\ H_{b3}(x) & 0_{2\beta\alpha} \end{bmatrix}, \quad H_{b1}(x) = -N^T \left(H_1(x, x) + H_2(x, x)\right) N, \\ H_{b2}(x) &= N^T H_1(x, 0) N_0, \quad H_{b3}(x) = N_1^T H_2(1, x) N, \\ \bar{H}(x, y) &= \frac{1}{2} N^T \left(\partial_y H_1(x, y) - \partial_x H_2(x, y)\right) N + \frac{1}{2}\left(N^T H_1(x, y) N_\partial - N_\partial^T H_2(x, y) N\right). \end{aligned}$$

The proof of this lemma is presented in Appendix 6.

We now present an integral expression derived using the definition of the boundary set $\mathscr{B}$ in (2c). The proof of the following lemma can be found in Appendix 6.

Lemma 3 *Given $F_1 \in \mathscr{R}^{\beta\alpha\times\beta(\alpha+1)}[x]$ and $F_2 \in \mathbb{R}^{\beta\alpha\times 2\beta\alpha}$ the following identity holds true for all $w \in \mathscr{B}$:*

$$\int_0^1 \bar{w}_\alpha(x)^T He\left(B_b(x)\right) \bar{w}_\alpha(x) dx + \int_\Delta w_\alpha(x)^T \Gamma\left[\bar{B}\right] w_\alpha(y) d\Delta = 0,$$

where

$$B_b(x) = \begin{bmatrix} 0_{\beta(\alpha+1)} & B_1(x)F_2 \\ B_2F_1(x) & B_2F_2 \end{bmatrix}, \quad \bar{B}(x,y) = \frac{1}{2}\left(B_1(x)F_1(y) + F_1(x)^T B_1(y)^T\right),$$

for any $B_1 \in \mathscr{R}^{\beta(\alpha+1)\times\beta\alpha}[x]$ *and* $B_2 \in \mathbb{R}^{2\beta\alpha\times\beta\alpha}$.

We now construct slack integrals using the results from Lemmas 1–3. Let us define

$$\begin{aligned} \mathcal{S}(w) = &\int_0^1 \bar{w}_\alpha(x)^T He\left(K_b(x) + H_b(x) + B_b(x)\right)\bar{w}_\alpha(x)dx \\ &+ \int_\Delta w_\alpha(x)^T \Gamma\left[\bar{H} + \bar{B}\right] w_\alpha(y)d\Delta, \end{aligned} \tag{20}$$

where K_b is parameterized by K_1 and K_2 as in Lemma 1, H_b and $\bar{H}$ are parameterized by H_1 and H_2 as in Lemma 2 and B_b and $\bar{B}$ are parameterized by B_1 and B_2 and the (polynomial) matrices F_1 and F_2 which define the set $\mathscr{B}$ as in Lemma 3.

We now state the main result of this section.

Theorem 3 *Given (polynomial) matrices* $F_1(x)$ *and* F_2 *which define the set* $\mathscr{B}$ *in (2c), the following identity holds true*

$$\mathcal{S}(w) = 0, \quad \forall w \in \mathscr{B}, \tag{21}$$

where $\mathcal{S}(w)$ *is parameterized by any* K_i, H_i *and* B_i, $i \in \{1, 2\}$, *as in (20).*

Proof We begin by considering the following decomposition

$$\begin{aligned} \mathcal{S}(w) = &\int_0^1 \bar{w}_\alpha(x)^T He\left(K_b(x) + H_b(x) + B_b(x)\right)\bar{w}_\alpha(x)dx \\ &+ \int_\Delta w_\alpha(x)^T \Gamma\left[\bar{H} + \bar{B}\right] w_\alpha(y)d\Delta = \sum_{i=1}^{3}\Theta_i, \end{aligned} \tag{22}$$

where

$$\begin{aligned} \Theta_1 &= \int_0^1 \bar{w}_\alpha(x)^T He\left(K_b(x)\right)\bar{w}_\alpha(x)dx, \\ \Theta_2 &= \int_0^1 \bar{w}_\alpha(x)^T He\left(H_b(x)\right)\bar{w}_\alpha(x)dx + \int_\Delta w_\alpha(x)^T \Gamma\left[\bar{H}\right] w_\alpha(y)d\Delta, \\ \Theta_3 &= \int_0^1 \bar{w}_\alpha(x)^T He\left(B_b(x)\right)\bar{w}_\alpha(x)dx + \int_\Delta w_\alpha(x)^T \Gamma\left[\bar{B}\right] w_\alpha(y)d\Delta. \end{aligned}$$

From Lemmas 1 and 2 we have that

$$\Theta_1 = 0 \quad \text{and} \quad \Theta_2 = 0, \quad \forall w \in \mathscr{H}^\alpha\left([0, 1]; \mathbb{R}^\beta\right). \tag{23}$$

From Lemma 3 we have that

$$\Theta_3 = 0, \quad \forall w \in \mathscr{B}. \tag{24}$$

Therefore, from (22) to (24) we conclude that the expression in (21) holds for all $w \in \mathscr{B}$. ■

5 Main Result

We now present the main result of the chapter. We use the results from the previous sections to devise a convex test for the verification of the conditions of Theorem 1.

Theorem 4 *Consider the PDE (2). For any given $d \in \mathbb{N}$, positive scalars ϵ, δ and polynomial $Y_{q(\alpha,\beta,d)} \in \mathscr{R}^{q(\alpha,\beta,d)\times\beta(\alpha+1)}[(x, y)]$ defined in (1), suppose there exist:*

$$\begin{aligned}
&\textit{(poly.) matrices defining} \quad \mathcal{R}(w) \textit{in (12.9) and } R(x) \textit{ in (12.11)}, && \text{(25a)}\\
&\textit{(poly.) matrices defining} \quad \mathcal{T}(w) \textit{ in (12.14) and } T(x) \textit{ in (12.16)}, && \text{(25b)}\\
&S_T \in \mathscr{S}^{\beta+2q(0,\beta,d)}[x], \quad S_R \in \mathscr{S}^{\beta(3\alpha+1)+2q(\alpha,\beta,d)}[x], && \text{(25c)}\\
&K_1 \in \mathscr{R}^{\beta\alpha\times\beta\alpha}[x], \quad K_2 \in \mathscr{R}^{\beta\alpha\times 2\beta\alpha}[x], && \text{(25d)}\\
&H_1, H_2 \in \mathscr{R}^{\beta\alpha\times\beta\alpha}[(x, y)], \quad S \in \mathscr{S}^{\beta(3\alpha+1)}[x], && \text{(25e)}\\
&B_1 \in \mathscr{R}^{\beta(\alpha+1)\times\beta\alpha}[x], \quad B_2 \in \mathscr{R}^{2\beta\alpha\times\beta\alpha}, && \text{(25f)}
\end{aligned}$$

such that

$$\begin{aligned}
&T(x) - \begin{bmatrix} \epsilon I_\beta & 0_{\beta,2q(0,\beta,d)} \\ 0_{2q(0,\beta,d),\beta} & 0_{2q(0,\beta,d)} \end{bmatrix} - S_T(x)\omega(x) \in \Sigma^{\beta+2q(0,\beta,d)}[x], && \text{(26a)}\\
&S_T \in \Sigma^{\beta+2q(0,\beta,d)}[x], && \text{(26b)}\\
&R(x) - S_R(x)\omega(x) \in \Sigma^{\beta(3\alpha+1)+2q(\alpha,\beta,d)}[x], && \text{(26c)}\\
&S_R(x) \in \Sigma^{\beta(3\alpha+1)+2q(\alpha,\beta,d)}[x], && \text{(26d)}\\
&He\,(U_b(x) + K_b(x) + H_b(x) + B_b(x)) - R_b(x) - S(x)\omega(x) \in \Sigma^{\beta(3\alpha+1)}[x], && \text{(26e)}\\
&S(x) \in \Sigma^{\beta(3\alpha+1)}[x], && \text{(26f)}\\
&\bar{U}(x, y) + \bar{H}(x, y) + \bar{B}(x, y) - \bar{R}(x, y) = 0_{\alpha+1}, && \text{(26g)}
\end{aligned}$$

where $\omega(x) = x(1 - x)$, K_b is defined using K_i as in Lemma 1, H_b and $\bar{H}$ are defined using H_i as in Lemma 2, B_b and $\bar{B}$ are defined using B_i and F_i as in Lemma 3 and polynomials U_b and $\bar{U}$ are defined in (4) with T_b and $\bar{T}$ defined in (14).

Then, (2) is exponentially stable.

Proof Since the polynomial matrix in (26a) is Sum-of-Squares (SOS), S_T is SOS and $\omega(x) \geq 0$, for all $x \in [0, 1]$, using the property that a SOS polynomial is non-negative for all $x \in \mathbb{R}$, we conclude that $T(x)$ satisfies (17). Therefore, using Corollary 1, we conclude that $\mathcal{T}(w) \geq \epsilon \|w\|_{\mathscr{L}_2}$, for all $w \in \mathscr{L}_2\left([0, 1]; \mathbb{R}^\beta\right)$. Moreover, from (3) we have that $\mathcal{V}(w) = \frac{1}{2}\mathcal{T}(w)$ and thus

$$\mathcal{V}(w) \geq \mu \|w\|_{\mathscr{L}_2}, \quad \forall w \in \mathscr{L}_2\left([0, 1]; \mathbb{R}^\beta\right), \tag{27}$$

with $\mu = \frac{1}{2}\epsilon$. Similarly, since (26c)–(26d) hold, we conclude from Theorem 2 that

$$\mathcal{R}(w) \geq 0, \quad \forall w \in \mathscr{H}^\alpha\left([0, 1]; \mathbb{R}^\beta\right), \tag{28}$$

where $\mathcal{R}(w)$ is defined in (9).

Now, let us define

$$\mathcal{P}(w) = \int_0^1 \bar{w}_\alpha(x)^T P_b(x) \bar{w}_\alpha(x) dx + \int_\Delta w_\alpha(x)^T \Gamma\left[\bar{P}\right] w_\alpha(y) \Delta, \tag{29}$$

where

$$\begin{aligned} P_b(x) =& He\left(U_b(x) + K_b(x) + H_b(x) + B_b(x)\right) - R_b(x), \\ \bar{P}(x, y) =& \bar{U}(x, y) + \bar{H}(x, y) + \bar{B}(x, y) - \bar{R}(x, y). \end{aligned}$$

Now, since (26e)–(26g) hold, we deduce that

$$P_b(x) \succeq 0, \quad \forall x \in [0, 1], \quad \text{and} \quad \bar{P}(x, y) = 0, \quad \forall (x, y) \in \Delta.$$

Therefore, we get

$$\mathcal{P}(w) \geq 0, \quad \forall w \in \mathscr{H}^\alpha\left([0, 1]; \mathbb{R}^\beta\right). \tag{30}$$

From the definition of $\mathcal{P}(w)$ in (29) it is clear that

$$\mathcal{P}(w) = \mathcal{V}_d(w) + \mathcal{S}(w) - \mathcal{R}(w),$$

where $\mathcal{V}_d(w)$ is defined in (4) and $\mathcal{S}(w)$ is defined in (20). Thus, using (30) we obtain

$$\mathcal{V}_d(w) + \mathcal{S}(w) - \mathcal{R}(w) \geq 0, \quad \forall w \in \mathscr{H}^\alpha\left([0, 1]; \mathbb{R}^\beta\right).$$

Using (28) the previous expression may be reduced to

$$\mathcal{V}_d(w) + \mathcal{S}(w) \geq 0, \quad \forall w \in \mathscr{H}^\alpha\left([0, 1]; \mathbb{R}^\beta\right). \tag{31}$$

Now, from Theorem 3 we have that $\mathcal{S}(w) = 0$, for all $w \in \mathscr{B}$, thus, from (31) we deduce that

$$\mathcal{V}_d(w) \geq 0, \quad \forall w \in \mathscr{B} \subset \mathscr{H}^{\alpha}\left([0,1];\mathbb{R}^{\beta}\right). \tag{32}$$

Since (27) and (32) hold, we apply Theorem 1 to complete the proof. ■

We claim that the search of variables verifying the conditions of Theorem 4 can be cast as a convex optimization feasibility problem and thus, such a search can be performed algorithmically in a numerically efficient manner. Indeed, all the search variables are polynomials, and since the set of polynomials is closed under linear operations, including differentiation and integration, it is evident from the conditions in (26) that we are searching for SOS polynomials subject to affine constraints. Importantly, the problem of searching for SOS polynomials subject to affine constraints is a Semi-Definite Program (SDP), [3, Chap. 3], [29], a convex optimization problem that is cast as

$$\textbf{SDPProblem}: \quad \text{Find (12.25) subject to (12.26).} \tag{33}$$

The numerical implementation is performed by constructing the underlying SDP for (33) by using the freely available packages SOSTOOLS [24] or YALMIP [20]. Then, the associated SDP is solved, for example, using SeDuMi [27] or SDPA [31] solvers.

5.1 Numerical Example

In this section we apply Theorem 4 to numerically obtain the certificate of stability of the trivial solution of the following hyperbolic Partial Integro-Differential Equation (PIDE)

$$\begin{aligned}
\partial_t w_1(x,t) &= a(x)\partial_x w_1(x,t) + b(x)w_1(x,t) + c(x)w_2(x,t) + \textstyle\int_0^x d(x,y)w(y,t)dy,\\
\partial_t w_2(x,t) &= e(x)\partial_x w_2(x,t),\\
w_1(0,t) &= f w_2(0,t), \quad w_2(1,t) = 0.
\end{aligned} \tag{34}$$

The PIDE is coupled in-domain and on the boundary. The system data is the following set of polynomials

$$\begin{aligned}
&a(x) = -2x^2 - x - 3, \; b(x) = -x^2, && c(x) = -x^3 - x + 2,\\
&d(x,y) = xy, && e(x) = (1-\lambda)(x^2+10), \; f = 5.
\end{aligned}$$

Using [8, Lemma 3.1] we conclude that (34) is exponentially stable in the $\mathscr{L}_2$ sense for $\lambda < 1$.

In order to analyze the example system using our methodology, we first present (34) in the form of (2) by choosing

$$\alpha = 1, \quad \beta = 2, \quad w(x,t) = \begin{bmatrix} w_1(x,t) \\ w_2(x,t) \end{bmatrix}, \tag{35a}$$

$$A_1(x) = \begin{bmatrix} b(x) & c(x) & a(x) & 0 \\ 0 & 0 & 0 & e(x) \end{bmatrix}, \quad A_2(x,y) = \begin{bmatrix} 0 & d(x,y) & 0 & 0 \\ 0 & 0 & 0 & 0 \end{bmatrix}, \tag{35b}$$

$$A_3(x,y) = 0_{\beta,\beta(\alpha+1)}, \quad F_1 = 0_{\beta\alpha,\beta(\alpha+1)}, \quad F_2 = \begin{bmatrix} 0 & 1 & 0 & 0 \\ 0 & 0 & 1 & -f \end{bmatrix}. \tag{35c}$$

For the system (35), we solve the problem in (33) using SOSTOOLS with $\epsilon = 10^{-3}$ and $\delta = 10^{-4}$. We perform a bisection search with a resolution of 10^{-3} to search for the maximum value of λ for which (33) is feasible. For a degree of 8 for the polynomial variables, we are able to construct Lyapunov certificates of exponential stability for the system under consideration for a maximum λ of 0.999, therefore obtaining a certificate of exponential stability close to the stability limit.

6 Conclusions and Future Work

We provided a numerically tractable method for the exponential stability analysis for a large class of linear PDEs. We propose a structure of Lyapunov functionals which allow us to the cast the analysis problem in the form of construction/verification of two integral inequalities on (subsets of) Hilbert spaces. Furthermore, for the verification of integral inequalities on the subspaces of Hilbert spaces defined by the boundary conditions, we construct slack integrals using the fundamental theorem of calculus and the Green's theorem. Finally, we use sum-of-squares polynomials to verify the constructed conditions. We show the effectiveness of the proposed method by constructing certificates of exponential stability for the trivial solution of a partial integro-differential equation.

We aim to extend the presented methodology to distributed and boundary controller synthesis of linear PDEs. Additionally, since our method is Lyapunov based, we also aim to extend it to robust controller synthesis for uncertain and non-linear PDEs.

Appendix 1

The following technical lemmas are used throughout the chapter.

Lemma 6.1 *For any* $\alpha, \beta \in \mathbb{N}$ *and Lebesgue integrable functions* $K_1 : \overline{\Omega} \to \mathbb{R}^{\beta(\alpha+1)\times\beta(\alpha+1)}$ *and* $K_2 : \underline{\Omega} \to \mathbb{R}^{\beta(\alpha+1)\times\beta(\alpha+1)}$, *the following identity holds*

$$\int_0^1 \int_0^x w_\alpha(x)^T K_1(x, y) w_\alpha(y) dy dx + \int_0^1 \int_x^1 w_\alpha(x)^T K_2(x, y) w_\alpha(y) dy dx$$
$$= \frac{1}{2} \int_\Delta w_\alpha(x)^T \Gamma \left[K_1(x, y) + K_2(y, x)^T \right] w_\alpha(y) d\Delta,$$

for all $w \in \mathscr{H}^\alpha\left([0, 1]; \mathbb{R}^\beta\right)$.

Proof We have

$$\begin{aligned}\int_0^1 \int_0^x w_\alpha(x)^T K_1(x, y) w_\alpha(y) dy dx &= \int_0^1 \int_0^x w_\alpha(y)^T K_1(x, y)^T w_\alpha(x) dy dx \\ &= \int_0^1 \int_y^1 w_\alpha(y)^T K_1(x, y)^T w_\alpha(x) dx dy \\ &= \int_0^1 \int_x^1 w_\alpha(x)^T K_1(y, x)^T w_\alpha(y) dy dx,\end{aligned}$$

where we first transposed the integrand, followed by a change of order of integration and finally switched between the variables x and y. Thus

$$\begin{aligned}&\int_0^1 \int_0^x w_\alpha(x)^T K_1(x, y) w_\alpha(y) dy dx \\ &= \frac{1}{2} \int_0^1 \int_0^x w_\alpha(x)^T K_1(x, y) w_\alpha(y) dy dx + \frac{1}{2} \int_0^1 \int_x^1 w_\alpha(x)^T K_1(y, x)^T w_\alpha(y) dy dx \\ &= \frac{1}{2} \int_\Delta w_\alpha(x)^T \Gamma \left[K_1(x, y) \right] w_\alpha(y) d\Delta.\end{aligned}$$

Following the same steps for

$$\int_0^1 \int_x^1 w_\alpha(x)^T K_2(x, y) w_\alpha(y) dy dx,$$

then completes the proof. ■

Lemma 6.2 *For any* $v \in \mathscr{L}_2\left([0, 1]; \mathbb{R}^\beta\right)$, $\alpha, \beta \in \mathbb{N}$, *and polynomials* $F_1, G_1, F_2, G_2 \in \mathscr{R}^{\beta \times \beta}[(x, y)]$, *the following identity holds*

$$\begin{aligned}&\int_0^1 \left(\int_0^x F_1(x, y) v(y) dy + \int_x^1 F_2(x, y) v(y) dy \right)^T \\ &\times \left(\int_0^x G_1(x, y) v(y) dy + \int_x^1 G_2(x, y) v(y) dy \right) dx = \frac{1}{2} \int_\Delta v(x)^T \Gamma \left[K \right] v(y) d\Delta, \end{aligned} \tag{36}$$

where

$$K(x,y) = \int_0^y \left(F_2(z,x)^T G_2(z,y) + G_2(z,x)^T F_2(z,y)\right) dz + \int_y^x \left(F_2(z,x)^T G_1(z,y) + G_2(z,x)^T F_1(z,y)\right) dz + \int_x^1 \left(F_1(z,x)^T G_1(z,y) + G_1(z,x)^T F_1(z,y)\right) dz.$$

Proof We begin by observing that the left hand side of (36) may be written as

$$\langle \mathscr{F} v, \mathscr{G} v\rangle_{\mathscr{L}_2} = \langle v, \mathscr{F}^\star \mathscr{G} v\rangle_{\mathscr{L}_2}, \tag{37}$$

where the linear bounded operators on $\mathscr{L}_2\left([0,1];\mathscr{R}^\beta\right)$ are defined as

$$(\mathscr{F} v)(x) = \int_0^x F_1(x,y)v(y)dy + \int_x^1 F_2(x,y)v(y)dy,$$
$$(\mathscr{G} v)(x) = \int_0^x G_1(x,y)v(y)dy + \int_x^1 G_2(x,y)v(y)dy,$$

and where the Hilbert adjoint of the operator $\mathscr{F}$ is given by

$$\left(\mathscr{F}^\star v\right)(x) = \int_0^x F_2(y,x)^T v(y)dy + \int_x^1 F_1(y,x)^T v(y)dy.$$

Now, using the fact that

$$(\mathscr{G} v)(y) = \int_0^y G_1(y,z)v(z)dz + \int_y^1 G_2(y,z)v(z)dz,$$

we get

$$\left(\mathscr{F}^\star \mathscr{G} v\right)(x) = \int_0^x F_2(y,x)^T \left(\mathscr{G} v\right)(y)dy + \int_x^1 F_1(y,x)^T \left(\mathscr{G} v\right)(y)dy = \sum_{i=1}^4 \Theta_i(x), \tag{38}$$

where

$$\Theta_1(x) = \int_0^x \int_0^y F_2(y,x)^T G_1(y,z)v(z)dzdy,$$
$$\Theta_2(x) = \int_0^x \int_y^1 F_2(y,x)^T G_2(y,z)v(z)dzdy,$$
$$\Theta_3(x) = \int_x^1 \int_0^y F_1(y,x)^T G_1(y,z)v(z)dzdy,$$
$$\Theta_4(x) = \int_x^1 \int_y^1 F_1(y,x)^T G_2(y,z)v(z)dzdy.$$

In each of the $\Theta_i(x)$ we change the order of integration and switch between the variables y and z to obtain

$$\Theta_1(x) = \int_0^x \int_y^x F_2(z,x)^T G_1(z,y)dzv(y)dy,$$
$$\Theta_2(x) = \int_0^x \int_0^y F_2(z,x)^T G_2(z,y)dzv(y)dz + \int_x^1 \int_0^x F_2(z,x)^T G_2(z,y)dzv(y)dy,$$
$$\Theta_3(x) = \int_0^x \int_x^1 F_1(z,x)^T G_1(z,y)dzv(y)dy + \int_x^1 \int_y^1 F_1(z,x)^T G_1(z,y)dzv(y)dy,$$
$$\Theta_4(x) = \int_x^1 \int_x^y F_1(z,x)^T G_2(z,y)dzv(y)dz.$$

Substituting into (38) and consequently in (36) we get

$$\langle \mathscr{F}v, \mathscr{G}v \rangle = \int_0^1 \int_0^x v(x)K_1(x,y)v(y)dydx + \int_0^1 \int_x^1 v(x)K_2(x,y)v(y)dydx, \tag{39}$$

where

$$K_1(x,y) = \int_0^y F_2(z,x)^T G_2(z,y)dz + \int_y^x F_2(z,x)^T G_1(z,y)dz$$
$$+ \int_x^1 F_1(z,x)^T G_1(z,y)dz,$$
$$K_2(x,y) = \int_0^x F_2(z,x)^T G_2(z,y)dz + \int_x^y F_1(z,x)^T G_2(z,y)dz$$
$$+ \int_y^1 F_1(z,x)^T G_1(z,y)dz.$$

Then, applying Lemma 6.1 to (39) completes the proof. ■

Remark 1 Lemma 6.2 also holds for any $v \in \mathscr{H}^\alpha\left([0,1];\mathbb{R}^\beta\right)$, $F_1, G_1, F_2, G_2 \in \mathscr{R}^{\beta(\alpha+1)\times\beta(\alpha+1)}[(x,y)]$ and with $v(x)$ replaced by $v_\alpha(x)$ in (36).

Appendix 2

We now present the proofs of the various results stated in the chapter. We begin with the proof of our assertion that (4) can be written in the form presented in (5).

Let us write the integral expression in (3) as[1]

$$\mathcal{V}(w) = \frac{1}{2}\langle \Xi w, w\rangle_{\mathcal{L}_2}, \tag{40}$$

where the self-adjoint operator Ξ on $\mathcal{L}_2\left([0,1];\mathcal{R}^{\beta}\right)$ is defined as

$$(\Xi w)(x) = T_b(x)w(x) + \int_0^x \bar{T}(x,y)w(y)dy + \int_x^1 \bar{T}(y,x)^T w(y)dy.$$

Since the operator Ξ is self-adjoint, we may use (2) to obtain

$$\begin{aligned}
\partial_t \mathcal{V}(w) =& \frac{1}{2}\langle \Xi \partial_t w, w\rangle_{\mathcal{L}_2} + \frac{1}{2}\langle \Xi w, \partial_t w\rangle_{\mathcal{L}_2} \\
=& \frac{1}{2}\langle \Xi w, \partial_t w\rangle_{\mathcal{L}_2} + \frac{1}{2}\langle \Xi w, \partial_t w\rangle_{\mathcal{L}_2} \\
=& \langle \Xi w, \partial_t w\rangle_{\mathcal{L}_2} \\
=& \int_0^1 \left(T_b(x)w(x)dx + \int_0^x \bar{T}(x,y)w(y)dy + \int_x^1 \bar{T}(y,x)^T w(y)dy \right)^T \\
& \times \left(A_1(x)w_\alpha(x) + \int_0^x A_2(x,y)w_\alpha(y)dy + \int_x^1 A_3(x,y)w_\alpha(y)dy \right) dx.
\end{aligned}$$

Therefore,

$$\begin{aligned}
\mathcal{V}_d(w) =& -\partial_t \mathcal{V}(w) - 2\delta \mathcal{V}(w) \\
=& -\int_0^1 \left(T_b(x)w(x)dx + \int_0^x \bar{T}(x,y)w(y)dy + \int_x^1 \bar{T}(y,x)^T w(y)dy \right)^T \\
& \times \left(A_1(x)w_\alpha(x) + \int_0^x A_2(x,y)w_\alpha(y)dy + \int_x^1 A_3(x,y)w_\alpha(y)dy \right) dx \\
& - \delta \int_0^1 w(x)^T T_b(x)w(x)dx - \delta \int_\Delta w(x)^T \Gamma\left[\bar{T}(x,y)\right] w(y)d\Delta = -\sum_{i=1}^4 \Phi_i,
\end{aligned} \tag{41}$$

where,

[1] For brevity we have dropped the temporal dependency of w.

$$\Phi_1 = \int_0^1 \left(T_b(x)w(x) + \int_0^x \bar{T}(x,y)w(y)dy + \int_x^1 \bar{T}(y,x)^T w(y)dy \right)^T A_1(x)w_\alpha(x)dx,$$

$$\Phi_2 = \int_0^1 (T_b(x)w(x))^T \left(\int_0^x A_2(x,y)w_\alpha(y)dy + \int_x^1 A_3(x,y)w_\alpha(y)dy \right) dx,$$

$$\Phi_3 = \int_0^1 \left(\int_0^x \bar{T}(x,y)w(y)dy + \int_x^1 \bar{T}(y,x)^T w(y)dy \right)^T$$
$$\times \left(\int_0^x A_2(x,y)w_\alpha(y)dy + \int_x^1 A_3(x,y)w_\alpha(y)dy \right) dx,$$

$$\Phi_4 = \delta \int_0^1 w(x)^T T_b(x)w(x)dx + \delta \int_\Delta w(x)^T \Gamma \left[\bar{T}(x,y) \right] w(y)d\Delta.$$

The term Φ_1 may be written as

$$\begin{aligned}\Phi_1 = &\int_0^1 \bar{w}_\alpha(x)^T \begin{bmatrix} T_b(x)A_1(x) & 0_{\beta,2\beta\alpha} \\ 0_{3\beta\alpha,\beta(\alpha+1)} & 0_{3\beta\alpha,2\beta\alpha} \end{bmatrix} \bar{w}_\alpha(x)dx \\ &+ \int_0^1 \int_0^x w_\alpha(x)^T \begin{bmatrix} \bar{T}(x,y)^T A_1(x) \\ 0_{\beta\alpha,\beta(\alpha+1)} \end{bmatrix}^T w_\alpha(y)dydx \\ &+ \int_0^1 \int_x^1 w_\alpha(x)^T \begin{bmatrix} \bar{T}(y,x)^T A_1(x) \\ 0_{\beta\alpha,\beta(\alpha+1)} \end{bmatrix}^T w_\alpha(y)dydx.\end{aligned}$$

Then, applying Lemma 6.1 to the double integrals and writing the single integral kernel in a symmetric form produces

$$\begin{aligned}\Phi_1 = &\int_0^1 \bar{w}_\alpha(x)^T He \left(\begin{bmatrix} T_b(x)A_1(x) & 0_{\beta,2\beta\alpha} \\ 0_{3\beta\alpha,\beta(\alpha+1)} & 0_{3\beta\alpha,2\beta\alpha} \end{bmatrix} \right) \bar{w}_\alpha(x)dx \\ &+ \frac{1}{2} \int_\Delta w_\alpha(x)^T \Gamma \left[\begin{bmatrix} \bar{T}(x,y)^T A_1(x) \\ 0_{\beta\alpha,\beta(\alpha+1)} \end{bmatrix}^T + \begin{bmatrix} \bar{T}(x,y)A_1(y) \\ 0_{\beta\alpha,\beta(\alpha+1)} \end{bmatrix} \right] w_\alpha(y)d\Delta.\end{aligned} \tag{42}$$

The term Φ_2 may be written as

$$\begin{aligned}\Phi_2 = &\int_0^1 \int_0^x w_\alpha(x)^T \begin{bmatrix} T_b(x)A_2(x,y) \\ 0_{\beta\alpha,\beta(\alpha+1)} \end{bmatrix} w_\alpha(y)dydx \\ &+ \int_0^1 \int_x^1 w_\alpha(x)^T \begin{bmatrix} T_b(x)A_3(x,y) \\ 0_{\beta\alpha,\beta(\alpha+1)} \end{bmatrix} w_\alpha(y)dydx.\end{aligned}$$

Applying Lemma 6.1 produces

$$\Phi_2 =\frac{1}{2}\int_{\Delta} w_\alpha(x)^T \Gamma\left[\begin{bmatrix} T_b(x)A_2(x,y) \\ 0_{\beta\alpha,\beta(\alpha+1)}\end{bmatrix} + \begin{bmatrix} T_b(y)A_3(y,x) \\ 0_{\beta\alpha,\beta(\alpha+1)}\end{bmatrix}^T\right] w_\alpha(y)d\Delta. \quad (43)$$

The term Φ_3 may be written as

$$\Phi_3 = \int_0^1 \left(\int_0^x \begin{bmatrix} \bar{T}(x,y)^T \\ 0_{\beta\alpha,\beta}\end{bmatrix}^T w_\alpha(y)dy + \int_x^1 \begin{bmatrix} \bar{T}(y,x) \\ 0_{\beta\alpha,\beta}\end{bmatrix}^T w_\alpha(y)dy\right)$$
$$\times \left(\int_0^x A_2(x,y)w_\alpha(y)dy + \int_x^1 A_3(x,y)w_\alpha(y)dy\right)dx. \quad (44)$$

Then, applying Lemma 6.2 to (44) with

$$F_1(x,y) = \begin{bmatrix} \bar{T}(x,y)^T \\ 0_{\beta\alpha,\beta}\end{bmatrix}^T, \quad F_2(x,y) = \begin{bmatrix} \bar{T}(y,x) \\ 0_{\beta\alpha,\beta}\end{bmatrix}^T,$$
$$G_1(x,y) = A_2(x,y), \quad G_2(x,y) = A_3(x,y), \quad v(y) = w_\alpha(y),$$

produces

$$\Phi_3 = \frac{1}{2}\int_{\Delta} w_\alpha(x)^T \Gamma\left[\underline{U}\right] w_\alpha(y)d\Delta, \quad (45)$$

where $\underline{U}(x,y)$ is defined in (5).

Finally, the term Φ_4 may be written as

$$\Phi_4 = \int_0^1 \bar{w}_\alpha(x)^T He\left(\begin{bmatrix} \delta T_b(x) & 0_{\beta,3\beta\alpha} \\ 0_{3\beta\alpha,\beta} & 0_{3\beta\alpha}\end{bmatrix}\right)\bar{w}_\alpha(x)dx$$
$$+ \frac{1}{2}\int_{\Delta} w_\alpha(x)^T \Gamma\left[\begin{bmatrix} 2\delta\bar{T}(x,y) & 0_{\beta,\beta\alpha} \\ 0_{\beta\alpha,\beta} & 0_{\beta\alpha}\end{bmatrix}\right] w_\alpha(y)d\Delta. \quad (46)$$

Substituting (42), (43), (45) and (46) into (41) produces (5).

We now provide the proofs of previously stated results.

Proof (Theorem 2) Throughout this proof, for notational brevity, we write q in place of $q(\alpha,\beta,d)$. If we define

$$f(x) = \begin{bmatrix} \bar{w}_\alpha(x) \\ \int_0^x Y_q(x,y)w_\alpha(y)dy \\ \int_x^1 Y_q(x,y)w_\alpha(y)dy \end{bmatrix}, \quad w \in \mathscr{H}^\alpha\left([0,1];\mathbb{R}^\beta\right),$$

then

$$\int_0^1 f(x)^T R(x) f(x)dx \geq 0, \quad \forall w \in \mathscr{H}^\alpha\left([0,1];\mathbb{R}^\beta\right), \quad (47)$$

since $R(x) \succeq 0$, for all $x \in [0, 1]$. Simplifying the expression we obtain

$$\int_0^1 f(x)^T R(x) f(x) dx = \int_0^1 \bar{w}_\alpha(x)^T R_b(x) \bar{w}_\alpha(x) dx + \sum_{i=1}^{3} \Theta_i, \tag{48}$$

where

$$\begin{aligned}
\Theta_1 =& 2\int_0^1 \int_0^x w_\alpha(x)^T R_{12}(x) Y_q(x, y) w_\alpha(y) dy dx \\
&+ 2\int_0^1 \int_x^1 w_\alpha(x)^T R_{13}(x) Y_q(x, y) w_\alpha(y) dy dx, \\
\Theta_2 =& \int_0^1 \left(\int_0^x Y_q(x, y) w_\alpha(y) dy\right)^T \\
&\times \left(\int_0^x R_{22} Y_q(x, y) w_\alpha(y) dy + \int_x^1 R_{23} Y_q(x, y) w_\alpha(y) dy\right) dx, \\
\Theta_3 =& \int_0^1 \left(\int_x^1 Y_q(x, y) w_\alpha(y) dy\right)^T \\
&\times \left(\int_0^x R_{23}^T Y_q(x, y) w_\alpha(y) dy + \int_x^1 R_{33} Y_q(x, y) w_\alpha(y) dy\right) dx.
\end{aligned}$$

Applying Lemma 6.1 to Θ_1 with $K_1(x, y) = 2R_{12}(x) Y_q(x, y)$ and $K_2(x, y) = 2R_{13}(x) Y_q(x, y)$ produces

$$\begin{aligned}
\Theta_1 =& \int_\Delta w_\alpha(x)^T \Gamma\left[R_{12}(x) Y_q(x, y)\right] w_\alpha(y) d\Delta \\
&+ \int_\Delta w_\alpha(x)^T \Gamma\left[Y_q(y, x)^T R_{13}(y)^T\right] w_\alpha(y) d\Delta,
\end{aligned} \tag{49}$$

Applying Lemma 6.2 to Θ_2 with

$$\begin{aligned}
F_1(x, y) =& Y_q(x, y), \quad F_2(x, y) = 0_{q,\beta(\alpha+1)}, \quad v(y) = w_\alpha(y), \\
G_1(x, y) =& R_{22} Y_q(x, y), \quad G_2(x, y) = R_{23} Y_q(x, y),
\end{aligned}$$

produces

$$\begin{aligned}
\Theta_2 =& \int_\Delta w_\alpha(x)^T \Gamma\left[\int_y^x Y_q(z, x)^T \frac{R_{23}^T}{2} Y_q(z, y) dz\right] w_\alpha(y) d\Delta, \\
&+ \int_\Delta w_\alpha(x)^T \Gamma\left[\int_x^1 Y_q(z, x)^T R_{22} Y_q(z, y) dz\right] w_\alpha(y) d\Delta.
\end{aligned} \tag{50}$$

Similarly, applying Lemma 6.2 to Θ_3 with

$$\begin{aligned}F_1(x, y) =& 0_{q,\beta(\alpha+1)}, \quad F_2(x, y) = Y_q(x, y), \quad v(y) = w_\alpha(y),\\ G_1(x, y) =& R_{23}^T Y_q(x, y), \quad G_2(x, y) = R_{33} Y_q(x, y),\end{aligned}$$

produces

$$\begin{aligned}\Theta_3 =& \int_\Delta w_\alpha(x)^T \Gamma \left[\int_0^y Y_q(z, x)^T R_{33}^T Y_q(z, y) dz \right] w_\alpha(y) d\Delta,\\ &+ \int_\Delta w_\alpha(x)^T \Gamma \left[\int_y^x Y_q(z, x)^T \frac{R_{23}^T}{2} Y_q(z, y) dz \right] w_\alpha(y) d\Delta. \end{aligned} \tag{51}$$

Substituting (49)–(51) into (48) produces

$$\int_0^1 f(x)^T R(x) f(x) dx = \int_0^1 \bar{w}_\alpha(x)^T R_b(x) \bar{w}_\alpha(x) dx + \int_\Delta w_\alpha(x)^T \Gamma \left[\bar{R}\right] w_\alpha(y) d\Delta$$

Then, (47) completes the proof. ■

Proof (Corollary 1) Following the same steps as for the proof of Theorem 2, it can be shown that

$$\mathcal{T}(w) - \epsilon \|w\|_{\mathcal{L}_2} = \int_0^1 f(x)^T \left(T(x) - \begin{bmatrix} \epsilon I_\beta & 0_{\beta,2q(0,\beta,d)} \\ 0_{2q(0,\beta,d),\beta} & 0_{2q(0,\beta,d)} \end{bmatrix} \right) f(x) dx,$$

where

$$f(x) = \begin{bmatrix} w(x) \\ \int_0^x Y_{q(0,\beta,d)}(x, y) w(y) dy \\ \int_x^1 Y_{q(0,\beta,d)}(x, y) w(y) dy \end{bmatrix}, \quad w \in \mathcal{L}_2\left([0, 1]; \mathbb{R}^\beta\right).$$

Then, (18) holds since (17) holds. ■

Proof (Lemma 2) Consider the vector field

$$\begin{bmatrix} \phi_1(x, y) \\ \phi_2(x, y) \end{bmatrix} = \begin{bmatrix} w_{\alpha-1}(x)^T H_1(x, y) w_{\alpha-1}(y) \\ w_{\alpha-1}(x)^T H_2(x, y) w_{\alpha-1}(y) \end{bmatrix},$$

Then, by Green's theorem

$$\oint_{\partial\overline{\Omega}} (\phi_1(x, y) dx + \phi_2(x, y) dy) + \int_{\overline{\Omega}} \left(\partial_y \phi_1(x, y) - \partial_x \phi_2(x, y)\right) d\overline{\Omega} = 0,$$

where $\partial\overline{\Omega}$ denotes the boundary of the domain $\overline{\Omega}$. Therefore, we obtain

$$\int_0^1 [\phi_1(x, 0) + \phi_2(1, x) - \phi_1(x, x) - \phi_2(x, x)] dx + \int_0^1 \int_0^x [\partial_y \phi_1 - \partial_x \phi_2] dy dx = 0. \tag{52}$$

Using the definitions of the projection matrices described in subsection on notation used in the chapter, we obtain

$$\int_0^1 [\phi_1(x,0)+\phi_2(1,x)-\phi_1(x,x)-\phi_2(x,x)]\,dx = \int_0^1 \bar{w}_\alpha(x)^T H_b(x)\bar{w}_\alpha(x)dx$$
$$= \int_0^1 \bar{w}_\alpha(x)^T He\,(H_b(x))\,\bar{w}_\alpha(x)dx. \tag{53}$$

Similarly, we also obtain the following identity

$$\int_0^1\int_0^x \left[\partial_y\phi_1(x,y)-\partial_x\phi_2(x,y)\right]dydx = \int_0^1\int_0^x w_\alpha(x)^T 2\bar{H}(x,y)w_\alpha(y)dydx.$$

Then, applying Lemma 6.1 with $K_1 = 2\bar{H}$ and $K_2 = 0$ produces

$$\int_0^1\int_0^x w_\alpha(x)^T 2\bar{H}(x,y)w_\alpha(y)dydx = \int_\Delta w_\alpha(x)^T \Gamma\left[\bar{H}(x,y)\right] w_\alpha(y)dydx. \tag{54}$$

Substituting (53) and (54) into (52) completes the proof. ■

Proof (Lemma 3) Since for all $w \in \mathscr{B}$ (see (2c)),

$$\int_0^1 \left[\, F_1(x)\;\; F_2 \,\right]\bar{w}_\alpha(x)dx = 0_{\beta\alpha,1},$$

we get for all $w \in \mathscr{B}$

$$0 = \int_0^1 \bar{w}_\alpha(y)^T \begin{bmatrix} B_1(y) \\ B_2 \end{bmatrix} dy \cdot \int_0^1 \left[\, F_1(x)\;\; F_2 \,\right]\bar{w}_\alpha(x)dx$$
$$= \int_0^1\int_0^1 \bar{w}_\alpha(y)^T \begin{bmatrix} B_1(y) \\ B_2 \end{bmatrix}\left[\, F_1(x)\;\; F_2 \,\right]\bar{w}_\alpha(x)dydx$$
$$= \int_0^1\int_0^1 \bar{w}_\alpha(y)^T \begin{bmatrix} B_1(y)F_1(x) & B_1(y)F_2 \\ B_2F_1(x) & B_2F_2 \end{bmatrix}\bar{w}_\alpha(x)dydx.$$

Switching between the variables x and y and changing the order of integration produces

$$0 = \int_0^1\int_0^1 \bar{w}_\alpha(x)^T \begin{bmatrix} B_1(x)F_1(y) & B_1(x)F_2 \\ B_2F_1(y) & B_2F_2 \end{bmatrix}\bar{w}_\alpha(y)dydx,$$

for all $w \in \mathscr{B}$. Recall that

$$\bar{w}_\alpha(x) = \begin{bmatrix} w_\alpha(x) \\ w_\alpha^b \end{bmatrix}.$$

Therefore, we simplify the previous expression to obtain for all $w \in \mathscr{B}$

$$
\begin{aligned}
0 &= \int_0^1 \int_0^1 \begin{bmatrix} w_\alpha(x) \\ w_\alpha^b \end{bmatrix}^T \begin{bmatrix} B_1(x)F_1(y) & B_1(x)F_2 \\ B_2F_1(y) & B_2F_2 \end{bmatrix} \begin{bmatrix} w_\alpha(y) \\ w_\alpha^b \end{bmatrix} dydx \\
&= \int_0^1 \bar{w}_\alpha(x)^T B_b(x)\bar{w}_\alpha(x)dx + \int_0^1 \int_0^1 w_\alpha(x)^T B_1(x)F_1(y)w_\alpha(y)dydx.
\end{aligned}
$$

Writing the single integral kernel in symmetric form and dividing up the double integral gives us

$$
\begin{aligned}
0 = &\int_0^1 \bar{w}_\alpha(x)^T He\left(B_b(x)\right)\bar{w}_\alpha(x)dx \\
&+ \int_0^1 \int_0^x w_\alpha(x)^T B_1(x)F_1(y)w_\alpha(y)dydx + \int_0^1 \int_x^1 w_\alpha(x)^T B_1(x)F_1(y)w_\alpha(y)dydx,
\end{aligned}
$$

for all $w \in \mathscr{B}$. Then, applying Lemma 6.1 with $K_1(x, y) = K_2(x, y) = B_1(x)F_1(y)$ completes the proof. ∎

References

1. A. Papachristodoulou, J. Anderson, G. Valmorbida, S. Prajna, P. Seiler, P.A. Parrilo, *SOSTOOLS: Sum of Squares Optimization Toolbox for MATLAB* (2013). arXiv:1310.4716, http://www.eng.ox.ac.uk/control/sostools
2. F. Argomedo, C. Prieur, E. Witrant, S. Brémond, A strict control Lyapunov function for a diffusion equation with time-varying distributed coefficients. IEEE Trans. Autom. Control **58**, 290–303 (2013)
3. G. Blekherman, P.A. Parrilo, *And R* (SIAM, Thomas. Semidefinite Optimization and Convex Algebraic Geometry, 2012)
4. F. Bucci, I. Lasiecka, Optimal boundary control with critical penalization for a PDE model of fluid-solid interactions. Calc. Var. Partial Differ. Equ. **37**, 217–235 (2010)
5. F. Castillo, E. Witrant, C. Prieur, L. Dugard, Boundary observers for linear and quasi-linear hyperbolic systems with application to flow control. Automatica **49**, 3180–3188 (2013)
6. R. Curtain, H. Zwart, *An Introduction to Infinite-Dimensional Linear Systems Theory*, vol. 21. (Springer Science & Business Media, 2012)
7. R. Datko, Extending a theorem of A.M. Liapunov to Hilbert space. J. Math. Anal. Appl. **32**, 610–616 (1970)
8. F. Di Meglio, R. Vazquez, M. Krstic, Stabilization of a system of $n + 1$ coupled first-order hyperbolic linear PDEs with a single boundary input. IEEE Trans. Autom. Control **58**, 3097–3111 (2013)
9. N. El-Farra, A. Armaou, P. Christofides, Analysis and control of parabolic PDE systems with input constraints. Automatica **39**, 715–725 (2003)
10. W. Evans, *Partial Differential Equations* (Wiley Online Library, 1988)
11. E. Fridman, Y. Orlov, An LMI approach to H_∞ boundary control of semilinear parabolic and hyperbolic systems. Automatica **45**, 2060–2066 (2009)
12. A. Gahlawat, M. Peet, A convex sum-of-squares approach to analysis, state feedback and output feedback control of parabolic PDEs. IEEE Trans. Autom. Control **62**, 1636–1651 (2017)

13. A. Gahlawat, G. Valmorbida, A semi-definite programming approach to stability analysis of linear partial differential equations, in *Proceedings of the 56th IEEE Conference on Decision and Control*, pp. 1882–1887 (2017)
14. P. Goulart, S. Chernyshenko, Global stability analysis of fluid flows using sum-of-squares. Physica D: Nonlinear Phenomena **241**, 692–704 (2012)
15. A. Halanay, V. Rasvan, Stability radii for some propagation models. IMA J. Math. Control Inf. **14**, 95–107 (1997)
16. C. Jin, *Analyse de stabilité de systèmes à coefficients dépendant du retard*. Ph.D. thesis, Université Paris-Saclay (2017)
17. C. Johnson, *Numerical Solution of Partial Differential Equations by the Finite Element Method* (Courier Corporation, 2012)
18. M. Krstic, A. Siranosian, A. Smyshlyaev, Backstepping boundary controllers and observers for the slender Timoshenko beam: Part I-Design, in *Proceedings of the American Control Conference*, pp. 2412–2417 (2006)
19. M. Krstic, A. Smyshlyaev, Adaptive boundary control for unstable parabolic PDEs—Part I: Lyapunov design. IEEE Trans. Autom. Control **53**, 1575–1591 (2008)
20. J. Lofberg, YALMIP: a toolbox for modeling and optimization in MATLAB, in *IEEE International Symposium on Computer Aided Control Systems Design*, pp. 284–289 (2005)
21. T. Meurer, A. Kugi, Tracking control for boundary controlled parabolic PDEs with varying parameters: combining backstepping and differential flatness. Automatica **45**(5), 1182–1194 (2009)
22. A. Movchan, The direct method of Liapunov in stability problems of elastic systems. J. Appl. Math. Mech. **23**, 686–700 (1959)
23. A. Paranjape, J. Guan, S. Chung, M. Krstic, PDE boundary control for flexible articulated wings on a robotic aircraft. IEEE Trans. Robot. **29**, 625–640 (2013)
24. S. Prajna, A. Papachristodoulou, P. Parrilo, Introducing sostools: a general purpose sum of squares programming solver, in *Proceedings of the 41st IEEE Conference on Decision and Control*, vol. 1, pp. 741–746 (2002)
25. A. Smyshlyaev, M. Krstic, Adaptive boundary control for unstable parabolic PDEs—Part II: estimation-based designs. Automatica **43**, 1543–1556 (2007)
26. A. Smyshlyaev, M. Krstic, Adaptive boundary control for unstable parabolic PDEs—Part III: output feedback examples with swapping identifiers. Automatica **43**, 1557–1564 (2007)
27. J. Sturm, Using SeDuMi 1.02, a MATLAB toolbox for optimization over symmetric cones. Optim. Methods Softw. **11**, 625–653 (1999)
28. G. Valmorbida, M. Ahmadi, A. Papachristodoulou, Stability analysis for a class of partial differential equations via semidefinite programming. IEEE Trans. Autom. Control **61**, 1649–1654 (2016)
29. L. Vandenberghe, S. Boyd, Semidefinite programming. SIAM **38**, 49–95 (1996)
30. E. Witrant, E. Joffrin, S. Brémond, G. Giruzzi, D. Mazon, O. Barana, P. Moreau, A control-oriented model of the current profile in tokamak plasma. Plasma Phys. Control. Fusion **49**, 1075 (2007)
31. M. Yamashita, K. Fujisawa, M. Kojima, Implementation and evaluation of SDPA 6.0 (semidefinite programming algorithm 6.0). Optim. Methods Softw. **18**, 491–505 (2003)

Index

G. Valmorbida et al. (eds.), *Accounting for Constraints in Delay Systems*,
Advances in Delays and Dynamics 12,
https://doi.org/10.1007/978-3-030-89014-8

P

Q

R

S

U

CPSIA information can be obtained
at www.ICGtesting.com
Printed in the USA
LVHW082216060422
715570LV00003B/29

9 783030 890131